The Ecology, Exploitation, and Conservation of

RIVER TURTLES

A Massacre of Turtles (From St. Cricq, 1874).

DON MOLL AND EDWARD O. MOLL

The Ecology, Exploitation, and Conservation of

RIVER TURTLES

OXFORD
UNIVERSITY PRESS
2004

OXFORD
UNIVERSITY PRESS

Oxford New York
Auckland Bangkok Buenos Aires Cape Town Chennai
Dar es Salaam Delhi Hong Kong Istanbul Karachi Kolkata
Kuala Lumpur Madrid Melbourne Mexico City Mumbai Nairobi
São Paulo Shanghai Taipei Tokyo Toronto

Published by Oxford University Press, Inc.,
198 Madison Avenue, New York, New York 10016

www.oup.com

Library of Congress Cataloging-in-Publication Data
Moll, Don.
The ecology, exploitation, and conservation of river turtles / Doll Moll, Edward O. Moll.
p. cm.
Includes bibliographical references (p.).
ISBN 978-0-19-510229-1
1. Turtles–Ecology. 2. Stream animals–Ecology. 3. Endangered species. 4. Wildlife conservation. I. Moll, Edward O. II. Title.

QL666.C5M57 2004
597.92′1764–dc21 2003049881

This book is dedicated to the memory of our parents for their love and devotion.

Preface

The following account is a tribute to the diverse and often spectacular turtles which live in the world's rivers. Some are truly beautiful animals with finely chiseled and ornamented carapaces, and with the gaudiest colors imaginable adorning their shells and heads and limbs. Others, such as the softshells, are most unturtle-like in appearance with flattened, leathery shells and alien-strange faces. A number of soft-shelled and hard-shelled forms attain sizes which approach and sometimes exceed proportions more commonly associated with marine species. Many of these species are capable of attaining incredible abundance in prime habitat, and this fact, as well as data available from research concerning the life histories and ecology of several riverine species suggest that they play key ecological roles which affect the dynamics of their river ecosystems in important ways.

In spite of all these attributes the river turtles remain some of the least known vertebrate species on earth—ignored by aquatic biologists and herpetologists alike. To some extent this remains inexplicable to us, but lacking the charisma and cuteness of many endangered warm-blooded species, or even the widespread concern increasingly evident for the magnificent, but dwindling marine turtles (largely due to the late Archie Carr's eloquent writing on the subject), endangered river turtles have been slow to receive the public awareness, research funding, and researchers that their desperate circumstances demand. The fact that many species live secretive lives in great rain forest rivers of remote, often politically unstable, regions where study and law enforcement are difficult at best has not helped matters at all. Collectively, these factors have masked the magnitude of the destruction of these species without mitigating the tragedy of its reality, both for those who depend upon them as an important source of food and for the complex ecosystems in which they have played unique roles. Unfortunately, river turtles have never been ignored by people who value them and/or their eggs as food in many protein-hungry parts of the world, or as ingredients in their traditional medicines. As human populations have grown to their modern levels, and use and trade in river turtles has increased accordingly, many river turtle species have precipitously declined—a process accelerated in some species by river habitat degradation and alteration, and their utilization in the international pet trade. There is mounting concern over the welfare of these species, but our ability to develop effective conservation and management strategies for many of them is often hampered as much by our ignorance of their needs as it is by legislative and cultural concerns.

In chapters 2 and 3 in this book we have presented an overview of both the identities of the world's river turtles and that which is known of their ecology. It will be obvious that although there has been a surge in interest and research concerning river turtles in recent years, we still know very little about them. We hope that these gaps in our knowledge stimulate the investigations of younger turtle enthusiasts and researchers in the near future. In chapters 4–6 we explore the threats that river turtles face from direct exploitation for various uses, and from indirect factors related to human alteration and degradation of their river ecosystems. In chapter 7 we provide an overview of considerations related to the conservation and management of river turtles, and provide our opinions concerning the best methodology and approaches available for the task at hand. We consider river turtles' future prospects in the epilogue.

This book is written for all those interested in river turtles, whether they are amateur turtle enthusiasts, conservationists, river managers, aquatic biologists, or professional turtle biologists. It incorporates both information gleaned from the research of a legion of dedicated biologists and conservationists, and original information contributed by the authors that is presented for the first time in this book. We have written the book to be as "user friendly" as possible to accommodate this broad intended readership with a minimum of jargon and widespread use of common names (most adopted from the *Checklist of the Turtles of the World with English Common Names* [Iverson, 1985]).

We wish to thank all of the turtle biologists and conservationists who have conducted the research upon which this book is based. Often frustrating (e.g., when the traps remain empty), physically difficult, and conducted far from home and families, it is for most truly a labor of love. Many of these people have generously shared their extensive knowledge of river turtles, provided insights into the mysteries of turtle lives, and encouraged our efforts to complete this book. We especially mention the following: Eric Bollinger, Donald Broadley, Arthur Georges, John Iverson, Dale Jackson, Christopher Phillips and Scott Thomson reviewed portions of the manuscript. Christophe Arvy, John Behler, Michael Ewert, Oscar Flores-Villela, Scott Gritters, John Legler, Anders Rhodin and John Tucker provided difficult to obtain and unpublished information. Michael Dreslik, Robert Fischer, and Kipp Kruse advised on statistical matters. Indraneil Das, Richard Funk, James Harding, Kevin Jansen, Wachira Kitimasak, Russell Mittermeier, and Lynn Robbins generously allowed us to use their photographs in this book. We are deeply grateful to our wives Judy (of EOM) and Janice (of DM) for their assistance and emotional support and encouragement during all phases in the development of this book. Janice especially deserves our gratitude for her time and effort devoted to typing and proofreading of our final drafts of the manuscript. Southwest Missouri State University provided a Faculty Summer Research Fellowship to DM which assisted in the preparation of this book. We both wish to specifically acknowledge our major professors for our Ph.D. studies—Dr. J. M. Legler (for EOM) and Dr. Lauren E. Brown (for DM). Although their contributions to this work have perhaps been more indirect than on our dissertation work, their influence on our professional careers has been profound. Finally we wish to pay tribute to our late parents to whom this book is dedicated. Their love, patience, and tolerance of our proclivities for critters—especially turtles—helped shape the direction of our lives long before we became professional biologists.

Contents

The Ecology, Exploitation, and Conservation of

RIVER TURTLES

This page intentionally left blank

Introduction

1

Profile 1. *It was tea time and very hot as we jolted into the river terrapin hatchery at Bota Kanan. The stillness and beauty of the well-manicured grounds were a relief after the roar of the old tour bus and the driver's kamikaze tactics in ruthless adherence to our schedule. Tea had been set in the shade by the hatchery staff and Louis Ratnam, our amiable host and then-director of research in Malaysia's Department of Wildlife and National Parks lectured us on the objectives and progress of the hatchery operations as we sampled the diverse fare of sweets and sandwiches, and drank tea and iced coconut milk from crystal cups. As we toured the facilities the mingled hopes and frustrations that are representative of all the attempts to bring river turtles back from the brink become evident. The large olive drab females have been in captivity since 1967 yet they have laid few eggs. Why? Will the beautiful wild caught males, jet back with bright white eyes, successfully mate with the females? No one knows. Large concrete tanks hold thousands of brown, saucer-sized juveniles of indeterminate sex, almost certainly more than all the wild river terrapins that still exist across their vast range from Indochina to India. Hatched in protected sites nearby, they are head-started to foil small turtle predators before their release. Many have been incubated at relatively high temperatures in styrofoam containers, but it has been determined that incubating temperatures determine the sex of hatchling turtles in many species (usually higher temperatures produce all females and slightly lower produce all males). What will be the effect of releasing huge numbers of juveniles with badly skewed sex ratios into natural populations? How do females find their way back to a nesting beach when they mature? If they imprint on the site how and when does it occur? There is so much we do not know that only time and further research can solve, but time is running out for the river terrapin and many other species.*

Later we walked down to the nearby Perak River and stood on a broad sandy beach that represented one of western Malaysia's last major river terrapin nesting sites. Sweating profusely, as thunderheads built on the horizon for the daily downpour, we waited for the hatchery jeep to pull up and unload a large yellow box filled to the brim with yearling terrapin. Set on the sand, one side of the box was opened and the little turtles tumbled out and flooded down the sloping beach into the river. We watched until all were safely in the water and hundreds of small heads peered back at us from the shallows near shore. Eventually they all disappeared, heading out for who-knows-where. It was good to watch them go, and to know that Malaysia would see that many more would follow them, but it was hard to be overly confident that any of them would someday haul back up on that beach, dig a hole, and fill it with their future. There are just too many trials that they must face in the meantime. The same could be said of any baby river turtles that defy the odds by surviving long enough to enter the water's edge. The responsibility of survival is a vastly greater weight on each of these tiny backs than it once was simply because there are so many fewer now to share the load. It is our sincere hope that they can bear the burden.

Don Moll

The turtle species that dwell in the world's rivers, play important, but incompletely understood and largely unappreciated, roles in both the ecology of their respective ecosystems, and in the economy and sociology of the human cultures through which their rivers flow. Accordingly, the precipitous decline of many populations from former levels, and the complete extirpation of some species from large areas of their former ranges is cause for alarm by those concerned, not only with turtles, but with the health and welfare of rivers and humans as well. The causes of these declines are virtually all the result of human activities—including those which involve direct predation on the turtles themselves (and their eggs), removal for the pet trade, and those which result in unfavorable changes in their habitats. The fact that no known modern species has yet been driven to total extinction is testimony to the resiliency and fecundity of river turtle species. This is no cause for complacency—many are almost certainly teetering on the brink now and are facing ever-increasing odds against long-term survival. Some species could be even now beyond hope of recovery, their populations sustained for a time by a relatively few long-lived adults without sufficient collective reproductive "power" to thwart the normal and various mortality sources which plague all turtles, and which their sheer numbers had formerly been able to overcome. However, we are still so ignorant of the population dynamics of these species that we know not which, if any, are doomed. It seems preferable, therefore, as we strive to conserve river turtle stocks that we assume the working philosophy that none is in this category, and focus our efforts and resources toward research and conservation action most likely to return even the most endangered species to robust, sustainable population levels.

The roots of the current crisis predate history, beginning with our hominid ancestors, and particularly hunter-gather *Homo sapiens*' unparalleled capacity to learn and adapt in order to maximize the harvest of all available food resources. The often large-bodied, relatively slow swimming river turtles, some of which concentrated in virtually helpless tens of thousands at nesting areas must have often served as a reliable source of protein for early humans. Certainly their abundant remains in the middens of many prehistoric cultures—some of which date back over 20,000 years before present (Nicholls, 1977), and the written accounts of early explorer naturalists such as von Humboldt (1852) and Bates (1863) document their importance in this regard. The observations by these two men and others of the capture techniques and magnitude of harvest of adults and eggs of the South American river turtle (*Podocnemis expansa*) in particular provide insight into the efficacy of aboriginal collection methods. Given their vulnerability in the face of such predatory zeal and effectiveness, the survival of river turtle species through the Paleolithic was probably due as much to low human population densities as to their own naturally selected survival strategies. They also had the advantage of living in a medium alien to that of their principal persecutor. Many of their terrestrial relatives, the giant tortoises, which did not enjoy this additional protection suspiciously went extinct during this period.

With the rise of agricultural economies and the development of cities over the last 10,000 years human populations have continued to increase and concentrate in many locations that increasingly threaten river turtles. Ever since the rise of the great civilizations of Mesopotamia, Egypt, the Indus Valley, and the Shang culture of northern China (Yellow River) river flood plains have continued to serve as prime sites for the development of cities. The rivers serve a host of utilitarian functions required by concentrations of people, of course, including supplying water for drinking and sanitation, irrigation, a means of transportation, and power for industry. They also were increasingly exploited as a source of food. With the occasional exception of cultures with religious taboos against the utilization of turtles, the harvest of available species and their eggs has traditionally been a part of this fishery. First undoubtedly collected for subsistence use, then a component of local commerce, turtles often became a part of intercity trade which reached national and international scale as efficient transportation networks developed (e.g., see Clark & Southall, 1920; Cahn, 1937; Das, 1990; Le Dien Duc & Broad, 1995; Jenkins, 1995; van Dijk et al., 2000a). Today as we write this book, river turtles, as well as species from many other biotopes, are threatened by international export trade (both legal and illegal) with East Asian and other lesser markets of such vast magnitude collectively that it nearly defies description. As a direct result of this trade the number of critically endangered freshwater turtles has more than doubled in just the last four years, with three-fourths of Asia's freshwater turtles now listed as threatened, and over half considered endangered (van Dijk et al., 2000a).

While exact numbers are often difficult to come by in the turtle trade a characteristic marker of its unsustainability, seen previously in so many other heavily exploited species, is the gradual widening of the collection area required to supply the market. As rivers adjacent to human population centers with a taste for turtle, or which act as staging centers for shipment, empty their waters of turtles, more distant stocks are tapped (e.g., see van Dijk et al., 2000a). Corollaries of this pattern also include the increased harvest of less desirable species or size classes as the earlier target species or adults are decimated, and the rise in prices of marketed turtle products when they are available at all (e.g., Johns, 1987). All of these warning signals are apparent in river turtle stocks in many parts of the world and in some cases have also attained an international global scale. When Arkansas and Louisiana rivers are being depleted of their turtles to supply Asian markets, as is now occurring, the time for evaluating the ramifications of the industry for river turtle conservation has certainly come.

Of course the other major consequence of human's close association with rivers for resident turtles is that we change them in various ways for our own specialized purposes, degrading their quality as an environment for life as a direct or indirect result. River turtles have not been immune to the effects of these alterations of their habitat. While many species are extremely resilient as long as certain basic requirements for nesting, feeding and cover are met, numerous others have been seriously threatened by the construction of dams and tidal barrages, channelization, flood plain drainage, and myriad forms of pollution.

In light of the nearly cosmopolitan distribution of river turtle species, their importance to humans from cultural, subsistence, and economic perspectives, and their significant ecological roles in riverine environments, the relatively limited attention that they have received from conservationists, limnologists, fisheries biologists, and government regulating agencies is peculiar. Interest in their ecology and biology, and appreciation of their importance in river ecosystems and human economics have surged in recent years however, even as concern for their survival has grown. Inevitably, with river turtles viewed from so many different perspectives, the literature which has developed in relation to these issues is diverse and scattered through numerous forms of publications and in other "gray" outlets. Although there are still huge gaps, both intellectual and geographical, in what we know about river turtles, there is a sufficient base of knowledge, and a desperate enough situation to encourage an attempt to synthesize what we do know—hence this book. In it we first provide an overview of the world's river turtles from taxonomic and ecological perspectives. Second, we describe and evaluate the severity of specific threats to the welfare of riverine species. Finally, we discuss the methodology and techniques for conservation, management, and rehabilitation of the world's river turtle stocks. We hope this book will serve as a useful reference for all of those persons who are concerned, as we are, with the lives and survival of the world's river turtle species.

River Turtle Diversity, Adaptations, and Roles in the River Ecosystem

2

Profile 2. *The painted terrapin male, one of the most beautiful turtles on Earth when in breeding coloration, basked twenty yards away on a palm trunk in Malaysia's Perak River estuary. His fully extended head and neck, white with trim of red, black, and cream and carapace of pearl and black glistened wetly in the tropical sun. His visage was amazingly clownlike as he turned his face toward us, warily assessing our degree of threat and measuring the critical safety cushion of distance necessary for escape should we approach nearer. Fiddler crabs peered out from the entrances to their burrows while mud skippers clambered up the muddy bank at our feet, then in the distance, through the heat haze of mid-day we could make out the black, fist-sized heads of river terrapin, another estuarine species, coursing powerfully through the swiftly flowing current of the turbid river channel.*

It was an exhilarating moment for those of us watching the scene from the riverbank at Durian Sabatang. Members of the Tortoise and Freshwater Turtle Specialist Group of the World Conservation Union (IUCN), we had traveled to Malaysia from around the world to discuss the plight of turtles such as these at the IUCN's 58th Species Survival Commission meeting in Kuala Lumpur. To observe closely some of the world's most endangered river turtles living wild in their natural habitat was at once incredibly lucky under the circumstances, and a moving and memorable experience of contact with animals which may not swim the waters of Earth much longer. The rivers in which they play key ecological roles and the world at large would be the poorer for their passing, and after admiring them from afar, we were more determined than ever, that we must not let this happen.

Don Moll

There are river turtles and then there are turtles that live in rivers; a statement that suggests that the definition of a river turtle may not be as straightforward as it might first appear. It is an issue that must be resolved, however, in determining the scope of the treatment of the group in this book. Complications ensue from both the context in which the term "river" is applied, and from the habitat catholicity of many chelonian species.

Our resolution of the first issue is to focus upon the chelonian inhabitants of larger, higher order lotic environments, and to place less emphasis on the treatment of those species that inhabit smaller streams. With a few notable exceptions (e.g., the flattened musk turtle, *Sternotherus depressus* due to attention by Mount, 1981; Dodd, 1988, 1990, and in press; U.S. Fish and Wildlife Service, 1990; and Bailey & Guyer, 1998), there is so little information available concerning most of these stream dwelling species that there is relatively little sacrificed with this approach. Nevertheless, we shall refer to the limited literature concerning such species when we consider it important to our treatment of river species.

As river ecosystems increase in size, however, they also increase in ecological complexity and habitat diversity. While a complete analysis of these ecosystems is well beyond the scope of this book, Welcomme (1979) distinguished between two broad categories of larger rivers that provide habitat for riverine chelonians—the reservoir river and the flood plain (or flood) river.

Reservoir rivers are those with generally stable flow throughout the year as a result of even annual precipitation, or the presence of marginal lakes and swamps that store water that may be slowly released into the channel throughout the year. Such rivers (e.g., the lower Zaire) overflow their banks only rarely during periods of exceptional rainfall. Many flood plain rivers have been altered to such an extent by water flow regulation due to dams, dikes, and impoundments that they now more closely resemble rivers of the reservoir type (Welcomme, 1979). The fish communities of reservoir rivers are diverse, trophically specialized, and structured into well-defined food webs (Roberts, 1973). They resemble lacustrine communities in behavioral and dynamic characteristics to a much greater extent than does the fish fauna of the flood plain river (Welcomme, 1979). These rivers also provide habitats for turtles in their marginal lakes, swamps, and flowing sections. Although based upon less evidence, their chelonian communities could probably be described similarly as more lacustrine in character compared to flood plain river turtle communities also. The greater regularity of environmental conditions in the reservoir river would permit proportionately more generalized species also associated with lentic habitats to be included in its fauna. This viewpoint would seem to be supported by our analysis of the turtle community of the Zaire River described in the next chapter, as most of its pelomedusid (side-necks) and trionychid (softshell) inhabitants are relatively generalized forms that inhabit both lotic and lentic habitats within the system (Ernst & Barbour, 1989; chapter 3, this volume). Also, there are examples of altered systems such as the Illinois River of the central United States and the Mae Klong system of western Thailand where lotic specialists such as the smooth softshell (*Apalone mutica*), and the Thai narrow-headed softshell (*Chitra chitra*) respectively, have declined at least in part in response to the changed habitats which resulted from flood-controlling "improvements" on these rivers (D. Moll, 1980; van Dijk & Thirakhupt, 1995). In the Illinois River, at least, the more generalized turtle species, such as common snapping turtles (*Chelydra serpentina*), red-eared sliders (*Trachemys scripta elegans*), painted turtles (*Chrysemys picta*), and spiny softshells (*Apalone spinifera*) now dominate the turtle fauna also (D. Moll, 1980).

Most natural rivers are flood plain rivers which, in contrast, are those that periodically respond to large seasonal changes in precipitation within their watersheds by overflowing the main channel and inundating the adjacent plain on a regular basis. The great fluctuations in water level result in seasonal cycles of flood and drought and an ever-changing array of ecological conditions and habitats (Welcomme, 1979). Leopold et al. (1964) and Welcomme (1979) listed and described eight typical features of flood plain ecosystems, all of which may be utilized by resident turtle species. These include: (1) the main channel and its anabranches which normally retain water throughout the year (although surface flow may cease or even disappear in dry periods in extreme environments such as those of East Africa and Australia thereby forcing its turtle fauna to seasonally move to permanent water or into subterranean estivation); (2) oxbows and disconnected oxbow lakes; (3) point bars; (4) meander scrolls and shallow depressions; (5) sloughs; (6) natural levees adjacent to the channel; (7) swamps and back-swamp deposits; and (8) coarse sand splays deposited by flooding. This mix of permanent, semi-permanent, and ephemeral aquatic habitats and associated terrestrial habitats are constantly changing in extent and character. Welcomme (1979) stated that the organisms inhabiting these flood plain river ecosystems have had to adapt to spatial and temporal fluctuations which are perhaps unique among aquatic environments. Predictably, many of the turtles of such ecosystems have responded with morphological, ecological, and behavioral specializations that allow them to cope with life in these extremely variable habitats, and flood plain river turtle communities are composed of mixtures of specialized and generalized species which fill niches associated with these biotopes. In general, the more extreme the differences in environmental characteristics encountered from the rivers' channel laterally toward the edges of its flood plain, the more distinctive the turtles of its various biotopes will be. Large, fast-flowing rivers with largely undisturbed flood plains will usually host the most diverse turtle faunas, for example (e.g., see Bury, 1979; Obst, 1986 and chapter 3, this volume). Similarly, rivers of considerable gradient change and length, culminating in broad estuaries will usually also have a greater diversity of resident turtles as a result of the increased number of habitats available along their course compared to shorter and more homogeneous systems (e.g., see Obst, 1986; chapter 3 in this volume).

Obst (1986) has categorized the turtles associated with larger bodies of water, including rivers, into four distinct groups. These include: (1) the semi-terrestrial inhabitants of river-bank regions which are not very strong swimmers and which may wander a considerable distance from water to feed and nest (e.g., Asian forest turtles [*Heosemys*], stream and pond turtles [*Clemmys*], and Blanding's turtle [*Emydoidea*]); (2) amphibious inhabitants of river-bank regions which usually feed in water and are good swimmers, usually leaving the water regularly only to bask and nest (e.g., European pond turtles [*Emys*], map turtles [*Graptemys*], African side-necks [*Pelusios*]); (3) aquatic inhabitants of river-bank regions that feed exclusively in water and seldom bask (e.g., common snapping turtles [*Chelydra*], alligator snapping turtles [*Macroclemys*], big headed side-necks [*Peltocephalus*]); and (4) aquatic turtles of the open water (river turtles) that are excellent swimmers and divers that can cover extensive areas of great rivers and lakes. Some of these species will bask, while others leave the water only to nest. All of the latter groups are mainly or totally aquatic feeders (e.g., giant softshells [*Chitra*], Asian river turtles [*Kachuga*], and Central American river turtles [*Dermatemys*]).

It is in the flood plain biotopes that the distinction between river turtles and lentic forms becomes particularly murky as flood plain-adapted forms from Obst's first three categories may mingle here with generalized species which range into lentic habitats beyond the flood plain (e.g., painted turtles [*Chrysemys*], sliders [*Trachemys*], Asian box turtles [*Cuora*]), and juveniles of river forms often use flood plain lentic habitats as nurseries during early life. Further complicating the issue is the fact that normally lentic forms may move into the channels of rivers where more typical lotic taxa are absent, essentially filling the "river turtle" niche. The large sliders (*Trachemys scripta*) and black turtles (*Rhinoclemmys funerea*) of Costa Rica's Tortuguero River exemplify this phenomenon (Pritchard & Trebbau, 1984; Moll, 1994). A considerable number of the generalist species may at least occasionally enter slower flowing portions of river habitats also. A perusal of Ernst and Barbour's (1989) habitat descriptions of the world's approximately 257 turtle species then recognized (the exact number is arguable due to the questionable specific status of some and more recent recognition of other species) indicates that some 169 (65%) of the world's chelonians (or approximately 72% if the confirmedly terrestrial tortoises [Testudinidae] are excluded) will sometimes enter flowing water habitats.

We fear that attempting to bring all of the species that occasionally enter running water into our treatment of river turtles would result in losing focus on our primary objectives. Therefore, we shall emphasize those species in our treatment that are river specialists—mainly associated with flowing portions of the river ecosystems, but also including those species which mainly inhabit other aquatic flood plain habitats, but not lentic habitats beyond the flood plain borders (e.g., the alligator snapper [*Macroclemys temminckii*] and its South American podocnemid analog the big-headed side-neck [*Peltocephalus dumerilianus*]). Nevertheless, generalist species play important roles in many rivers and, where relevant, we shall discuss them in this context here and in chapter 3 in particular, where the characteristics of such "combined" river turtle communities are considered.

While there may not be an ideal solution to this rather subjective issue this approach will allow us to place our emphasis upon the ecology, exploitation, conservation, and management of a much smaller group of chelonians (about 99 species or approximately 35% of the world's chelonians) which are primarily or completely river dwellers. This group would encompass all of Obst's category 4 species and the more aquatic forms associated with his 2nd and 3rd categories.

Unfortunately, many of these species also share endangered status—largely as a result of their restriction to river habitats where they are concentrated and vulnerable to overexploitation and habitat destruction. Although, these species range greatly in size, many are impressively large, and powerful swimmers. All are well adapted for life in dynamic and difficult river ecosystems. The larger turtles are often residents of broad channels and estuaries where strong currents and considerable depths preclude the presence of all but the most highly adapted forms. Indeed, some of these species defy the general rule that the channels of major rivers are more correctly considered occasionally used dispersal "highways" rather than "homes" (which are more often along the margins and in the flood plains) for resident life forms (Galat et al., 1996). For these river turtles the channel is often an integral component of their habitat where feeding and breeding regularly occur and, in low water when sand bars and islands emerge, where nesting will take place. Other river dwellers may seek out other biotopes within the river when it suits their purposes, such as the shallow riffles (e.g., Australia's Fitzroy River turtle [*Rheodytes leukops*]), or the more forgiving backwaters, scours, and tributaries along the margins where they mingle with the lentic generalists.

Taxonomic Composition of the World's River Turtles

The world's extant turtles are well represented in riverine environments. Both chelonian suborders (Pleurodira or "side-necked turtles" and Cryptodira or "hidden-necked turtles"—named for the retraction plane of the head and neck) contain numerous lotic representatives (table 2.1). Ten of thirteen living families are also represented, with only the marine families Cheloniidae and Dermochelyidae, and the terrestrial Testudinidae excluded. The exact number of genera and species, which could be categorized as river-dwelling specialists, is hard to pin down exactly (as is the number of extant turtle species as a whole). Again, this is because new genera and species are regularly being described as taxonomic relationships are reconsidered and new species are occasionally discovered, and some of these designations are controversial—accepted by some but not by other turtle specialists. In addition, some species are so poorly known ecologically, or known only from a few old museum specimens or older literature references that placement in the river specialist group entails a certain amount of guesswork. The 39 genera and 104 species listed in table 2.1, therefore represent our best estimate of the modern taxa that are typically associated with larger river environments. The specific composition of riverine turtle communities varies greatly in different geographic areas, and is considered in depth in chapter 3.

Adaptations to River Life

River turtles have evolved numerous ecological, behavioral, and morphological adaptations which allow them to cope with river life, and which set them apart

Table 2.1
Turtles usually associated with larger rivers with summaries of distribution and status of species of concern as indicated by the most recent World Conservation Union's (IUCN) Red List Major categories 1996–2001 (Baillie & Groombridge, 1996, 2001 revisions), and in reference to Asian species the recommendations of van Dijk et al. (2000a) are provided in parentheses if different from the Red List recommendation.

Family	Genus and Species	Common name	Distribution	IUCN Red List	TCF	CITES Appendix Designation
Bataguridae						
	Batagur baska	River terrapin	Tropical South Asia, Indian Sunderbans to Vietnam	CR	CR	I
	Callagur borneoensis	Painted terrapin	Thailand through Malaysia to Sumatra	CR	CR	II
	Orlitia borneensis	Malayan giant turtle	Malaysia and Western Indonesia	EN	EN	
	Kachuga kachuga	Red-crowned roofed turtle	Ganges Drainage	CR	CR	
	Kachuga tecta	Indian roofed turtle	Widespread over the northern Indian subcontinent	NE (LR)		I
	Kachuga trivittata	Burmese roofed turtle	Irawaddy and Salween River Systems	EN*	EN	
	Kachuga dhongoka	Three-striped roofed turtle	Ganges and Brahmapurta River Drainages	EN	EN	
	Kachuga smithi	Brown-roofed turtle	Ganges and Indus River System	NE (LR)		
	Kachuga tentoria	Indian tent turtle	Bangladesh and India (except extreme south and northwest)	NE (LR)		
	Hardella thurjii	Crowned river turtle	Pacific Drainages of Western Columbia, NW Ecuador	LR		
	Rhinoclemmys nasuta	Large-nosed wood turtle	Pacific Drainages of Western Colombia, NW Ecuador	LR		

	Morenia petersi	Indian eyed turtle	Northeastern India, Bangladesh	LR (VU)	
	Morenia ocellata	Burmese eyed turtle	Southern Myanmar	LR (VU)	
Emydidae					
	Pseudemys texana	Texas river cooter	Central and South Central, Texas		
	Pseudemys concinna	River cooter	Southern and Eastern North America	NE	
	Pseudemys alabamensis	Alabama red-bellied turtle	Baldwin and Mobile Counties, Alabama	EN	EN
	Pseudemys gorzugi	Rio Grande cooter	Rio Grande River Drainage	LR	
	Trachemys gaigeae	Big Bend slider	Big Bend region of Rio Grande and Conchos River Drainage – U.S. and Northern Mexico	VU	
	Graptemys geographica	Common map turtle	Pecos River Drainage of West Texas and New Mexico, Eastern North America	NE	
	Graptemys barbouri	Barbour's map turtle	Apalachicola River System, Florida and Georgia	LR	
	Graptemys ernsti	Escambia map turtle	Rivers Feeding Escambia Bay, Alabama and Florida	LR	
	Graptemys pulchra	Alabama map turtle	Mobile Bay Drainage, Alabama and Georgia	NE	
	Graptemys gibbonsi	Pascagoula map turtle	Pascagoula and Pearl River Systems, Mississippi and Louisiana	LR	
	Graptemys pseudogeographica	False map turtle	Mississippi River Drainage, Central U.S.	NE	
	Graptemys ouachitensis	Ouachita map turtle	Central U.S. Mississippi River Drainage	NE	

(continued)

Table 2.1
(*continued*)

Family	Genus and Species	Common name	Distribution	IUCN Red List	TCF	CITES Appendix Designation
	Graptemys kohnii	Mississippi map turtle	Southern and Western Mississippi River Drainage	NE		
	Graptemys versa	Texas map turtle	Colorado River Drainage, Texas	LR		
	Graptemys caglei	Cagle's map turtle	Guadelupe and San Antonio River Drainages, Texas	VU		
	Graptemys oculifera	Ringed map turtle	Pearl River System, Mississippi and Lousiana	EN	EN	
	Graptemys flavimaculata	Yellow-blotched map turtle	Pascagoula River System Mississippi	EN	EN	
	Graptemys nigrinoda	Black-knobbed map turtle	Alabama, Tombigbee and Black Warrior Rivers, Alabama and Mississippi	LR		
	Malaclemys terrapin	Diamondback terrapin	Estuaries and coastal habitats from Northeastern U.S. to Texas Gulf Coast	LR		
Pelomedusidae						
	Pelusios bechuanicus	Okavango mud turtle	Zambezi System of South Central Africa	NE		
	Pelusios adansonii	Adanson's mud turtle	Savannah Rivers of North-Central and West Africa	NE		III (Ghana)

	Pelusios gabonensis	African forest turtle	Rainforest Rivers of Central and West Africa	NE		III (Ghana)
	Pelusios sinuatus	East African serrated mud turtle	East African Rivers from Somalia Southward to South Africa	NE		
	Pelusios castaneus	West African mud turtle	Rivers and Streams of West Africa	NE		III (Ghana)
	Pelusios carinatus	African keeled mud turtle	Congo basin of West Central Africa			
	Pelusios upembae	Upemba mud turtle	Tributaries of Fungwe River, Shaba Province, Congo	DD		
Podocnemidae						
	Erymnochelys madagascariensis	Madagascan big-headed side-neck turtle	Madagascar	EN	EN	II
	Podocnemis expansa	Giant South American river turtle	Amazon and Orinoco Drainage, South America	LR		II
	Podocnemis erythrocephala	Red-headed Amazon river turtle	Rio Negro and Rio Casiquiare Drainages, South America	VU		II
	Podocnemis lewyana	Magdalena river turtle	Rio Magdalena and Rio Sinu Drainages, Columbia	EN	EN	II
	Podocnemis sextuberculata	Six-tubercled Amazon river turtle	Amazon Drainage, Brazil, Peru, Columbia	VU		II
	Podocnemis unifilis	Yellow-spotted Amazon river turtle	Carribbean Drainages and Upper Amazon Tributaries, South America	VU		II
	Peltocephalus dumerilianus	Big-headed Amazon River turtle	Amazon and Orinoco Drainage, South America	VU		II
Chelidae**	*Phrynops geoffroanus*	Geoffroy's side-necked turtle	Tropical and Temperate South America	NE		

(*continued*)

Table 2.1
(continued)

Family	Genus and Species	Common name	Distribution	IUCN Red List	TCF	CITES Appendix Designation
	Acanthochelys pallidipectoris	Chaco side-necked turtle	Chaco region of Argentina, probably Southern Paraguay, Bolivia	VU		
	Chelodina expansa	Giant snake-necked turtle	Eastern Australia	NE		
	Chelodina burrungandjii	Arnhem Land snake-necked turtle	Northern Australia			
	Chelodina longicollis	Common snake-necked turtle	Eastern Australia			
	Chelodina steindachneri	Steindachner's snake-necked turtle	Western Australia			
	Chelodina oblonga	Narrow-breasted snake-necked turtle	Southwestern Australia	LR		
	Chelodina rugosa	Northern Australian snake-necked turtle	Northern Australia			
	Chelodina novaeguineae	New Guinea snake-necked turtle	Northern Australia, New Guinea			
	Chelodina pritchardi	Pritchard's snake-necked turtle	Southern Papua, New Guinea (PNG)	EN		
	Chelodina parkeri	Parker's snake-necked turtle	Irian Jaya and PNG	VU		
	Chelodina siebenrocki	Siebenrock's snake-necked turtle	Irian Jaya and PNG	LR		

Chelodina kuchlingi	Kuchling's snake-necked turtle	Northwestern Australia		
Emydura macquarii	Murray River turtle	Southeastern Australia		
Emydura krefftii	Krefft's River turtle	Northeastern Australia		
Emydura signata	Brisbane short-necked turtle	Northeastern Australia		
Emydura subglobosa	Red-bellied short-necked turtle	Extreme northern Australia, Irian Jaya, PNG		
Emydura victoriae	Victoria short-necked turtle	Northern Australia		
Emydura australis	Australian big-headed turtle	Northwestern Australia		
Emydura tanybaraga	Northern yellow-faced turtle	Northern Australia		
Emydura worrelli	Worrell's short-necked turtle	Northern Australia		
Elseya dentata	Northern Australian snapping turtle	Northern and northeastern Australia		
Elseya lavarackorum	Nicholson River snapping turtle	Northern Australia		
Elseya irwini	Irwin's snapping turtle	Northeastern Australia		
Elseya latisternum	Saw-shelled snapping turtle	Northern and northeastern Australia		
Elseya bellii	Bell's snapping turtle	Southeastern Australia	EN	EN
Elseya purvisi	Purrvis' snapping turtle	Southeastern Australia	DD	
Elseya georgesi	Georges' snapping turtle	Southeastern Australia	DD	
Elseya novaeguineae	New Guinea snapping turtle	Irian Jaya, PNG		
Elseya branderhorstii	Branderhorst's snapping turtle	Irian Jaya	VU	
Elusor macrurus	Mary River turtle	Eastern Australia	EN	EN
Rheodytes leucops	Fitzroy River turtle	Eastern Australia	VU	

(*continued*)

Table 2.1
(continued)

Family	Genus and Species	Common name	Distribution	IUCN Red List	TCF	CITES Appendix Designation
Trionychidae						
	Trionys triunguis	Nile softshell turtle	Most of Africa, Eastern Mediterranean Sea to Greece and Turkey	NE	CR***	III (Ghana)
	Rafetus euphraticus	Euphrates softshell turtle	Tigris and Euphrates River	EN	EN	
	Rafetus swinhoei	Shanghai softshell	Lower Reaches of Chan Jiang and Tai Hai Drainages in Shanghai and Jiangsu Provinces (possibly Yunnan, Province, China and Vietnam)	CR	CR	
	Apalone spinifera	Spiny softshell turtle	Widespread in Central and Southern North America, also represented by populations in Northeastern and Southwestern U.S.	NE		
	Apalone mutica	Smooth softshell turtle	Central and Eastern U.S.	NE		
	Aspideretes nigricans	Black softshell turtle	Chittagong, Bangladesh (captive colony)	EW	CR	I
	Aspideretes leithii	Leith's softshell turtle	Bhavani, Godaveri, and Moyer Rivers of India	VU		
	Aspideretes hurum	Indian peacock softshell turtle	Brahmaputra and Ganges River	VU		I

Aspideretes gangeticus	Indian softshell turtle	Ganges, Indus and Mahanadi River Systems	VU		I
Palea steindachneri	Wattle-necked softshell turtle	China (introduced in Hawaii)	EN	EN	
Nilssonia formosa	Burmese peacock softshell turtle	Irrawaddy, Sittang, and Salween Rivers, Myanmar	EN	EN	
Amyda cartilaginea	Asiatic softshell turtle	Widespread in South and Southeast Asia	VU		
Pelochelys cantorii	Asian giant softshell turtle	Widespread in South and Southeast Asia	EN	EN	
Pelochelys bibroni	New Guinea giant softshell turtle	Southern New Guinea	VU		
Pelochelys signifera	Variegated giant softshell turtle	Northern New Guinea			
Chitra vandijki ****	Myanmar narrow-headed softshell turtle	Ayeyarwady and Salween Rivers, Myanmar			
Chitra indica	Indian narrow-headed softshell turtle	Ganges, Godavari, Mahhanadi, Sutlej and Indus Drainages to Myanmar	EN	EN	
Chitra chitra	Thai narrow-headed softshell turtle	Khwae, Mae Klong and Mae Ping Rivers of Western Thailand, Malaysia and Indonesia (Java)	CR	CR	
Cycloderma frenatum	Zambezi flapshell turtle	Rufiji River Basin, Tanzania to Save River Basin, Mozambique and Zimbabwe	LR		
Cycloderma aubryi	Aubry's flapshell turtle	Rainforest rivers of the Congo (Zaire) River basin			
Cyclanorbis elegans	Nubian flapshell turtle	Larger rivers of Western and Central Africa	LR		

(*continued*)

Table 2.1
(continued)

Family	Genus and Species	Common name	Distribution	IUCN Red List	TCF	CITES Appendix Designation
Dermatemyidae						
	Dermatemys mawii	Central American river turtle	Southern Mexico, Belize and Guatemala	EN	EN	II
Carettochelyidae						
	Carettochelys insculpta	Pig-nosed turtle	New Guinea and Northern Australia	VU		
Chelydridae						
	Macroclemys temminckii	Alligator snapping turtle	Southeastern U.S. Northward to Midwest	VU		
Kinosternidae						
	Staurotypus salvinii	Chiapas giant musk turtle	Mesoamerica, Pacific Slope	LR		
	Staurotypus triporcatus	Giant Mexican musk turtle	Mesoamerica, Atlantic Slope	LR		

Red list code: CR = critically endangered; EN = endangered; EW = extinct in the wild; VU = vulnerable; LR = lower risk; NE = not evaluated; and DD = data deficient. Convention on International Trade in Endangered Species of Wild Fauna and flora (CITES, 1973) or also provided are Turtle Conservation Fund (TCF) status listings (code same as in Red List) from TCF Action Plan (2002) and "CITES" appendix designations are also provided for listed species (I, II, or III). For elaboration of the causes of these status designations, the reader is referred to the above references and chapter 7, this volume.

* This species had not been recorded since 1935 until Kuchling rediscovered it in 2002 in the upper Chindwin and Dokthawady Rivers, and in a Mandalay Buddhist temple pond in Myanmar (Kuchling, 2003a).

** The taxonomy of the Australian and New Guinea chelid genera *Chelodina*, *Elseya*, and *Emydura* is currently in flux and some species listed here may actually be more correctly considered as a complex of closely related species (see Georges & Adams, 1992, 1996; Legler & Georges, 1993a,b; Cann, 1998; Georges et al, 1998; van Dijk et al., 2000a; Rhodin & Genorupa, 2000; Samedi & Iskandar, 2000; Thomson et al. 2000; Iverson et al., 2001). Suspected, but yet undescribed new species have not been included.

*** Mediterranean population only.

**** Another name *Chitra burmanica*, has also been applied to this species (McCord & Pritchard, 2002; Jaruthanin 2001–2002). It has been ruled an unavailable name by McCord et al. (2003).

from more generalized lentic forms (table 2.2). Many of these adaptations have evolved independently in distantly related riverine taxa, and collectively these characteristics help to define the group. These adaptations and representative species which exemplify them (as well as some notable exceptions) will be examined below.

Optimum Habitat

River turtles live in higher-order (i.e., formed from several tributaries) fluviatile systems, which are often characterized by considerable depth and strong currents. As mentioned above, however, there is also a mosaic of ever-changing, usually less rigorous habitats nearer shore, in tributaries, and on the flood plain. While we can provisionally categorize river-dwelling turtles into the four biotopes of Obst (1986) described above, recent research suggests that even the most lotic adapted forms are seasonally and ontogenetically dependent on a wide variety of the other habitats represented in the flood plain river ecosystem (e.g., Anderson et al., 2002). The reasons for movement into these other habitats are encompassed in Gibbons et al.'s

Table 2.2
Common characteristics of river turtles
(not every species will possess every character).

- Optimum habitat: large, flowing bodies of water.
- Mainly strong swimmers; rather than bottom-walkers (*Macroclemys* and *Peltocephalus* are exceptions) (Zug, 1971).
- Medium to large species (maximum size 25→ 100 cm CL; *Graptemys* and *Pangshura* group of *Kachuga* are exceptions). Most of the largest species are tropical.
- Sexually dimorphic with females often exceeding males in size. At least three species seasonally sexually dichromatic.
- Type I reproduction (i.e., laying large numbers of relatively small eggs at ancestral nesting beaches during a discrete nesting season) (Moll, 1979).
- Nest on sand banks.
- Diverse foraging styles; with some groups tending toward increased herbivory.
- Elongated snout or proboscis, sometimes elongated neck.
- Bony lung chambers, sometimes encased by large shell buttresses in diving species.
- Cutaneous, buccopharyngeal and cloacal respiration.
- Limbs flattened and oar-like; feet extensively webbed.
- Reduced shell lip (skin of neck attaches near leading edge of shell). Semi-aquatic and terrestrial forms have a prominent shell lip.
- Streamlined body shape.
- Body more dense than water (river turtles sink, bottom walkers and tortoises float) (Zug, 1971).
- Tendency for plica media of penis to sport elaborate flaps (Trionchids/Bataguridae).

Adapted from Moll and Moll (2000)

(1990) list of population objectives accomplished through relatively long range (i.e., extrapopulational) movements. These include: (1) searches for habitats supplying seasonally available food resources; (2) migration to suitable nesting locations by adult females; (3) mate seeking by adult males; (4) migration to hibernation and aestivation sites; (5) migration from nest sites by hatchlings to juvenile "nursery" habitats; and (6) departure from habitats which have become unsuitable.

In the mid-1990s, turtle biologists became concerned about the conservation implications of such extrapopulational movements (e.g., Burke & Gibbons, 1995; Burke et al., 1994, 1995). Only by becoming familiar with the types of habitats required to meet these needs, and the areal extent of land and water encompassing these required habitats could effective critical habitat recommendations be formulated, and appropriate boundaries for refuges be established for threatened species. With these objectives in mind, Bodie and Semlitsch (2000a) and Bodie et al. (2000), have conducted the most detailed studies since Vogt's (1980a) investigations of Mississippi River map turtles to determine the extent to which river turtles utilize the various habitats available to them in their river and its flood plain. Bodie and Semlitsch (2000a) specifically examined the spatial and temporal use of the lower Missouri River and its flood plain by a representative lotic species, the false map turtle (*Graptemys pseudogeographica*), and a representative lentic generalist, the red-eared slider. They found that the river proper, and connected scours (pond-like habitats formed on the flood plain by concentrated flow through levee breaks) were important habitat for both species throughout the year, but especially during the cooler months (September–February). During this period the river especially provided hibernacula in near-shore locations, and mating mainly occurred in fall and spring while the sexes are concentrated there. Similar hibernation and mating patterns were also described by Vogt (1980a) in false map turtles, common map turtles (*Graptemys geographica*), and Ouachita map turtles (*G. ouachitensis*) in his Wisconsin study of Mississippi River map turtles. Specifically, he found that Wisconsin *Graptemys* would move into wing dams along sloughs and channels in October where they would wedge themselves into rock piles in large concentrations to spend the winter. Mating occurred in April, October and November. Bodie and Semlitsch (2000a) found that in the warmer months (March–August) both species utilized all of the seven habitats they differentiated in the river and its flood plain (including the river, levees, scours, sloughs, flooded forests less than four years old, flooded forests more than four years old, and flooded agricultural land). They also found sexual differences in usage patterns, with males being consistently more sedentary and restricted in movements, and ontogenetically, with juveniles usually found further from the river (often near nesting areas and associated with highly productive, aquatic predator-free ephemeral habitats) than adults. Bodie and Semlitsch (2000a) found that the turtles they studied annually utilized several kilometers of habitat along one side of the river (which was not crossed), and that a 449 m riparian zone (minimum estimate) encompassed 95% of the populations' movements laterally from the river proper. Movements from riverine habitats into seasonally flooded and connected flooded forests and marshes has also been described for Wisconsin false map turtles (although closely related Ouachita map turtles, especially males, tend to remain in the flowing parts of the river [Vogt, 1980a]), for giant South American river turtles, (*Podocnemis expansa*) (Pritchard & Trebbau, 1984), yellow-blotched map turtles (*Graptemys flavimaculata*) in Mississippi (Jones, 1996), sliders in Panama (Moll & Legler, 1971) and for the western pond turtle (*Clemmys marmorata*) in the western United States (Reese, 1996) to take advantage of the additional resources offered by these productive environments. Similarly, tropical estuarine river terrapin (*Batagur baska*) and painted terrapin (*Callagur borneoensis*) in some rivers move daily into much smaller tidal creeks to feed on river bank vegetation during the rising tide, then retreat with the ebb tide back into the river and estuary (E. Moll, 1980 a,b). D. Moll (1990) also noted that Belizean scorpion mud turtles (*Kinosternon scorpioides*), and white-lipped mud turtles (*K. leucostomum*) would move overland from drying ephemeral habitats such as ditches into a permanent stream as the spring dry season progressed. Few lotic species are as willing or able to migrate terrestrially to reach seasonally important resources and habitats as are the semi-terrestrial kinosternids, however. One such example is provided by Jones (1996), who observed two female yellow-blotched map turtles that moved from the Pascagoula River into a nearby cypress pond in early spring—presumably to take advantage of the pond's warmer water and foraging opportunities. As Bodie and Semlitsch (2000a) noted, their research and the other cited work underscores the critical importance of natural, seasonal flooding of flood plains to provide access to seasonally important habitats for adult river turtles, to replenish these habitats for adult river turtles, and to replenish the highly productive ephemeral aquatic habitats that serve as important nurseries for juvenile turtles.

Bodie et al. (2000) also examined the diversity and structure of turtle assemblages in Missouri River flood plain wetlands. They found that turtle diversity peaked in sloughs, oxbows, and connected and isolated scour habitats, while abundances of fewer species of turtles were higher in remnant sloughs. Both diversity and evenness are positively correlated with degree of permanence of water within these habitats and factors which influence it (e.g., depth, prevalence of flooding, etc.). They found that river-connected scours supported proportionately more lotic adapted species (including smooth softshells, spiny softshells, and false map turtles, while oxbows supported the highest proportion of lentic species (including sliders, painted turtles, and common snapping turtles). Wetland habitats which were closer to the river, turbid, and relatively insect-poor contained higher proportions of lotic species (with the first criterion probably the most influential), while wetlands which were frequently flooded, relatively distant from the river and insect-rich contained the highest proportion of lentic species. This study also underscores the importance of regular seasonal flooding in accordance with natural rhythms, and flood plain habitat diversity in supporting the ecological requirements of the river turtle community as a whole.

The flooding period itself may have varied effects on resident river turtles depending on its extent and timing. Increased current velocity in heavy floods could potentially dislodge and displace turtles from their normal home ranges (e.g., painted turtles—MacCulloch & Secoy, 1983; sliders—Moll & Legler,

1971), but Plummer and Shirer's (1975) study of Kansas River smooth softshell movements suggest that highly adapted lotic species such as these would have little problem returning upstream if they so wish. Probably if displacement due to flooding does occur it would mainly affect generalist species and smaller juveniles. Tucker et al. (1998a) noted that Illinois River red-eared sliders had reduced reproductive output in 1994 and 1996, the years following the largest and third-largest floods in river history. While other influences cannot be ruled out, the extensive flooding may have adversely affected slider feeding by making aquatic vegetation inaccessible, or eliminating it from large areas, resulting in the accumulation of lower energy reserves for reproduction in the following season. As environmental influences affect growth rates in young turtles (e.g., Gibbons, 1967, 1968; D. Moll, 1976a; Dunham & Gibbons, 1990; J. K. Tucker et al., 1995a), the loss of food resources via flooding would probably negatively affect growth as well. Certainly, greater than average flooding or mistimed floods can and do disrupt successful nesting and hatching of young in such a year. Many river turtles' reproductive cycles are adapted to the rhythms of the normal seasonal hydrologic patterns of flooding and drying as discussed above, but early and prolonged periods of flooding may cover flood plain nesting sites or prevent channel sand bars and islands from emerging during the critical season, thereby reducing or eliminating reproductive output for that year. Later than normal flooding may inundate and silt over nesting sites causing the deaths of the developing embryos, or in the latter case preventing their successful emergence from the nests (Plummer, 1976). Roze (1964) reported the deaths of several million eggs and embryos of the colonial nesting giant South American River turtle due to nest flooding caused by premature rains—just one spectacular example of a fairly regular and predictable catastrophe endured by flood plain river turtle species. Also predictably, human tampering with normal flooding rhythms will adversely affect river turtle species by preventing the replenishment of flood plain habitats as discussed above, or more directly by inundating nesting sites before and/or after nesting. An example of the latter is provided by recent research in Thailand's Mae Klong River system concerning the conservation and ecology of the Thai narrow-headed softshell. Studies in progress indicate that the release of large amounts of water during *Chitra*'s spring dry season nesting period from an upstream reservoir periodically inundates the sand bar nesting sites after oviposition, killing the eggs within the nests (W. Kitimasak, unpubl.). Regardless of other threats this species is facing, the lack of reproduction will surely doom this critically endangered, spectacular species if this situation is not rectified.

We have been dwelling on the negative here, however, and wish to re-emphasize that normal, seasonal flooding is a beneficial phenomenon on the whole. It is by definition, a part of life on the flood plain river to which its resident species are remarkably well adapted. River-dwelling sliders are benefited directly as flooding normally reduces the distances that both females and hatchlings must travel to find a suitable nesting site in the spring, and, upon emergence from the nest to find suitable aquatic, hatchling nursery habitat respectively (Tucker, 1997). The reproductive cycles of these, and other river turtle species such as the giant South

American river turtle are surely adapted to take advantage of these seasonal phenomena. Unlike sliders, foraging success may actually be enhanced during flood periods for some species also. Moll (1976b) found that Mississippi River false map turtles were more likely to have found food, and were able to expand their diet to include more terrestrial animal foods during flood periods when these inundated terrestrial resources became accessible to them. As their sand-bottomed aquatic habitat was relatively sterile and apparently offered few benthic or other food resources in abundance, it is probable that the flood periods were extremely important in sustaining the very large population of false map turtles (and perhaps the other very common species in the study area, the smooth softshell) through the year. In contrast to Tucker et al.'s (1998a) findings concerning post-flooding reproduction in the more herbivorous sliders, Mississippi River's false map turtles were found to grow faster than in normal flood years during the great 1993 flood year, and to have higher than normal reproductive output in 1994 (D. Moll, unpubl.).

Within the river itself various ecological requirements, and the adaptations and capabilities of the species themselves influence the locations and extent of the habitats utilized. These requirements include foraging areas where food sources are available and concentrated, suitable basking sites for those species so inclined, sites which serve as hibernacula, sites suitable for nesting, sites suitable for nocturnal resting areas such as those with submerged logs, root tangles, or undercut banks, and areas which provide safety from predators. Obviously some locations will provide more than one, or even all of these requirements and species' home ranges are usually of relatively modest proportions here (e.g., Plummer & Shirer, 1975; Gordon & MacCulloch, 1980; Vogt, 1980a; MacCulloch & Secoy, 1983; Pluto & Bellis, 1986, 1988; Jones, 1996; Plummer & Burnley, 1997). Others may be met in locations at some distance from the others, requiring daily or seasonal movements to and from the areas to satisfy the particular need. The daily movements of Asian river terrapin and painted terrapin from estuaries into tidal creeks to feed, and the annual nesting migrations of these species in opposite directions (river terrapin upstream to sand bars, and painted terrapin downstream to the sea and sea beaches) provide examples of such patterns (E. Moll, 1980 a,b). Similarly the annual movements of giant South American river turtles from dispersed feeding areas to concentrated nesting areas between wet and dry seasons (Pritchard & Trebbau, 1984), and those of spiny softshells in Vermont's Lamoille River, and Missiquoi Rivers, and Quebec's Riviére aux Brochets, where movement occurs between summer nesting and feeding sites in the rivers' lower reaches and in Lake Champlain, then back upstream to river hibernacula in autumn provide additional examples of seasonal patterns (Graham & Graham, 1997; Galois et al., 2002). Pluto and Bellis (1988) and Jones (1996) also studied seasonally changing river usage patterns in map turtles in Pennsylvania (Raystown Branch of Juniata River), and yellow-blotched map turtles in Mississippi (Pascagoula River) respectively. Sloan and Taylor (1987), Shipman (1993), Harrell et al. (1996a), and Trauth et al. (1998) found that alligator snapping turtles moved extensively within their aquatic habitats. Stream dwellers moved both upstream and downstream, and males often, but not always (Trauth et al., 1998) moved further distances than females. Harrell et al. (1996a) found that water temperature and cover were important factors influencing movements in subadults.

As suggested by Bodie et al. (2000), poorly understood biotic interactions between species (e.g., competition, predation, symbioses, etc.) may also influence the ways that river turtles utilize and divide up available habitats and resources (e.g., see Vogt, 1980a; Fuselier & Edds, 1994; Lindeman, 2000; and chapter 3 this volume). Alternatively, phylogeny may be more important in influencing resource use patterns than biotic interactions in some turtle communities (Lindeman, 2000).

Many of the relationships between ecological requirements and distributions and densities (a reflection of habitat optimality) are probably too obvious to elaborate upon further, but others have only recently been clarified by relevant studies. The importance of suitable basking sites in influencing basking species' distributions and densities is one such relationship.

Basking provides important thermoregulatory effects (Boyer, 1965; Moll & Legler, 1971; Auth, 1975; Vogt, 1980a) and perhaps other benefits related to parasite removal (e.g., Vogt, 1979). Even some normally non-basking species such as common snapping turtles (D. Moll, pers. obs.) and alligator snapping turtles (Ewert, 1976) may occasionally bask, and species have been recorded to bask on such unusual substrates as on snapping turtle's carapaces (Legler, 1956), the head of a water buffalo (MacGeorge, 1928 in Boyer, 1965), loon nesting platforms, and even the backs of loons (Gelatt & Kelley, 1995). Although different species sometimes prefer different types of basking sites (e.g., smaller mud turtles [Kinosternidae] may climb out onto small skinny snags inaccessible to larger baskers [Lindeman, 1996]; smooth softshells prefer sand bars and beaches over snags and stumps [Lindeman, 2001a]), most species utilize emergent deadwood for this purpose. It not only provides a platform for basking in the sun, but also below water level it often serves as substrate for algal growth and invertebrates which serve as food for turtles (Waters, 1974; Shively & Jackson, 1985; Kofron, 1991; Seigel & Brauman, 1994; Lindeman, 1997). It may also serve as a subsurface shelter against the current and a sleeping area for species which may or may not bask (Chaney & Smith, 1950; Polisar, 1990a). Studies by Fuselier and Edds (1994) in Kansas, Pluto and Bellis (1986) in Pennsylvania, Jones (1996) in Mississippi, Shively and Jackson (1985) in Louisiana, and Kramer (1995) in Florida all found that deadwood basking substrates were important habitat variables influencing the abundance and distribution of riverine emydid turtles (principally map turtles [*Graptemys*]). Lindeman (1999) provided the most convincing evidence of the importance of the relationship with his surveys of basking turtles and deadwood abundance in the Pearl and Pascagoula river drainages of Mississippi and Louisiana, and in the lower Tennessee River in Kentucky. He found positive correlations of *Graptemys* densities, and total turtle species densities with deadwood densities in all cases, and that high basking turtle densities were never associated with low deadwood densities. Although all three drainages had roughly similar quantities of emergent deadwood available, he found that total turtle densities in the Tennessee River were three times higher than in the Pearl River, and six times higher than in the Pascagoula River. These data indicate that deadwood is necessary for supporting large populations of map turtles and other river species, but other factors such as current speed and channel morphology may also influence habitat quality for these species as well (Lindeman, 1999). Shively and Jackson (1985) examined factors that limited the upstream distribution of Sabine map turtles (*G. ouachitensis sabinensis*) in Whiskey Chitto Creek, Louisiana. They considered stream width, but not depth, to be the ultimate determinant of turtle density and the upstream limit of the distribution through a variety of direct and indirect effects influencing map turtle ecology (also noted as an important habitat influence by Fuselier & Edds, 1994). They concluded that range termination finally occurs because suitable habitats become progressively smaller and more isolated upstream, and eventually the turtle's dispersal capabilities over the larger areas of unsuitable habitat in between are inadequate to sustain successful colonizations of suitable habitat patches. Competition with false map turtles and sliders, which do range considerably farther upstream in Whisky Chitto Creek, may also help to limit the Sabine map turtle's upstream progress. Finally, Flaherty and Bider (1984) have suggested that social factors may influence habitat use in the common map turtles (*Graptemys geographica*) at the northern extent of its range in Quebec. They noted that apparently suitable habitats in some bays (e.g., good feeding, basking and nesting sites available) were unused while other similar or less well-endowed habitats supported large populations of turtles. Hopefully these North American studies of the ecology of basking, and the significance of basking sites will provide insights and stimulate similar studies of basking relationships in other parts of the world also. Many non-North American riverine genera such as *Batagur*, *Callagur*, *Kachuga*, *Hardella*, *Podocnemis*, *Emydura*, *Elseya*, *Pelomedusa*, *Pelusios*, *Aspideretes*, and *Trionyx* are also confirmed baskers. Doody et al. (2001) described an unusual form of aquatic "basking" in small isolated thermal springs in river beds by pig-nosed turtles (*Carettochelys insculpta*) in Australia. Turtles, especially gravid females, spend many hours in the vicinity of the warm springs during the cooler months, apparently practicing a form of behavioral thermoregulation when river temperatures were lower than those in the springs.

Aquatic substrate type also plays an important role as a habitat determinant of many basking and non-basking river turtles (e.g., Fuselier & Edds, 1994; Plummer & Shirer, 1975; D. Moll, 1980, 1985; Das, 1995). This preference may be directly due to the physical characteristics of the substrate itself that provide advantages for turtles associated with it. Indirect influences of the habitats' characteristics may also be beneficial. For example, smooth softshells and giant Asian softshells, Indian narrow-headed and Thai narrow-headed softshells can bury themselves rapidly in clean sand substrates to protect themselves and to ambush prey which is visible to them due to the water clarity in such environments. Conversely, turtles such as South American mata matas (*Chelus fimbriatus*) and Australian snapping turtles (*Elseya latisternum*), which may inhabit silt-bottomed, very turbid environments (Pritchard & Trebbau, 1984; Legler & Winokur, 1979), may benefit from the associated turbidity of the habitat which helps to shield them from their intended prey as they slowly stalk it in the murky water. Interestingly, both of these species are well equipped with non-visual sensory, "mechanoreceptor" structures (flaps, fringes, barbels, and perhaps even tympana which are adapted for sensing and "hearing" prey movements in mata matas, and elongated sensory follicular tubercles on the neck of the Australian snappers), which provide them with a hunting advantage in dark turbid waters. Many species use submerged logs, brush piles, boulders, crevices, and other benthic features to their advantage as they seek shelter or hunt prey, and turtles are seldom seen on clear unobstructed bottoms or in open water (e.g., Legler & Winokur, 1979; Mount, 1981; Polisar, 1992; Pritchard, 1984; Legler & Cann, 1980). Particular substrate types may benefit turtles also by providing appropriate habitat for growth of favored food plants and algae or animal prey. D. Moll found Tortuguero River sliders in dense population concentrations only where their aquatic food plants grew in particularly dark organic substrates patchily distributed through the waterway. In the Illinois River in 1973 and 1974, three species of map turtles were only abundant in locations where dense benthic concentrations of dipteran and trichopteran insect larvae, their principal prey, were present (D. Moll, 1980, 1985).

Water depth may not be an important habitat determinant in some locations (Shively & Jackson, 1985), but Lindeman (2000) did find different depth usage patterns in four emydid species in Kentucky Lake, an impoundment of the Tennessee River, and different sexes of sexually dimorphic smooth softshells live and forage mainly in habitats of different depths, with larger females preferring deeper waters and smaller males shallower sections of the Kansas River nearer shore (Plummer & Shirer, 1975; Plummer & Farrar, 1981). The Australian Fitzroy River turtle is a specialized, very fast-swimming form that lives in shallow, fast-flowing riffles with sand and gravel bottoms studded with rocks. These turtles forage for freshwater sponges and insect larvae in these habitats, and use rocks and deadwood snags (in slower sections) for shelter (Legler & Cann, 1980). Width of habitat was found to be important in influencing habitat use in *Graptemys*; however, as described above (Shively & Jackson, 1985; Fuselier & Edds, 1994) Shively and Jackson, in particular stated that they never observed Sabine map turtles closer to the banks than 2 m, presumably to avoid predation by terrestrial mammals, and Fuselier and Edds found false and Ouachita map turtles to prefer wider habitats than common map turtles in Kansas. Almost certainly though, especially when considering big rivers and big river turtles, the extremes in depth and width and perhaps current velocity present in these habitats collectively provide a very effective envelope of safety for those river turtles highly adapted enough to cope with them themselves.

Big Waters, Big (and not so Big) Turtles

Lotic turtles display a large range of sizes, and moderate-sized to larger rivers often provide habitat for small to medium-sized species as well as larger forms. This is true of both temperate and tropical rivers. Examples of small to moderate-sized species which have diversified into numerous forms which are successful in many rivers are the map turtles, genus *Graptemys* of North America and the morphologically similar *Pangshura* subgenus of *Kachuga*, the smaller roofed turtles, in tropical Asian rivers (Ernst et al., 1994; E. Moll, 1987a). The most striking ecological and behavioral differences between the smaller and larger forms (and their juveniles) relate to the ways they utilize their home rivers. Most freshwater turtles are not very strong swimmers (Pritchard & Trebbau, 1984), and the smaller species, smaller sex of dimorphic species, and juveniles utilize habitats which are characterized by less extreme conditions such as those which predominate in smaller rivers, and protected areas in larger rivers. They spend most of their time near shore and in protected bays, connected backwaters and tributaries, and they seldom venture into the deeper, wider expanses where current velocity and other ecological factors are less favorable for them. As discussed above, the Missouri River's modest-sized map turtles and sliders were never observed to venture as far as a quarter of the way across the channel by Bodie and Semlitsch (2000a); and Bodie et al. (2000) and Tucker et al. (1998b) found the connected scours, sloughs, and other backwater areas to be important habitats for several smaller lotic adapted forms in the Missouri and upper Mississippi Rivers. We have observed the importance of similar near shore habitats with modest currents and shallow depths along the central Mississippi River for false map turtles and softshell males and juvenile females (mainly smooth but a few spiny softshells also) (D. Moll, 1976a,b). Hundreds of heads of these species could often be observed emerging from the water in these protected habitats. Even in the much less expansive Kansas River, Plummer and Shirer (1975) observed very striking differences in habitat use by smaller male and usually larger female smooth softshells. Here, females tended to occupy both sides of the river, swimming freely back and forth through the greater current velocity of the channel, while males tended to reside on only one side (if a crossing was made the change was permanent). Females also moved more frequently than males and for longer distances within a day, utilizing the deep channel during their longer movements. During floods, the only turtles that shifted positions and moved away from shorelines were females. Vogt (1980a) also found that female false and Ouachita map turtles traveled extensively within the large pools of his Mississippi River study area. In contrast, MacCulloch and Secoy (1983) found that male painted turtles traveled further than females or juveniles in Saskatchewan's Qú Appelle River. In all three of these studies occasional movements of relatively longer distances than normal were accomplished by individuals, and there was little evidence of fidelity to a restricted home range. In general, larger turtles, and the larger sex of dimorphic species, utilize faster and deeper sections of river, and move and bask further from shore than do smaller forms (Plummer & Shirer, 1975; Pluto & Bellis, 1986, 1988; Vogt, 1980a; Jones, 1996).

The really large, powerfully swimming river turtles are all residents of the big tropical rivers (North America's giant alligator snapper is considered a bottom-walking form [sensu Zug, 1971; and Berry & Shine, 1980]) (table 2.3). The largest

Table 2.3

Maximum sizes attained by selected larger river turtles species in Temperate and Tropical zones. Greatest carapace lengths and weights recorded (when data are available).

Species	Carapace length (cm) max	Wt. (kg)	Range	Source
Temperate zone:				
*Macroclemys temminckii**	71.1	112.8	N. Am	Pritchard, 1980a
*M. temminckii**	80		N. Am.	Pritchard, 1980a
Apalone spinifera	54.0	11.7	N. Am.	Pritchard, 2001
A. mutica	35.6		N. Am.	Ernst et al., 1994
Pseudemys concinna (suwanniensis)	41.6		N. Am.	Carr, 1952
Graptemys barbouri	32.7		N. Am.	Cagle, 1952
Chelodina expansa	48.5		Australia	Cann & Legler, 1994
Elusor macrurus	40		Australia	Cann & Legler, 1994
Tropical zone:				
Chitra indica	183.0		India	Das, 1991
Chitra chitra	115.0	120	Thailand to Java	Nutphand, 1979
Pelochelys cantorii	90–100		SE Asia	Rhodin et al., 1993
P. bibroni	102		New Guinea	Rhodin et al., 1993
Rafetus euphraticus	68		Middle East	Taskavak & Atatür, 1998
R. swinhoei	104.1	150	Eastern China & North Vietnam	Pritchard, 2001

Nilssonia formosa	40		Myanmar	
Trionyx triunguis	95		Africa & Mediterranean	Branch, 1988
Aspideretes gangeticus	94 m	67 m	Greater India	Das, 1991
A. leithii	63.5		India	Das, 1991
*A. nigricans**	to 91		Bangladesh	Das, 1995
A. hurum	60		Greater India	Das, 1991
Amyda cartilaginea	est. to 83*		South East Asia	
Cycloderma frenatum	56		East & Southern Africa	
Batagur baska	60		Southern Asia	
Callagur borneoenis	50		South East Asia	
Kachuga trivittata	58		Myanmar	
K. kachuga	49		Greater India	
K. dhongoka	48		Greater India	
Dermatemys mawii	65		Central Am.	
Erymnochelys madagascariensis	50		Madagascar	
Peltochephalus dumerilianus	48		S. Am.	
Podocnemis unifilis	46.5		S. Am.	Pritchard & Trebbau, 1984
Podocnemis expansa	89	perhaps to 118	S. Am.	Pritchard & Trebbau, 1984
Elseya dentata	40		Australia	Cann & Legler, 1994
Carrettochelys insculpta	55		New Guinea & N. Australia	

*Captive specimen(s).
Data from Ernst and Barbour (1989) unless otherwise indicated.

temperate zone-swimming species are of more modest proportions as, with the notable exception of the Missouri–Mississippi River system, are the rivers they inhabit. An obvious question, though is why don't the larger turtles reside (at least currently) in the Missouri–Mississippi River or other large temperate zone rivers? It seems that large size either may confer a greater survival advantage in the tropics than in the temperate zones—perhaps due to more intense competitive interactions or other unknown factors—or that the tropical environment is more conducive toward the evolution of large size—perhaps due to greater annual productivity and year-around growth opportunities made possible by the absence of cold winters, or the existence of even larger river systems here. Perhaps both stronger tropical selection pressures and greater opportunity combine to explain why, for instance, Asian big river batagurids and trionychids attain larger body sizes than their Missouri–Mississippi River emydid and trionychid counterparts. Pritchard (2001) has also speculated that the environmental regimes which characterize big temperate zone riverine habitats like the Mississippi River and those which characterize the tropical equivalents are so different that they are very likely responsible for the different size patterns displayed by their resident turtle communities in as yet poorly understood ways.

The inhabitants of specific river systems and the link between big rivers, the tropics, and big turtles are further explored in the next chapter. Here, we will limit ourselves to more general observations concerning the advantages of large size in exploiting river resources and habitats.

Larger species become more frequent in evolutionary lineages over time (i.e., Cope's Rule) because of ecological and/or sexual advantages which are conferred upon such species (Brown & Maurer, 1986; Vogt & Villarreal-Benitez, 1997). Vogt and Villareal-Benitez (1997) concluded that similar phenomena influence size increases in chelonian lineages. Numerous hypotheses have been proposed to account for the attainment of large size in various aquatic turtles, a striking characteristic of many tropical river dwellers (table 2.3). These include: (1) a need to maximize reproductive potential and hatchling survivorship in predator-rich tropical environments (Moll & Legler, 1971; Valenzuela, 2001a); (2) a response to greater productivity and longer growing seasons in the tropics (Moll & Legler, 1971); (3) an adaptation to minimize crocodilian or other forms of predation (Moll & Moll, 1990); (4) an adaptive response which results from filling the herbivorous river turtle niche (Pritchard & Trebbau, 1984); (5) an adaptive response related to travel risks (Gibbons, 1990a); (6) an adaptive response to facilitate utilization of channel and estuarine and marine habitats (Pritchard & Trebbau, 1984; Moll, 1994); and (7) a form of character displacement which facilitates the partitioning of resources between species (Pritchard, 2001). One common explanation for large size in terrestrial species is the advantage it provides a predator in overcoming and killing large prey. This has not usually been considered seriously in regard to turtles because the larger forms are mainly herbivorous, or are considered to be adapted for ambush feeding (e.g., the softshells, alligator snapping turtle), where stealth and speed of strike, rather than power is employed to capture prey. Perhaps this view prevails because we know so little of the ecology of the larger carnivorous species—especially the Asian *Aspideretes* softshells, and the Nile softshell (*Trionyx triunguis*) which are big, fast, and armed with formidable jaws and aggressive temperament (see also section on Revenge of the River Turtles in chapter 5). These species are known to eat a variety of animal prey (as well as vegetation) including fish, frogs, waterfowl, other turtles, mollusks, and meat of other vertebrates usually thought to be taken as carrion (Loveridge & Williams, 1957; Villiers, 1958; Das, 1991, 1995; Akani et al., 2001). The Ganges softshell (*A. gangeticus*) is known to feed in large groups (Das, 1995), and one such group has been observed attacking, killing and eating a Nilgai antelope in a canal in India (Singh, 2000). Akani et al. (2001) observed more than 20 Nile softshells of various sizes feeding on a goat and a bush-pig carcass along the banks of Nigeria's Sombreiro River. Considering the Nilgai observation, it is conceivable that larger vertebrates entering river shallows to drink or forage may be attacked crocodile-style by submerged, waiting turtles (as has been recorded on a smaller scale in cases of bird predation by lurking pelomedusids in southern Africa [Boycott & Bourquin, 2000]), and then overpowered by a swarm of voracious turtles before escape to shore can be accomplished. With this type of ambush strategy, large size and strength coupled with aggressiveness would definitely be advantageous in obtaining prey.

It is beyond the scope of our treatment to tease out all of the pros and cons of these various hypotheses as they apply to specific riverine species—undoubtedly some complex combination of these factors affect all river turtle size to some degree. Although we may not completely understand all of the influential factors, there is certainly a link between big turtles and big waters, with the largest swimming turtles on earth being associated with the seas, and in fresh waters, with larger tropical rivers (and sometimes with both as discussed below).

Whichever specific selective forces have led to the development of large size in particular species, this key adaptation has pre-adapted them to expand the scope of their habitat utilization, and of their niche opportunities in the big river ecosystem. Larger river turtles are more heavily muscled, and hence more powerful swimmers than smaller turtles, a prerequisite for movement into and mastery of the turbulent channels and estuaries of major rivers. Once entry into these difficult environments has been attained, natural selection will continue to tinker with design to further "fine tune" these species to increase their efficiency and allow them to capitalize on the new opportunities presented. This process occurs inevitably whenever taxa enter new adaptive zones (e.g., the evolution of whales and birds). While smaller turtle species may be successful residents of big rivers by utilizing specific less rigorous habitats within them, as described above, they are restricted from mastering the entire river by the barriers imposed by its more difficult sections. As a result, many suitable habitats and their resources (e.g., along an opposite shore) remain unavailable to them. For the larger species, however, the channels and estuarine portions are no longer barriers. They may be crossed and entered repeatedly to access resources or favorable microenvironments, and may well become an integral component of their habitat used for dispersal, migration, foraging, mating and, when islands and sand bars emerge in low water periods, for nesting sites.

As discussed above, Plummer and Shirer (1975) pointed out different size-related habitat usage patterns by smaller males and larger female smooth softshells in the Kansas River. E. Moll's (1980a,b) studies of river terrapin and painted terrapin in Malaysian rivers also provide examples of the diversity of habitats utilized daily and seasonally by these large species, most of which are made accessible by their powerful swimming ability.

Both of these latter Asian species inhabit the wide and deep estuarine portions of rivers under tidal influence (which they share with the large softshells, the Asiatic softshell [*Amyda cartilaginea*] and the Asian giant softshell [*Pelochelys cantorii*]), through which they move freely, traversing the channels and moving upstream daily into small tidal creeks to feed on river bank vegetation (usually on a rising tide in the Perak River). Both species are omnivorous and behaviorally osmoregulate, moving throughout their habitats to forage in less saline portions, and to restore comfortable fluid and salt concentrations in their tissues after sojourns into more highly saline portions of their habitats (Dunson & Moll, 1980; E. Moll, 1980a,b; Davenport & Wong, 1986; Davenport et al., 1992). At the onset of their respective nesting seasons (which vary with species and river systems in Malaysia), females of both species accomplish relatively extensive migrations to suitable oviposition sites. River terrapin swim upriver to nest on emerging sand bars as far as 80 km upstream from the river mouth in the Perak River (E. Moll, 1980a), while painted terrapin swim downriver and out of the river mouths to nest on sea beaches at least as far as several kilometers away (Dunson & Moll, 1980; E. Moll, 1980b). In some rivers the estuarine and non-tidal influenced upstream portions are partitioned for use by separate species of big river turtles with relatively little overlap. Pritchard (2001) provided an example from the Ganges River where the Indian softshell (*Aspidertes gangeticus*) lives upstream, while the peacock softshell (*Aspideretes hurum*) dominates the delta. Again, these species exemplify a level of versatility in movements and habitat usage patterns that result from physical mastery of nearly all the challenging environments offered by their river ecosystem. In Malaysia and other tropical Asian countries, in addition to powerful swimming capabilities, large size probably also does confer protection against the saltwater crocodiles (*Crocodylus porosus*) that share these estuaries (especially in conjunction with the heavy shell buttressing found in these species), and helps to minimize or slow osmoregulatory imbalances during salt water travel (Pritchard & Trebbau, 1984; Dunson & Mazzotti, 1989; Davenport et al., 1992; Moll, 1994). In reference to the latter, Dunson and Mazzotti (1989) would consider the larger sizes, heavily calcified shells (see also, Dunson & Heatwole, 1986) and behavioral osmoregulatory activities practiced (e.g., limiting prolonged exposure to high salinity, cessation of eating, drinking, defecating and urinating in high salinity, and resumption of bodily functions in fresher water, etc. [Davenport & Wong, 1986]) to represent early stages of adaptation to life in salt water environments. It is instructive that a much smaller estuarine species, North America's diamondback terrapin (*Malaclemys terrapin*), represents a more advanced stage of adaptation to brackish habitats where the osmoregulatory advantages of larger size have been minimized by other adaptations (Dunson & Mazzotti, 1989). It copes with osmoregulatory problems with a functional salt gland and physiological and behavioral adaptations, which help it to compensate for its greater surface area to volume ratio (Dunson, 1970, 1976, 1979a, 1984; Gilles-Baillien, 1970, 1973a,b; Davenport & Macedo, 1990). It is particularly interesting that these same advantages associated with larger size and presumably evolved to better exploit big, tropical, river environments have resulted in the extensive use of an even larger aquatic medium by some riverine species—the sea itself.

Several larger riverine and estuarine species utilize sea beaches near river mouths and delta islands as nesting sites. These include the painted terrapin discussed above; river terrapin, which formerly nested in the Irrawady River delta in Myanmar (Maxwell, 1911), and still nest on island beaches in the Sunderbans of India and Bangladesh (E. Moll, 1990b; Bhupathy, 1995); the Nile softshell along the African coast (Carr & Carr, 1985); the giant *Pelochelys* softshells which wander through tropical Asia and Indonesia to Papua, New Guinea (Nair & Badrudeen, 1975; Vijaya, 1982a; Kar & Rao, 1985; and especially see Rhodin et al., 1993 for a thorough review of marine activity; Hussain, 2003); and the pig-nosed turtle (*Carettochelys insculpta*) which uses coastal beaches in New Guinea as nest sites (Georges & Rose, 1993). In Caribbean Costa Rica river dwelling sliders, which interestingly have the largest female body sizes of any populations of this wide-ranging species (to 44 cm carapace lengths) also swim out river mouths to nest on sea beaches as far as several kilometers away (Pritchard & Trebbau, 1984; Moll, 1994). The most thoroughly pelagic species are probably the *Pelochelys* softshells, and especially the Nile softshell, which ranges into the Atlantic Ocean off West Africa and is widespread throughout the eastern Mediterranean Sea. (Villiers, 1957; Atatür, 1979; Carr & Carr, 1985; Taskavak et al., 1999; Atatür et al., in press). *Pelochelys* softshells have often been reported swimming in deep water, and several kilometers offshore, as well as in larger rivers within their vast range (which very likely is a result of their saltwater dispersal abilities) (Rhodin et al., 1993). In one interesting account reported by these authors, 30–40 large (~90 cm) individuals believed to be New Guinea giant softshells (*Pelochelys bibroni*) were observed traveling together in the sea of about four fathoms depth and two or three miles from the New Guinea shore. The Nile softshell has been recorded nesting on African sea beaches as mentioned above (Carr & Carr, 1985), but it is also known to nest at upriver sites in Turkey and Israel in the eastern Mediterranean region. It has been recorded numerous times at sea off Turkey (Atatür, 1979; Taskavak et al., 1999), and more recently off the island of Kos, in Greece (Taskavak et al., 1999). This species had become more common than sea turtles in the by-catch of Turkish shrimp trawlers (Oruc et al., 1996). Most of the animals caught were large (78% over 50 cm in carapace length), and most were caught in very deep water (68% caught in nets set between 10 and 20 m and 30% from depths between 20 and 40 m). Unfortunately, very little information is currently available concerning the ecology and adaptations for marine life of these sea-going populations for comparison with the more typical riverine populations of Africa. Atatür (1979) recorded nest and egg data, and the presence of both a freshwater water snake and a marine anthozoan in the stomach contents of one individual, suggesting that both freshwater and marine resources may be utilized. The Nile softshells Mediterranean population has become critically endangered recently, perhaps as a result of the shrimp trawler by-catch referred to above (TCF, 2002). Presumably, these sea-going riverine forms are repeating the initial stages of ocean colonization accomplished by the true sea turtles in the Mesozoic Era (Dunson & Mazzotti, 1989; Kinneary, 1996).

Sexual Dimorphism

Many riverine species are sexually dimorphic in size and related shell proportions, with females of more nektonic forms usually attaining larger average dimensions than males. The North American map turtles (*Graptemys*), softshells (*Apalone*), and Asian roofed terrapin (*Kachuga*) exemplify extremes in this trait, but it is well represented—if sometimes less pronounced—in many other groups also (fig. 2.1a,b). Reverse dimorphism, with males consistently larger, or similar-sized sexes is more common in "bottom-walking" and semi-aquatic river dwellers such as the North American snapping turtles (*Chelydra, Macroclemys*), South American big headed side-neck (*Peltocephalus*), some North American and Neotropical mud and musk turtles (*Kinosternidae*), and perhaps some of the Asian softshells in the genera *Amyda*, *Dogania*, *Rafetus*, and *Pelodiscus*. Data concerning size and its sexual correlations in the giant Asian softshell species are particularly limited and therefore conclusions concerning patterns of dimorphism are very tentative (Pritchard, 2001).

The reasons for the independent evolution of dimorphism in so many groups have been presented in detail (and sometimes hotly debated) elsewhere (e.g., Berry & Shine, 1980; Vogt, 1980a; Gibbons & Lovich, 1990; Prichard, 2001). Berry and Shine (1980) suggested that species with larger or similar sized males are usually those in which male combat may frequently occur and/or where less mobile females may be forcibly inseminated by males (the latter explanation vehemently rejected by Gibbons & Lovich, 1990 and contradicted by studies of courtship and mating behavior in alligator snapping turtles [Harrell et al., 1996b]). Pritchard and Trebbau (1984) and Pritchard (2001) noted the phenomenon of "female size conservatism" in land tortoises, mud and musk turtles, and softshells in which females are usually larger in the smaller species within the lineage while males are larger in the larger forms. Pritchard (2001) suggested that once a species has evolved a large enough size, then adequate female body capacity for egg storage is no longer an evolutionary issue, and other selective forces may either favor increased male size (e.g., mate acquisition) or dissuade further increases in female size (e.g., lack of mobility during nesting forays). Certainly the primary advantage of larger size in females of the smaller more typical nektonic species is that it allows them to contain larger eggs and/or egg clutches before oviposition, but numerous secondary advantages may accrue from larger size such as greater safety from predators during oviposition and the opening of other niche opportunities, as discussed previously. The advantage of smaller size of males in these species is

(a)

(b)

Figure 2.1
(a) Average-sized adult male (smaller) and female spiny softshells (*Apalone spinifera*) displaying typical sexual dichromatism and size dimorphism. (b) Average-sized adult male (smaller) and female Indian tent turtles (*Kachuga tentoria*) displaying typical size dimorphism.

less obvious, but is probably a result of a balance being struck between sexual selection guiding the advantages of early entry into the reproductive component of the population and natural selection favoring growth and larger size and the advantages it confers (Gibbons & Lovich, 1990). Complex pre-mating behavioral patterns and related secondary sexual characteristics such as the long fore-claws of male emydids have evolved in these species to synchronize the sexes for copulation (obviously a necessity when larger, equally mobile females can easily break off the encounter), although only a few species such as the sliders (*Trachemys*), river cooters (*Pseudemys*), and some map turtles (*Graptemys*) have been carefully studied (e.g., Cagle, 1950; Jackson & Davis, 1972; Vogt, 1980a; [see summaries of pre-mating behavior in North American species in Ernst et al., 1994]). The lack of knowledge of many species precludes a current assessment of how taxonomically widespread and variable these behavioral patterns may be. There are indications that the level of complexity of these behavioral patterns may be related to the presence or absence of congeneric species within the range (Legler, 1990).

Distinctive sexual differences in other shell and non-shell body parts may frequently be observed in dimorphic species also. In some smaller batagurid turtles a kinetic plastron has evolved to facilitate oviposition of the large, hard-shelled eggs characteristic of this group (E. Moll, 1985a). The structure of the jaws and their triturating surfaces, and jaw musculature may also differ between sexes due to dietary differences. This phenomenon is exemplified by several of the map turtles (e.g., *G. barbouri*) where females may become mainly molluskivorous, and develop the appropriate apparatus for crushing the shells of their prey, while males remain mainly insectivorous with accordingly less powerful jaw architecture. Head shape between the sexes varies accordingly in these species, but may also be very distinctive in other species, though with less-obvious functional relationships. Mesoamerican sliders provide a striking example of the latter with males displaying narrower heads and larger snouts than the more bulbous heads and shorter snouts of females (fig. 2.2a,b). Legler (1990) suggested that such distinctive profiles may provide sexual recognition in populations where the sexes differ less in overall size dimorphism (i.e., tropical sliders compared to temperate sliders). Another sexually divergent feature usually explained by functional relationship would be the longer, thicker tails of males in many species. While clearly useful for housing the penis, the extreme divergence in some species such as the Australian Mary River turtle (*Elusor macrurus*) described by Cann and Legler (1994) could easily serve for sexual recognition by individuals approaching from behind or below another adult.

A few river turtles display sexual dichromatism (Moll et al., 1981). Male and female spiny and smooth softshells display different carapacial markings and are easily differentiated well before maturity is attained (e.g., Graham, 1991; Graham & Cobb, 1998). These species are size dimorphic as stated above and, as a consequence, the sexes inhabit and forage in divergent microhabitats in their home rivers (Plummer & Farrar, 1981; Ernst et al., 1994). The divergent markings may allow them to be more cryptic in their respectively favored haunts.

Ontogenetic melanism is displayed primarily in male sliders, usually developing progressively, well after sexual maturity has been attained. The causative

(a)

(b)

Figure 2.2
(a) Adult female Costa Rican slider (*Trachemys scripta venusta*). (b) Male displaying typical distinctive sexual dimorphism in snout shape.

mechanisms and possible social significance have been investigated by several researchers (e.g., Lovich et al., 1990; Garstka et al., 1991; Tucker et al., 1995b; Thomas et al., 1999; Thomas & Parker, 2000), but its function in wild populations is still not well understood. Thomas et al. (1999) and Thomas and Parker (2000) found correlations between male melanism and changes in mating behavior and movement patterns in red-eared sliders.

Central American river turtle males display very different head coloration in different, but interconnected river systems in Belize, ranging from bright orange (Belize River) and dull orange (Mussell Creek), to yellow (Whitewater Lagoon/ Freshwater Creek) to dull olive or yellow (Spanish Creek), while female heads remain consistently brownish-olive to gray in all populations (D. Moll, pers. obs.; Polisar, 1990b) (fig. 2.3a,b). The significance of these color differences, or their possible seasonality, are unknown and deserve further investigation.

Seasonal sexual dichromatism is known definitely to occur in Asian river terrapin and painted terrapin, and probably also in the Indian red-crowned roofed terrapin (*Kachuga kachuga*) (fig. 2.4a,b,c). The breeding males of these species are among the most beautiful and strikingly colored turtles on earth. The extremely rare Burmese roofed turtle (*K. trivittata*) is also sexually dichromatic, with males displaying striped carapaces and red heads versus the more uniformly brown females (Theobald, 1868a; Moll et al., 1981). Whether its color changes occurred seasonally or not is not known.

Moll et al., (1981) studied the mechanism of color change in Malaysian painted terrapin. They concluded the process to be hormonally controlled and correlated with the breeding period of the species. Dark brownish gray males slowly transformed into lighter colored animals characterized by a light gray or cream-colored carapace, and a white head with black eyes and a dorsal scarlet stripe (fig. 2.5). Females retained the same uniform dark brownish-gray coloration throughout the year (fig. 2.6). These authors suggested that the most likely function of the color change was to allow females to differentiate between males of their own species and those of sympatric river terrapin, the males of which undergo an analogous, but opposite seasonal color change characterized by a blackening of carapace, limbs, and head featuring white eyes (see fig. 2.4). Moll et al. (1981) pointed out that these four documented or suspected seasonally dichromatic species are members of the *Batagur* species complex (sensu McDowell, 1964), and are big river/estuary dwellers. They suggest that in such habitats visual signals may be more effective than olfactory ones, which would quickly dissipate and flow away in the strong currents of their home rivers.

Reproduction

An extensive literature concerning turtle reproduction in general has developed which is beyond the scope of our treatment focusing on river turtles to summarize adequately here. For the reader wishing to explore this topic in greater depth, we recommend Moll, 1979; Ewert, 1979, 1985; Wilbur & Morin, 1988; Elgar &

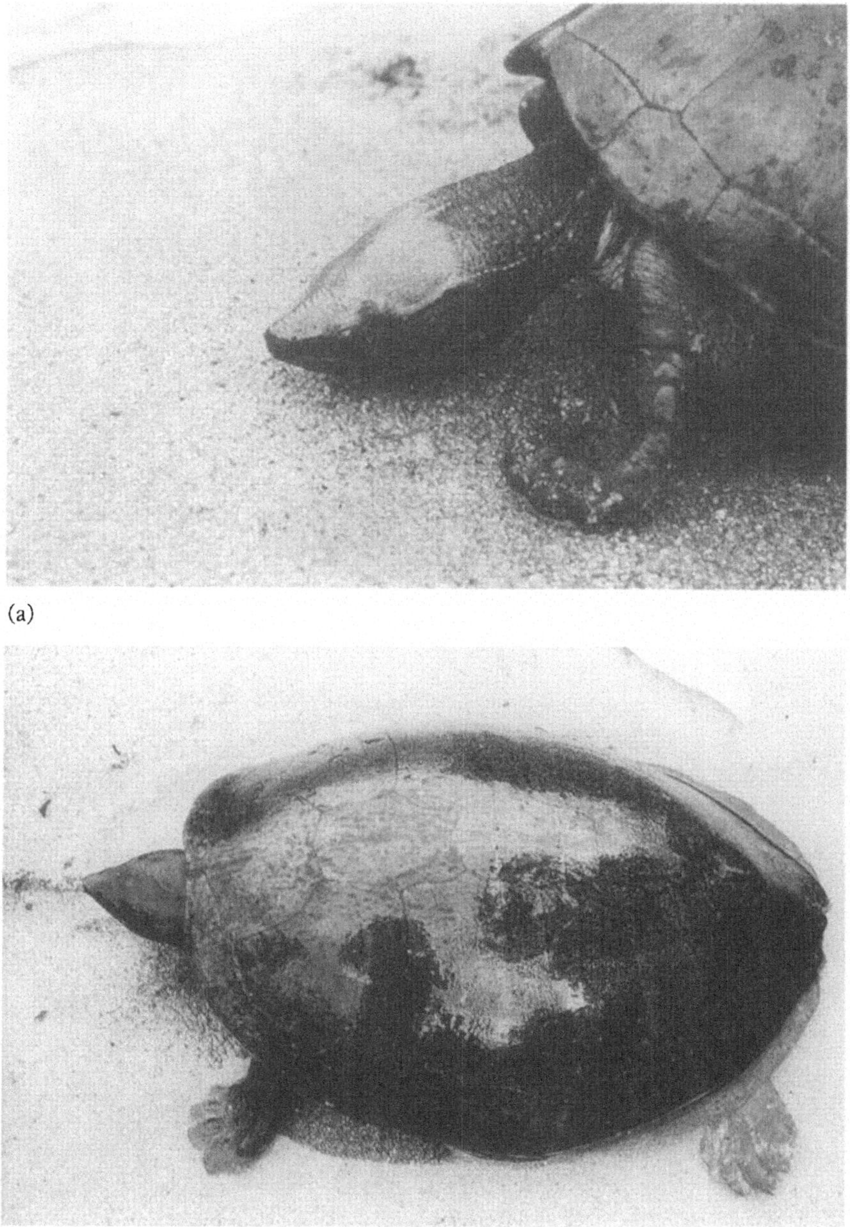

Figure 2.3

(a) Adult male Central American river turtle (*Dermatemys mawii*) from Belize displaying typical sexual dichromatism. Males in different river systems display dorsal head coloration ranging from tan to day-glo orange while females' are always dark gray.
(b) Adult female Central American river turtle (*Dermatemys mawii*) from Belize displaying typical sexual dichromatism. (See color insert.)

(a)

(b)

Figure 2.4
(a) Adult male river terrapin (*Batagur baska*) in breeding coloration. (b) Adult male painted terrapin (*Callagur borneoensis*) in breeding coloration. (See color insert.)

(c)

Figure 2.4 (cont.)
(c) Adult male red-crowned roofed terrapin (*Kachuga kachuga*) in breeding coloration. (See color insert.)

Heaphy, 1989; Gibbons, 1990a (several chapters); and Kuchling, 1999 for further reading.

General Patterns

Since the assemblage of turtles which dwell in rivers is composed of diverse evolutionary lineages, a diversity of reproductive adaptations designed for successfully coping with the vagaries of these highly variable, often-unpredictable environments would also be expected. Much variation does indeed exist, but perhaps even more striking is the fact that in spite of our very incomplete knowledge of almost all of these species, enough data are available to suggest that general patterns have developed independently across taxonomic lines which equip chelonians for life in lotic ecosystems.

Most specialized riverine turtles lay relatively large clutches of small, leathery shelled eggs with fairly short incubation periods, many (probably) produce multiple clutches during a well-defined nesting season, nest communally in well-defined ancestral nesting areas, and deposit egg clutches in nest burrows which are methodically dug (sometimes after a preliminary nest "pit" has been scraped out), and subsequently resealed after the egg clutch is deposited within. Moll (1979) termed this a type I pattern, and he considered it a primitive reproductive pattern shared

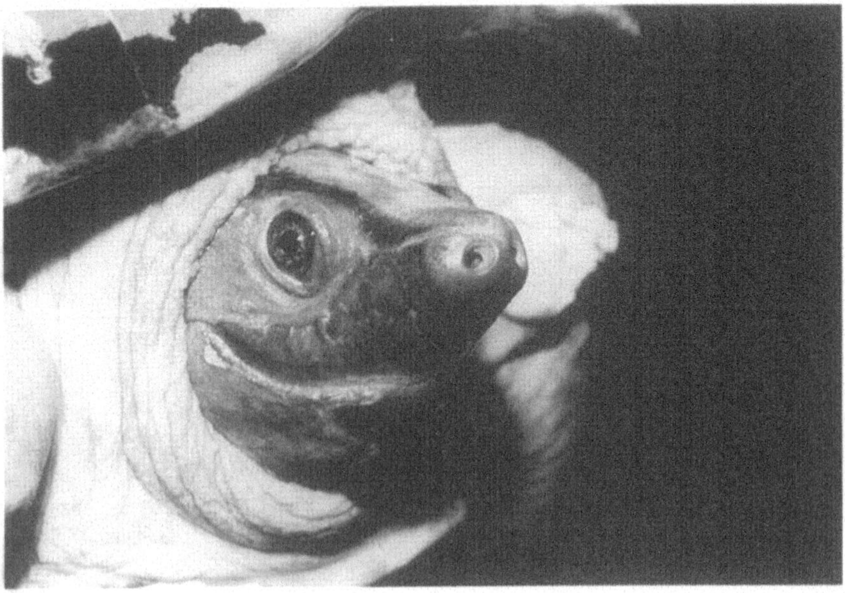

Figure 2.5
Adult male painted terrapin (*Callagur borneoensis*) in transition between non-breeding and breeding coloration. (Photo by Indraneil Das)

Figure 2.6
Typically drab-colored adult female painted terrapin (*Callagur borneoensis*).

with river turtles by all other temperate zone turtles and sea turtles. In contrast, most other tropical turtles more closely approach Moll's type II or advanced reproductive pattern where small clutches of large, brittle-shelled eggs with long incubation periods are produced acyclically or continuously by isolated individuals which lay them in scattered locations, and little or no nest construction is attempted (e.g., eggs are simply covered by leaf litter or laid beneath the edge of a fallen log). These two reproductive patterns, as described above, represent the extremes of a continuum that may be observed in various lineages and geographic areas, and numerous exceptions to the rule are notable in one or more defining characteristics in numerous river-dwelling (and other) chelonians. Moll (1979) suggested that the extreme patterns of both are best exemplified in the tropics, and that most temperate zone species fall somewhere in between (but usually more toward the type I end of the spectrum). When temperate origin turtles invade the tropics (and tropical rivers), such as the slider and common snapping turtle in the Neotropics (Moll & Moll, 1990; Vogt, 1990; D. Moll, 1997a), and the chelid genera *Emydura* and *Elseya* in Australia (Legler, 1985; but see comments by Kuchling, 1999), they tend to maintain the temperate zone reproductive habits rather than evolving toward patterns which have evolved and proven successful in long-term tropical residents (e.g., Neotropical and Oriental Bataguridae, and the Australian genera *Chelodina* and *Elseya*) (Legler, 1985; Moll & Moll, 1990: Kuchling, 1999). As Legler (1985) suggested, perhaps these temperate reproductive adaptations (i.e., type I-like characteristics), have preadapted these species for success in any environment in which their reproductive potential would not be significantly reduced (e.g., tropical environments with distinct wet and dry seasons in which species adapted for reproduction in the temperate summer could successfully reproduce in the latter). We suggest that these same type I-like characteristics in conjunction with other generalized habits pre-adapt these species for successfully colonizing tropical rivers as well, especially where river turtle niches have not been filled by other resident groups (e.g., podocnemids). Available data suggest that extreme type II reproductive patterns are not successful reproductive strategies for living in rivers. Even batagurid river dwellers such as river, painted, and roofed terrapin display much more type I-like reproductive characteristics (e.g., larger egg clutches, seasonal nesting, nest burrows dug and sealed) than their semi-aquatic and terrestrial close relatives which epitomize the type II pattern described above (see Moll & Moll, 1990 for examples of riverine and non-riverine batagurid reproductive characteristics). Viewed from this perspective, the type I-like reproductive patterns characteristic of many tropical river-dwelling species of diverse lineages may be more correctly considered as parallel developments which result from convergent evolution with temperate zone species, and other river dwellers rather than as primitive traits *per se*. Perhaps the greater variability and unpredictability of tropical river ecosystems compared to other more stable tropical environments mirrors the variability and unpredictability of higher latitude ecosystems compared to tropical ecosystems in general. The collective selective forces associated with these more rigorous environments result in the dominance of similar (type I-like) reproductive patterns in their respective chelonian faunas.

It is perhaps instructive to note that tropical riverine batagurids, although more similar to other river-dwelling lineages in reproductive patterns than to other semi-aquatic and terrestrial batagurids, still tend to have lower fecundities and bigger eggs relative to their body sizes than do the other groups (e.g., riverine emydids, trionychids, and podocnemids) (Moll & Moll, 1990). A plausible explanation for this phenomenon would be the former group's subsequent radiation from small, specialized, type II reproducing semi-aquatic batagurid ancestors. Larger size, greater fecundity, and other type I-associated reproductive traits (it is interesting that the only batagurids known to produce leathery [i.e., parchment] shelled eggs rather than typical type II brittle-shelled eggs are the riverine genera *Batagur*, *Callagur*, and *Kachuga*) would necessarily evolve in order to master the river environment (Pritchard & Trebbau, 1984; Moll, 1994), but the genetic constraints imposed by "being batagurid" would continue to limit relative fecundity while maintaining successful traits such as larger egg size (and the presumably greater survivorship of the larger hatchlings which emerge from them). The combination represents a batagurid "twist" on the standard type I prescription for being a successful river turtle (E. Moll, 1980a).

Egg Characteristics, Clutch size and Number of Clutches

All river turtles lay whitish eggs, which may be either spherical to ovate or symmetrically elongated in form. Spherical eggs are usually produced by large-bodied species, which typically lay many eggs in a given clutch (Kuchling, 1999) (fig. 2.7). Examples of riverine forms, which produce spherical (or nearly so) eggs, include the snapping turtles (Chelydridae), the softshells (Trionychidae), the pig-nosed turtle (Carettochelyidae), and the larger side-necks such as the South American river turtle and Madagascan big-headed turtle (*Erymnochelys madagascariensis*). All Emydidae, Batguridae, and Kinosternidae and many of the smaller chelid and podocnemid side-necks produce elongated, symmetrical eggs in clutches which range widely in number.

Chelonian eggs may be leathery or parchment-shelled that indent easily when laid and expand due to water uptake during incubation; harder-shelled but expansible eggs which behave much like parchment eggs during incubation; and brittle-shelled eggs which do not indent when laid and usually take up little or no external water during incubation (Ewert, 1979, 1985; Kuchling, 1999). Vogt et al. (1994) described what may be another type of egg in the big-headed Amazon River turtle (*Peltocephalus dumerilianus*)—a brittle egg when laid, becoming 25% heavier through water uptake and pliable by the 18th day of incubation. Egg shell characteristics may differ between species of different evolutionary lineages (e.g., parchment and hard-expansible in Emydidae versus brittle in Trionychidae), between species with different ecological characteristics within one lineage (e.g., most riverine batagurids and podocnemids produce parchment or hard-expansible eggs while some or all non-riverine forms produce brittle-shelled eggs in these families), and even within a species (e.g., river terrapin have been reported to produce both parchment and hard-expansible eggs) (Loch, 1950; Khan bin

Figure 2.7
Clutch of spherical eggs of the common snapping turtle (*Chelydra serpentina*) with a U.S. $0.25 coin in the foreground for scale.

Momin Khan, 1964; E. Moll, 1980a; Ewert, 1979, 1985; Kuchling, 1999). Generally, parchment eggs develop more rapidly than the other categories (the realized rate is temperature controlled [Páez & Bock, 1998]), and provide a definite advantage for beach- and sand bar-nesting species of flood plain rivers which often have very limited windows of opportunity for the incubation of their clutches before rising water levels inundate and destroy them (Ewert, 1979; Pezzuti & Vogt, 1999; Kuchling, 1999; Pritchard, 2001). The giant South American river turtle, for example lays parchment eggs, which can hatch within only 42–47 days (Kuchling, 1999). The softshells are an exception to the rule. Although often utilizing similar sand beach nesting habitat they produce a fast-developing, brittle-shelled egg (Ewert, 1979, 1985; Kuchling, 1999). The Chinese softshell (*Pelodiscus sinensis*), for example, has the shortest known incubation period of any turtle at 28 days (Kuchling, 1999).

Alternative solutions to the problem of flooding are exemplified by two riverine brittle egg species, which lay in flood periods, the Australian northern long-necked turtle (*Chelodina rugosa*) and the Central American river turtle. The northern long-necked turtle is the only chelonian known to nest and lay its eggs under water (Kennett et al., 1993a,b). It buries its eggs underground in shallow water near the end of the wet season, and embryonic development remains arrested (for as long as twelve weeks under experimental conditions) until water levels recede, the ground dries, and development commences. Hatchlings may not emerge for many weeks until the following wet season, perhaps with the aid of embryonic estivation, when heavy rains and/or rising water levels again soften the dried mud and flood the nest (Kennett et al., 1993a,b). The Central American river turtle nests on land (probably) during the latter part of the fall wet season in northern Belize in locations that may often be subsequently and intermittently flooded (Polisar, 1996). Embryos are able to withstand over thirty days of flooding and, unlike the long-necked turtle embryos which will die if inundated after post-flooding development has began, they are able to survive numerous oscillations of flooding and drying during their incubation period. Polisar (1996) recorded incubation times for Belizean eggs to range between 217 and 300 days, allowing hatchlings to emerge during the subsequent rainy season.

As Moll (1979) noted for turtles in general, egg size, clutch size, and number of clutches in river turtles may be regarded as an adaptive compromise for survival achieved over millions of years. Similarly, certain other chelonian reproductive generalities largely apply to river turtles as well. These generalities include: (1) the observation that (after adjusting for body size) a significant inverse correlation between egg size and latitude, and a positive correlation between clutch size (but not clutch mass) and latitude exists (e.g., Iverson, 1992a; Iverson et al., 1993); and (2) at both the generic and family levels, egg mass and clutch size show positive correlations with adult carapace lengths, but when the effects of body size are removed there is a negative correlation between clutch size and egg mass (e.g., Iverson, 1992a; Iverson et al., 1993; Kuchling, 1999). These conclusions generally support the belief that an evolutionary trade-off is struck between clutch size and egg mass by chelonians (Elgar & Heaphy, 1989; Kuchling, 1999), although evidence for this in common snapping turtles and giant South American river turtles two of the best studied species remains elusive (Iverson et al., 1997; Valenzuela, 2001a). Complicating this picture at the population level, however, is the fact that local environmental conditions may often influence the specific characteristics of reproductive output by affecting resource availability (e.g., Tucker et al., 1998a; Kuchling, 1999).

The specific reproductive characteristics of most river turtle populations are very poorly known. Available data reflect variability influenced by both the evolutionary/genetic heritage of the resident species, and by myriad environmental influences alluded to above. Spherical turtle eggs range from a minimum of about 20 mm in diameter in the Chinese softshell to 40–55 mm in giant South American river turtles (Moll, 1979; Pritchard & Trebbau, 1984). Much larger eggs have occasionally been reported in the latter species (called "oil eggs" by Alho & Padua, 1982), but they do not hatch and their significance is unknown (Pritchard & Trebbau, 1984). Elongated eggs over 70 mm in length are produced by larger batagurids like the river, painted, and red-crowned roofed terrapins. E. Moll (1980a) referred to another large Asian batagurid, the Malaysian giant turtle's (*Orlitia borneensis*) mean egg size as 80 × 43 mm (fig. 2.8). Moll (1979) noted that the relatively modest-sized Central American black turtle (*Rhinoclemmys funerea*) (a batagurid with carapace length usually under 250 mm) produces relatively gigantic eggs, which attain dimensions as great as 76 × 39 mm. Clutch sizes

Figure 2.8
The Malaysian giant turtle (*Orlitia borneensis*), a batagurid turtle that produces very large eggs.

range from as high as 178 eggs in the giant Indian narrow-headed softshell (*Chitra indica*) (Bhadauria et al., 1990) and 150 eggs in the giant South American river turtle (Mittermeier, 1978) to fewer than a dozen in many of the smaller species (see Moll, 1979 for a review). Of the larger river dwellers the batagurids, big-headed Amazon River turtle, and the Central American river turtle typically produce the smallest clutches (often fewer than 20 very large eggs) (Moll, 1979; Moll & Moll, 1990; Polisar, 1996; Vogt et al., 1994). Another surprising example of a large species, superficially similar to *Chitra* but which in contrast to it usually lays much smaller clutches of eggs (20–28 per clutch; Vijaya, 1982a; Rhodin et al., 1993) is the Asian giant softshell. Kar and Rao (1985) identified the latter species as a sea beach-nesting species in India, while *Chitra* nests inland on upriver sand bars (fig. 2.9). Pritchard (2001) suggested that *Chitra* is probably constrained to production of one large clutch of eggs by the ephemeral nature of its nesting habitat, while *Pelochelys* has the "luxury" of spreading out its reproductive effort over multiple, smaller clutches on the more stable sea beach habitat.

In general, larger turtles produce heavier clutches and greater annual clutch mass than do smaller ones (e.g., *P. expansa* versus smaller podochemids; Valenzuela, 2001a), but if relative clutch mass is considered (i.e., clutch mass compared to adult body mass) then smaller species usually produce relatively heavier individual clutches and total annual clutch masses (Iverson, 1992a). The difference between individual clutch masses and total annual clutch masses (both

Figure 2.9
Thai narrow-headed softshell (*Chitra chitra*) sand beach nesting habitat, Mae Klong River System, Thailand. (Photo by Wachira Kitimasak)

absolute and relative) depends, of course, on the fact that many riverine species are known or suspected to produce more than one clutch annually. Because nesting seasons are shorter at higher latitudes, high-latitude species and higher latitude populations of wide-ranging species typically have lower clutch frequencies than lower-latitude species and populations (Kuchling, 1999). Moll (1979) reviewed the estimated number of clutches produced annually for chelonians, including riverine species (which range from 2 to 6) for which data were available, but there are few or none for many species, especially the larger tropical riverine forms. There is evidence that some of these species produce multiple clutches. Kuchling (1999) argued, based upon ovarian analyses, that the Madagascan big-headed side-neck may lay several clutches of eggs in a given year, and Kitimasak (pers. com.) has observed hatchery-confined Thai narrow-headed softshells to lay four clutches annually under semi-natural conditions. E. Moll (1980a) confirmed that Asian river terrapin nested at least twice in one season in Malaysia by capturing three females twice while laying in 1986. In contrast, other studies suggest single annual nesting events for several Neotropical side-neck species including giant South American river turtles, red-headed side-neck turtles (*Podocnemis erythrocephala*), six-tubercled Amazon River turtles (*P. sextuberculata*), and yellow-spotted Amazon River turtles (*P. unifilis*) (Pritchard & Trebbau, 1984; Thorbjarnarson et al., 1993; Junk & Silva, 1997; Vogt & Soini, in press). Pardo (1969) and Ramo (1982) observed the Savanna side-necked turtle (*P. vogli*) species to nest multiple times

(2–4 times) annually. The snapping turtles (*Chelydra* and *Macroclemys*) are confirmed single annual nesters in almost all of North America (Ernst et al., 1994; Tucker & Sloan, 1997), but the former genus is known to lay multiple clutches in south Florida (Ewert, 2000), and possibly throughout its range in the Neotropics (Alvarez del Toro, 1960; Medem, 1977; D. Moll, 1997a). River turtles are usually considered to reproduce annually, but there are very few data available to support this hypothesis for many larger tropical species in particular. Kuchling (1993, 1999) in fact, after a detailed ovarian analysis, concluded that the Madagascan big-headed turtle is a biennial breeder, and Junk and Silva (1997), without presenting supporting evidence, stated that giant South American river turtles lay eggs every 2–4 years. Pritchard and Trebbau (1984) considered the latter to lay one clutch every year, however. Kuchling (1993, 1999) considered those riverine species with the most sea turtle-like reproductive ecologies, including extensive movements to nesting areas (see below) (e.g., the giant South American river turtle) to be most likely to exhibit multi-year reproductive cycles (the norm in sea turtles). Doody et al. (2003) also confirmed that the pig-nosed turtle exhibits biennial reproduction, producing two egg clutches every other year in the wet–dry tropics of northern Australia. As turtles produced larger, heavier, and more eggs per clutch following "big" wet seasons than they did after "small" wet seasons, they concluded that the biennial pattern and within-nesting year reproductive potentials were probably influenced by the low energy content of the dry season diet. Data from Obbard (1983) and Congdon et al. (1987, 1994) suggest that some female common snapping turtles may not nest every year, and Dobie (1971) suggested that some female alligator snappers may skip a year also.

Nesting Movements, Timing, and Sites

Freshwater turtles often move between hibernacula, foraging, and basking sites and, in females at least, to nesting areas within the course of a year, though usually the distances traveled are relatively modest (usually a few kilometers or less) (e.g., Obbard & Brooks, 1980; Graham & Graham, 1997; Plummer et al., 1997), compared to the hundred or thousands of kilometers often traveled by sea turtles. In some river turtles, however, where suitable nesting sites are some distance from foraging areas, the movements are more extensive. A number of river dwellers are known to nest on sea beaches (e.g., painted terrapin, some river terrapin populations), some Costa Rican riverine sliders, some pig-nosed turtles, some Asian giant softshells) and require relatively extensive movements from their river foraging habitats in some cases (Maxwell, 1911; Dunson & Moll, 1980; E. Moll, 1980a,b; Georges & Rose, 1993; D. Moll, 1994). The longest movements known are by the giant South American river turtles and Asian river terrapin. E. Moll (1980a) recaptured a tagged female river terrapin nesting about 80 km upstream from its original capture site at the mouth of Malaysia's Perak River (fig. 2.10). He noted that most suitable nesting sites and female concentrations were between 65 and 96 km upstream from the primarily estuarine foraging habitat of this species in the Perak. Moll also studied the river terrapin's movements

Figure 2.10
Adult female river terrapin (*Batagur baska*) from the Perak River, Malaysia, affixed with radio transmitter (rear of carapace) for movement studies.

and nesting in Malaysia's sandier Terengganu River, where suitable nesting sites were more widespread and located closer to the estuarine foraging habitat (beginning at about 19 km from the estuary and extending for approximately 32 km upriver). Here, the females' movements were more modest and nesting tended to be less concentrated than on the Perak. Distances moved between foraging and nesting areas in the giant South American river turtle are not known, as actual migration has not been observed (Pritchard & Trebbau, 1984), but they may be extensive. These turtles are thought to spread out widely to forage through vast areas of flooded forest associated with the Orinoco and Amazon drainages, while only a limited number of sites are suitable for nesting. Large concentrations of females appear at these sites (or did so before extensive exploitation reduced populations) during the nesting season, presumably from widespread foraging locations. Vogt (in press) radio-tracked post-nesting females in the Rio Trombetas for up to 45 km (in just two days) from the nesting site before their signals were lost after they entered the main channel of the Amazon. In the Orinoco River, tagged females of this species demonstrated extensive movements upstream (as far as about 100 km), and downstream (as far as a few hundred kilometers in some cases) from their nesting site, as well as displaying considerable lateral movement into smaller tributaries (Ojasti, 1967). Sites et al.'s (1999) study of mitochondrial DNA sequences within and between giant South American river turtle populations inhabiting the Araguai and Tapajós river systems in the Brazilian Amazon Basin suggested extensive within system gene flow (up to 275 km), although minor genetic differences between nesting rookeries were observed on the Rio Araguaia. In contrast, they observed very little evidence of gene flow between the two river systems (~2400 km). Another genetic study by Bock et al. (2001) found significant allele frequency differences between South American river turtle populations from Peru and Brazil, as expected from Sites et al.'s (1999) results, but that significant allele frequency differences also occurred between individuals from Brazilian nesting beaches only 80 km apart. They concluded in the latter case that either adults in some populations may not migrate great distances from their nesting areas, or that adults may wander considerable distances but return to breed at their natal beaches in sea turtle fashion. Given the contrasting findings by Sites et al. (1999), and Bock et al. (2001) concerning the extent of within-river gene flow, and the lack of evidence that males accompany females back to their natal beaches for mating during nesting movements (although the possibility was discussed by Pritchard & Trebbau, 1984), it is obvious that more research will be necessary to resolve this issue. Kuchling (1999) has used the issue of accompanying males as a criterion for differentiating between true nesting "migrations" as practiced by sea turtles (see Bowen & Karl, 1997 for a review) and the nesting "movements" characteristic of freshwater turtles. Unless evidence to the contrary is presented, all river turtle nesting travel—including that of Asian river terrapins (E. Moll, 1980a)—must currently be considered "movements" according to Kuchling's definition.

Bock et al. (2001) also examined inter-population genetic differentiation in yellow-spotted Amazon River turtles. This species showed much more inter-population differentiation than did the giant South American river turtles, with demes (reproductive units) separated by as little as 60 km exhibiting as much genetic differentiation as those seen between populations separated by hundreds of kilometers. Bock et al.'s (1998) preliminary radiotelemetry studies also indicated that females' dispersal away from their nesting sites in the non-breeding season was relatively limited.

Timing of nesting is extremely uniform in temperate zone turtles and extremely variable in tropical species. With the exception of a very few species (which may have tropical origins) such as North American chicken turtles (*Deirochelys reticularia*) and Australian broad-shelled turtles (*Chelodina expansa*), neither of which are specialized riverine species (although the latter does enter slow-flowing Australian rivers), almost all nesting activity occurs in the early part of the warm season (late spring and early summer) (Ernst et al., 1994; Cann, 1998; Kuchling, 1999). Although temperature is the ultimate controlling factor influencing the general timing of nesting in temperate turtles (perhaps augmented by photoperiod changes in some species (see Kuchling, 1999 for a review), precipitation patterns and related flooding cycles may play a role in fine-tuning its specific timing in many river species (Tucker 1997; Tucker et al., 1997).

In the tropics the nesting period for riverine species is more often influenced by precipitation patterns than by temperature or season, with considerable variation resulting geographically, between populations, and between species (e.g., Pritchard & Trebbau, 1984; Kuchling, 1999). Vogt (1997) recognized five broad categories of nesting patterns in the neotropics: (1) the rainy season; (2) the end of the rainy season; (3) initiation of the lowering of river depth; (4) lowest river depth; and (5) spring nesters.

The species which nest with the onset or culmination of declining water levels are usually in synchrony with the seasonal availability of specific suitable nesting areas such as clean sand bars, beaches, and islands which are inundated during the wetter times of the year as discussed above (Pritchard & Trebbau, 1984; Vogt, 1997; Kuchling, 1999). The specific proximate cues which trigger nesting activity in these species (and those with other patterns) are very poorly known however. Kuchling (1999) has speculated that pre-nesting sand beach basking behavior may trigger such cues in giant South American river turtles (also recorded in Burmese sea beach nesting river terrapin [Maxwell, 1911], but not in Malaysian populations [E. Moll, 1980a]) by stimulating the final development of the egg clutch for oviposition (fig. 2.11) (see also Vogt [1980a] for the significance of pre-nesting basking behavior and nesting initiation in Mississippi River map turtles).

As discussed previously, clean sandy substrates usually with no or sparse vegetation cover are used as nesting substrates (required by many) in numerous species from a variety of taxonomic groups (table 2.4) (fig. 2.12). These substrates are provided by the bars, beaches, and low islands, which usually become available when spring floods begin to recede in temperate zone rivers and when minimum dry season river levels are attained in the tropics. Some human activities, such as

Figure 2.11
Adult female giant South American river turtles (*Podocnemis expansa*) displaying pre-nesting basking behavior. (Photo by R. A. Mittermeier)

Table 2.4
River turtles known to utilize sand substrate for nesting.

Chelidae	Chelydridae
Chelodina expansa	*Macroclemys temminckii*
Elseya	
dentata	Bataguridae
latisternum	*Batagur baska* (P)
Elusor macrurus (P?)	*Callagur borneoensis* (P)
Rheodytes leukops (P?)	*Kachuga*
	dhongoka (P)
Podocnemidae	*kachuga* (P)
Podocnemis	*smithi*
erythrocephala (P)	*trivittata* (P)
expansa (P)	*tentoria*
sextuberculata (P)	
unifilis	Emydidae
	Graptemys
Dermatemyidae	*barbouri* (P)
Dermatemys mawii	*ernsti* (P)
	geographica
Carettochelyidae	*gibbonsi* (P)
Carettochelys insculpta (P)	*nigrinoda* (P)
	ouachitensis (P)
Trionychidae	*oculifera* (P)
Apalone	*pseudogeographica* (P)
mutica (P)	*pulchra* (P)
spinifera	*Pseudemys concinna*
Amyda cartilaginea	*Trachemys scripta*
Chitra	*venusta* (Caribbean Costa
indica (P)	Rican Population) (P)
chitra (P)	
Pelochelys cantori (P)	
bibroni (P)	
Trionyx triunguis (P)	

(P) indicates those which use sand predominantly for nesting. The list has been assembled mostly from Ernst and Barbour (1989); Ernst et al. (1994); Pritchard (1979); Pritchard and Trebbau (1984); Lindeman (2001b); and the authors' own observations.

dredging and the accumulation of riparian sandy spoil mounds and beaches, may also provide sand substrates used by some nesting turtles such as alligator snapping turtles (Ewert & Jackson, 1994; see also chapter 6, this volume). In other locations and species such spoil banks may not be suitable nesting sites however (e.g., Mississippi River *Graptemys*; Vogt, 1980a). Such substrates heat rapidly (thus enhancing incubation rates), are porous and well-drained, and do not dry into concrete-hard consistencies which might entomb hatchlings within their nest

Figure 2.12
Smooth softshell (*Apalone mutica*) sandy nesting, feeding, and basking habitat on the Mississippi River, Lake Co., Tennessee.

chambers. These nesting sites are particularly important in densely forested areas since the forest substrates (in conjunction with canopy cover) are usually unsuitable for most species. Mass nesting species, such as giant South American river turtles and some river terrapin populations, use these nesting habitats—as do many species which normally nest in smaller groups and as solitary individuals. Where inland sandbars and beaches are unavailable, species or populations which require open sandy areas for nesting will sometimes utilize sea beaches as nest sites (e.g., Costa Rican sliders; Moll, 1994). In any case, because of the ephemeral nature of many of these habitats, nesting usually peaks soon after these areas are exposed. Other species such as the Amazon yellow-spotted side-necked turtle may use sand areas, but also use other substrate types for nesting (Pritchard & Trebbau, 1984; Vogt & Soini, in press). D. Moll (1997a) found dry season nesting Costa Rican snapping turtles (*Chelydra acutirostris*) to only use abandoned agricultural sites with lateritic soils near shorelines for nesting (fig. 2.13).

Rainy season nesting species such as the Indian softshell turtle (*Aspideretes gangeticus*), underwater-nesting, northern long-necked turtle and Central American river turtle do not require sand substrates for nesting either (Kennett et al., 1993a; Polisar, 1996; Vasudevan, 1998). Polisar (1996) found nests of the latter species in Belize in a variety of substrate types, and both with and without vegetation cover. The main requirement was that the placement of the nests of this highly aquatic species was extremely close to the water's edge (in fact, Polisar did not discount the

Figure 2.13
Tropical snapping turtle (*Chelydra acutirostris*) typical nest site—an abandoned agricultural plot or "milpa" on the Tortuguero River, Costa Rica.

possibility of Northern Australian snake-necked turtle [*Chelodina rugosa*]-like underwater nesting in *Dermatemys* as nest construction was never actually observed). The eggs and hatchlings of these species do become encased in hardened mud for long periods of time, a circumstance with which these species cope through having impermeable (brittle) egg shells, and periods of embryonic developmental arrest until subsequent rains and/or floods facilitate the hatchlings' eventual release. The black turtle is another rainy season nesting species, and presumably nests in closed-canopy rain forest in which it covers the eggs with leaves (E. Moll & Legler, 1971; Acuña-Mesen, 1993; D. Moll, 1997b; Acuña-Mesen & Moll, in press) (fig. 2.14). The big-headed Amazon River turtle nests in both open sites and under closed forest canopy in a diverse array of substrates and settings (e.g., sometimes in termite nests and ant mounds) near stream banks (Vogt et al., 1994). All of these species are presumed to be solitary nesters, although few—if any—actual nestings have been witnessed.

Some turtles are known to lay their eggs in alligator nest mounds—a common occurrence for the Florida red-bellied turtle (*Pseudemys nelsoni*) (see Kushlan & Kushlan, 1980; and Ernst et al., 1994 for a review). Cox (1984) has also observed nests of New Guinea giant softshells in freshwater crocodile nest mounds in Papua, New Guinea.

Most river turtles which have been observed nesting nest in a straight-forward turtle-like manner with the actual nest chamber dug with the hind legs (some

Figure 2.14
Typical rain forest river habitat (Tortuguero River, Costa Rica) of the black turtle (*Rhinoclemmys funerea*) and tropical slider (*Trachemys scripta venusta*). Black turtles nest in the surrounding forest, while the sliders swim out to sea to nest on sea beaches.

species such as the river terrapin may first dig a body pit before the egg chamber is constructed), the eggs laid within, and then covered, once again with the hind legs. Some species tamp down the loose soil, capping the nest chamber with their hind legs (e.g., the yellow-spotted Amazon river turtle). The river terrapin compacts the loose soil with its plastron, producing a characteristic sound which carries for some distance and is the source of its onomatopoeic Malay name "tuntong" (E. Moll, 1980a). E. Moll (1980a) suggested the possibility—believed by local egg collectors—that this behavior, which is often produced synchronously by more than one female, may serve a secondary function of attracting other females to a suitable nesting area. River terrapin may occasionally divide a clutch of eggs into more than one nest (up to three observed) in one night (E. Moll, 1980a). Fitch and Plummer (1975) suggested that smooth softshells might also sometimes divide their clutch into more than one nest. A related, odd nesting behavior is practiced by river cooters (*Pseudemys concinna*), which dig three-holed nests—one main egg chamber and two adjacent shallow-false nests (often left uncovered)—in which up to two eggs may be laid (Ashton & Ashton, 1985). Nesting activity usually takes place at night in colonial nesting species, but may occur in daylight (usually near dawn or dusk) in others. Some species, such as a number of Australian chelids, are stimulated to nest by rainy weather (Kuchling, 1999).

While genotypic sex determination (GSD) is the rule in most vertebrates (i.e., sex determined by maternal and paternal sex chromosomes), in most turtle species sex is dependent upon nest temperatures at critical periods during incubation (TSD). While many species remain to be examined, TSD seems to be the most common form of sex determination in river turtles. The trionychids, the subfamily *Staurotypini* of the Kinosternidae, probably all the Australian chelid turtles and certainly the common snake-necked (*Chelodina longicollis*) and Murray River turtles (*Emydura macquarii*), the American wood turtle (*Clemmys insculpta*) and South America's Geoffroy's side-necked turtle (*Phrynops geoffroanus*; another chelid), are known exceptions as GSD species (Vogt 1992; Kuchling, 1999) (Vogt et al., [1994] thought the big-headed Amazon River turtle was a likely GSD species based upon its often-shaded nesting sites within the rain forest, but this remains to be verified).

Two general patterns of TSD are known in turtles: one in which relatively low temperatures produce males and higher temperatures produce females (TSDIa); and a second one in which cool temperatures produce females, intermediate temperatures produce males, and warmer temperatures again produce females (TSDII) (Kuchling, 1999; Valenzuela, 2001a,b). The first pattern occurs mainly in sexually dimorphic species where adult females average larger sizes than males (i.e., most river turtles), and the second where males are larger or where dimorphism does not occur (e.g., bottom-walking forms like the snapping turtles) (Kuchling, 1999). Some river turtles, such as the giant South American river turtle and the red-headed side-necked Amazon river turtle may actually display the TSDII condition under laboratory conditions when only exposed to lower pivotal temperatures briefly during the thermosensitive period, but are functionally TSDIa species in nature as prolonged exposure to low enough temperatures to produce (cool) females are usually lethal (Valenzuela, 2001b). The thermosensitive period (i.e., the time during incubation when temperature characteristics actually determine sex) may be quite long in some riverine species like the Amazon yellow-spotted side-neck (Souza & Vogt, 1994), or quite short and specific to a particular time during incubation, as in the giant South American river turtle (Valenzuela et al., 1997; Kuchling, 1999; Valenzuela, 2001b). Valenzuela (2001b) found giant South American river turtles to have the highest pivotal temperature (the temperature which determines sex during the thermosensitive period), at 32.6°C, of any turtle yet examined. Souza and Vogt (1994) found yellow-spotted Amazon river turtles to have pivotal temperatures nearly as high at between 31°C and 32°C. Hatchling sex ratios in TSD species are therefore, determined by a complicated mixture of factors which affect nest temperatures such as female size and resulting nest depth, degree of shading, and nest orientation (i.e., north-facing slope versus south-facing, etc.), and some studies also suggest the possible influence of nesting female site choice in this regard (e.g., Janzen, 1994; Janzen et al., 1995; Roosenburg & Kelley, 1996; Tucker et al., 1998c; Kolbe & Janzen, 2002). Natural nests as a result of these variables may produce either all females, all males, or some mix dependent upon the temperature gradient from top to bottom in the nest chamber. Valenzuela (2001a) discussed the importance of female *P. expansa* population size structure in influencing the sex ratio of hatchling sex ratios. As larger females dig deeper, cooler nests they contribute more males to the population than do smaller females which dig shallower, warmer nests. For more information on this subject the reader is referred to the following publications: Bull & Vogt, 1979; Bull, 1980; Vogt & Bull, 1982, 1984; Alho et al., 1984, 1985; Vogt & Flores-Villela, 1986, 1992a; Paukstis & Janzen, 1990; Ewert & Nelson, 1991; Janzen & Paukstis, 1991; Vogt, 1994; and Kuchling, 1999.

Another population characteristic which predominates among reptiles with TSD is the prevalence of female-biased hatchling, and (probably) adult sex ratios (Bull & Charnov, 1988; Ewert & Nelson, 1991). This has been recorded to be the case in some river turtle populations (e.g., *Graptemys* in the upper Missouri River [Timken, 1968] and upper Mississippi River [Vogt, 1980a] and in Florida's Apalachicola River alligator snapping turtles [Ewert & Jackson, 1994]), but other studies have recorded 1:1 sex ratios (Gibbons, 1970a), and still others have reported male-biased sex ratios (e.g., White River, Arkanasas *Graptemys*, Vogt, 1980a, and Perak River, Malaysia river terrapins, E. Moll, 1980a). Gibbons (1990c) has listed four demographic factors which influence sex ratios within a turtle population. These include the sex ratio of hatchlings, differential mortality of the sexes, differential emigration and immigration rates of the sexes, and differences in age at maturity of the sexes. He considered the last to be the most probable cause of biased adult sex ratios in many turtle populations (see Gibbons, 1990, 1990c and Ernst et al., 1994 for reviews of sex ratios of species for which data are available). Clearly this issue requires further investigation.

Hatchling Ecology

While our overall knowledge of river turtle life histories is poor, we probably know less about hatchlings than any other phase. A case in point would be the behavior of neonates after hatching and before emergence from the nest chamber onto the surface. The process of accomplishing mass emergence from the nest chamber through proto-cooperative, collective digging and thrashing toward the surface (perhaps in response to a temperature gradient [Bleakney, 1963; Depari, 1996]) has been described in sea turtles (Carr & Hirth, 1961), but it is not known if similar phenomena are involved in the emergence of most river turtle offspring. Alho and Padua (1982) did observe that South American river turtle hatchlings might emerge from their eggs in two waves, with about 60% of the group hatching about three days before the rest of the clutch. The first group waits near the surface of the nest until joined by the later-hatching group before explosive nocturnal emergence of the entire group. Yellow-spotted Amazon River turtle hatchlings may remain in the egg for up to a week after pipping the shell, and may remain in the nest for two weeks or more before emerging to the surface (Soini, 1995; Paez & Bock, 1998). As in sea turtles, it is often advantageous for river turtle hatchlings to emerge at night, both for protection from diurnal predators and from the heating and desiccating effects of the sun. The actual emergence may be triggered by a drop in temperature at the nest's surface layers with the onset of darkness (Miller, 1985). Conversely, Tucker (1999) found that hatchling sliders which had overwintered in west central Illinois nests emerged when soil temperature gradients become warmer at the top of the nest cavity than at the bottom, and when external environmental temperatures were generally on the rise. Rainfall and nest flooding may also stimulate hatching and emergence in some species, regardless of time of day (Moll & Legler, 1971). Those riverine species that nest in ephemeral habitats (e.g., sand bar nesters) usually emerge as fast as developmental rates allow, as do those which live in climatically stable and benign locations (i.e., where there is no advantage obtained by delaying emergence). Conversely, the survivorship of the hatchlings of many species depends on emergence under specific environmental circumstances, and dormancy in the nest in some developmental stage (probably coupled with some final environmental triggers to initiate emergence under appropriate conditions) may prolong the period within the nest chamber considerably (Kennett et al., 1993a,b; Polisar, 1996). Ewert and Wilson (1996) describe the four types of embryonic developmental arrest and environmental factors which terminate the arrest, allowing active development to resume in turtles. As it behooves many tropical turtles to emerge during the rainy season, species such as the northern long-necked turtle and the Central American river turtle which nest during the previous rainy season must be prepared to employ one or more of these mechanisms to ensure that survival in the intervening weeks is possible, and that premature dry season emergence does not occur (the latter also enforced by dried hardened mud around the egg chamber before subsequent rain and flooding soften it). In the higher latitudes of both hemispheres hatchlings of some species may over-winter in the nest, emerging after temperatures warm the following spring, while others emerge in late summer, and spend the winter in water (e.g., Gibbons & Nelson, 1978; Clay, 1981; Obbard and Brooks, 1981; Packard & Packard, 1997; Tucker & Packard, 1998; Kuchling, 1999). One of the former species which may enter slower rivers and backwaters, the painted turtle (*Chrysemys picta*), is able to remain in the nest through the winter by the hatchling's ability to sustain a state of super-cooling, and prevent freezing due to the integumentary barrier that prevents the penetration of ice into body compartments from frozen soil (Packard & Packard, 1995; Packard et al., 1997a,b).

Once hatchlings do emerge from the nest they undoubtedly face a dangerous world of predators and unfavorable environmental conditions, but only a few studies have been conducted to provide details of neonate turtles once they have left their nests.

Hatchling size varies among different taxonomic groups (e.g., batagurid hatchlings hatch from relatively larger eggs and are usually larger than the hatchlings of similar-sized emydid species), and is also often positively influenced by female size (e.g., Tucker & Moll, 1997; Valenzuela, 2001a) and hydric conditions of the nest environment in some species (e.g., more moist substrates usually produce larger hatchlings) (Packard, 1999). Wetter substrates may also reduce the amount of residual yolk remaining, provide greater hydration of tissues, and increase locomotor performance of hatchlings produced in such substrates (Miller et al., 1987; Packard et al., 1988; Packard, 1999; Finkler et al., 2000).

A number of studies have indicated that larger hatchlings may have greater survivorship than smaller hatchlings (Janzen, et al., 2000a,b; Tucker & Paukstis, 1999; Tucker, 2000a,b), but there may be circumstances where greater size provides no survivorship advantage (e.g., Congdon et al., 1999; Filoramo & Janzen, 2002). Controlled experiments with sliders suggest that greater size and corresponding enhanced locomotor ability make it more likely that the hatchlings will escape predation by bird predators in the hatchlings' movement from nest to water after emergence (Janzen et al., 2000a,b). However, in drier conditions, hatchlings may bury themselves during the trip, and these advantages against predators may be negated as they more erratically move toward the water (Filoramo & Janzen, 2002).

The hatchlings of most species—especially those which nest near the water's edge—are assumed to enter the water directly after emergence (Cahn, 1937; Anderson, 1958; Butler & Graham, 1995; Burke et al., 2000). Differences in light intensities over water and land are used as an ocean-finding guide by hatchling sea turtles (Witherington, 1995), but whether similar or other directional cues are used by river turtles is not yet clear. Some turtles which nest further away from water, such as Blanding's turtles (*Emydoidea blandingii*), a species which may sometimes enter river sloughs, may use olfactory cues, (including scent trails of other hatchlings) to find their way to wetlands after emergence from the nest (Butler & Graham, 1995). Hatchlings of some species are known to spend a variable amount of time on land before assuming an aquatic existence. Moll and Legler (1971) observed that slider hatchlings in Panama probably spend as long as a month hidden in terrestrial ground cover before entering aquatic habitats. Jansen (1993) found that black turtles in Caribbean Costa Rica (river dwellers as adults) spent the first year of life in low second growth vegetation and swampy areas before entering sloughs and backwaters at about 80–90 mm carapace lengths, and eventually fluviatile portions of rivers between 104 and 160 mm in carapace length.

Several studies have suggested that cryptic behavioral patterns, coloration, and utilization of specific microhabitats are important components of the survival kit of hatchlings, both in travel to the water, and in their aquatic habitat choice (e.g., immobility, fossorial behavior, color patterns which blend into surroundings, residence in dense mats of vegetation) (Moll & Legler, 1971; Burger, 1976; Hart, 1983; D. Moll, 1990; Lovich et al., 1991; Congdon et al., 1992; Janzen et al., 1992; Pappas & Brecke, 1992; Ross & Lovich, 1992; Butler & Graham, 1985). We personally have observed a striking example of cryptic behavior, coloration, and unusual microhabitat utilized by hatchling snapping turtles in Wisconsin. The black-colored hatchlings burrow into the black mud filling in the sides of beaver dams at or just below water level, rendering them nearly invisible.

Another interesting, but essentially unknown phase of hatchling ecology involves those riverine species, which nest on sea beaches (e.g., painted terrapin, Dunson & Moll, 1980; E. Moll, 1980b; river terrapin, Maxwell, 1911; E. Moll,

1980a; pig-nosed turtles, Georges & Rose, 1993; Costa Rican sliders, Moll, 1994). How do their hatchlings, apparently ill-equipped for sea travel, return to their home rivers? Dunson and Moll (1980) thought it likely that painted terrapin hatchlings from sea beach nests probably enter the sea and follow declining salinity gradients from their hatching sites to the river mouth from which their mother probably emerged into the sea to swim to the coastal nesting site (a distance of up to several kilometers). Moll (1994) thought it more likely that Costa Rican slider hatchlings from beach nests entered the swampy, forested peninsula separating the sea beaches from the Torguguro River where their parent population resides—entering the river itself after some unknown period of time spent inhabiting the interior swamps (fig. 2.15). The lack of evidence of sea travel in hatchlings from this population, and the documented presence and growth patterns of hatchlings from the peninsular wetlands lends support for this hypothesis (Moll, 1994; Tucker et al., 1999). Perhaps different survival strategies suited to the particular populations and their ecological circumstances are employed to cope with the unusual requirements of sea beach nesting by freshwater species.

While there are many predators of hatchling and small juvenile turtles (see the next sections for more discussion of this subject), studies by Semlitsch and Gibbons (1989) and Britson and Gutzke (1993) suggest that some hatchling turtles are able to defend themselves by violently scratching and thrashing about when ingested by predatory fish. This behavior may cause the fish to spit them out in order to prevent

Figure 2.15

A hatchling tropical slider (*Trachemys scripta venusta*) in the cocoplum nesting habitat on Tortuguero Beach, Costa Rica.

Figure 2.16
Colorful plastron markings of a juvenile mata mata turtle (*Chelus fimbriatus*) suggest a possible aposematic (warning) function for hatchlings and juveniles in this species. (Photo by Lynn W. Robbins). (See color insert.)

damage to its gills, and it learns to avoid future encounters. The bright colors of hatchlings of some species may serve an aposematic (i.e., warning) function, especially those in which their colors fade as they age and grow (e.g., the pink plastrons of hatchling loggerhead musk turtles, the yellow and black plastrons of young sliders, the salmon pink and black plastrons of hatchling western painted turtles (*C. p. bellii*), or the bright red, pink and blacks of the plastron and throat of hatchling mata mata turtles) (fig. 2.16).

The Roles of Turtles in Riverine Food Webs

While there is a rapidly expanding body of herpetological literature which suggests that turtles are major components of riverine food webs, playing important roles in such vital functions as energy flow, nutrient cycling, dispersal of riparian vegetation, and maintenance of water quality, recognition of their value has been surprisingly slow to be incorporated into the general limnological literature. Even comprehensive treatments of river ecology, wetlands, and limnology (e.g., Mills et al., 1966; Hynes, 1970; Welcomme, 1979; Lock & Williams, 1981; Goldman & Horne, 1983; Payne, 1986; Maitland, 1990; Cole, 1994; Horne & Goldman, 1994) mention turtles only in passing, if they are mentioned at all. Goldman and Horne

(1983) even included a section concerning aquatic reptiles, which failed to mention turtles (only basilisk lizards, water snakes, and limnologist-attacking crocodilians were briefly discussed). Ironically, a recent treatment of wetland ecology (Keddy, 2000) even pictured a drawing of a common snapping turtle on its cover, but refers to turtles (species unspecified in text, but identifiable as referring to common snapping turtles from the listed reference) only once (as wetland consumers) in its 614 pages. Starretts' (1972) account of the effects of human activity on the Illinois River and its fauna, including turtles, and Odum's (1957) inclusion of turtles in a study of trophic structure and energy flow in a Florida riverine ecosystem exemplify notable exceptions to their general neglect. Our objective in emphasizing this state of affairs is not to denigrate these otherwise valuable works, but to underscore the necessity of including turtles in any consideration of aquatic ecosystems and their functional dynamics, for to neglect them will inevitably result in a less than complete understanding of the ecosystem in question.

Turtles are seldom as diverse as fishes in the rivers they share, but their capacity to attain very high densities and biomass in both lentic and lotic ecosystems (although particularly difficult to quantify in larger lotic systems—Vogt and Villareal-Benitez (1993) suggest their substantial contribution to energy flow and nutrient cycling patterns in these ecosystems (e.g., Odum 1957; Chessman, 1978; Iverson, 1982; Congdon et al., 1986; Congdon & Gibbons, 1989; D. Moll, 1990; Meylan et al., 1992; Thompson, 1993; Vogt & Villareal-Benitez, 1997; Moll & Moll, 2000; Souza & Abe, 2000). In one example of the extremely high biomass and density that a single turtle species can attain in rivers, Kennett estimated that the northern Australian snapping turtle contributed 105–170 kg/ha (from 49 to 78 turtles/ha or 19 to 31 turtles/100 m of river channel) in some northern Australian rivers (in Woinarski, 1993). Souza and Abe (1997) estimated the density and biomass of *Hydromedusa maximiliani* within their actual Brazilian stream microhabitats at 193.5 turtles/ha. and 41.6 kg/ha., respectively. In another study, Souza and Abe (2000) reported 170–230 Geoffroy's side-necked turtles/ha, or 255–345 kg/ha, in a badly polluted river in Brazil. Ewert and Jackson (1994) estimated the density of the alligator snapping turtle, a specialized predator, at around 18.4 adults per km^2 of open water in prime habitat of the lower Appalachicola River. Thompson (1993) published estimates of the density, biomass, and contribution to energy flow of a three species river turtle community (composed of the Murray River turtle, common snake-necked turtle, and giant snake-necked turtle [*Chelodina expansa*]) in Australia's Murray River. Using his own data and those of Chessman (1978), Thompson projected that the densities of the first two species ranged between 4 and 159 turtles per hectare and the rarer giant snake-necked turtle at only 1 per hectare. Using average body mass estimates, he then estimated that all three species collectively might contribute as low as 3050 metric tonnes of biomass or as high as a staggering 230,000 metric tonnes of biomass to the Murray's total standing crop (Thompson's own preference for the actual total turtle biomass in the Murray was between 10,000 and 100,000 metric tones). Based upon his preferred estimates, Thompson calculated that standard metabolic energy consumption of the Murray's turtles would be between 13.5×10^7 and 13.5×10^9 kJ per day (Thompson, 1993). Thompson is to be commended for his groundbreaking work in initiating investigations designed to quantify the density, biomass, and energetic contributions of an entire river turtle community. While more research will be necessary to refine his rough estimates, his contributions suggest that turtles are more than just irrelevant, archaic curiosities in their home ecosystems. Even if lower-end values in Thompson's range of estimates are eventually confirmed to be most accurate (and especially if higher), then river turtles should more accurately be considered as keystone species in food webs and in energy flow and nutrient cycling processes in lotic ecosystems. Souza and Abe (2000) and Akani et al. (2001) arrived at similar conclusions based on the key roles that river turtles play in rivers they studied in Brazil and Nigeria, respectively.

River turtles display a considerable range of foraging patterns and are well adapted for filling their various trophic roles in riverine ecosystems [see Ernst and Barbour (1989) and Ernst et al. (1994) for an overview of world and North American turtle diets]. The specific diets of riverine species are strikingly variable, and their composition may be influenced by ontogeny, geographical or seasonal availability of foods, microhabitat usage patterns by different species, sexes and/or sizes classes, sexual dimorphism, and female reproductive requirements, abundance of a resource in relation to other resources, competitive interactions, and perhaps individual preferences (e.g., Clark & Gibbons, 1969; Moll & Legler, 1971; Berry, 1975; Hart, 1983; D. Moll, 1976b, 1980, 1985, 1989, 1990; E. Moll, 1980a,b; Parmenter, 1980; Vogt, 1980a, 1981a; Plummer & Farrar, 1981; Georges, 1982; Das, 1991, 1995; Vogt & Guzman-Guzman, 1988; Parmenter & Avery, 1990; Fachín-Terán et al., 1995; Kennett & Tory, 1996; de Lima et al., 1997; Luiselli, 1998; Souza & Abe, 1995, 1998, 2000).

Many of the medium to larger-sized hard-shelled river turtles are herbivorous or predominantly so as adults (e.g., black turtle, Moll, 1997b; Northern Australian snapping turtles [*Elseya dentata*]; Kennett & Tory, 1996; river terrapin and painted terrapin; Moll, 1980a,b; Central American river turtle; Moll, 1989; giant South American river turtle and yellow-spotted Amazon River turtle; Fachín-Terán et al., 1995; river cooter; Fahey, 1987; Buhlmann & Vaughn, 1991; Turner, 1995). One small stream-dwelling species, the red Amazon side-necked turtle (*Phrynops rufipes*) is apparently a confirmed frugivore throughout its rain forest range, subsisting principally on palm fruits, which fall into the water (de Lima et al., 1997). While many species are more carnivorous or omnivorous as juveniles and become more herbivorous as adults (e.g., false map turtles, Moll, 1976b; sliders, Hart, 1983; D. Moll, 1990) some specialized herbivores feed exclusively on plant material throughout life (e.g., Central American river turtle, Vogt & Flores-Villela, 1992b; Vogt et al., in press; giant South American river turtle, Fachín-Terán et al., 1995; some river cooter populations, Fahey, 1987; Turner, 1995). The Central American river turtle and the river cooter (also a habitat generalist species, the Florida red-bellied turtle [*Pseudemys nelsoni*]) have been shown to have symbiotic associations with gut microorganisms, which assist in the breakdown of plant structure (Bjorndal & Bolten, 1990, 1992; Vogt & Flores-Villela, 1992b;

Thomas et al., 1994). Some of the herbivores have serrated jaws and large alveolar plates with serrated ridges that enhance their ability to crop vegetation (e.g., the river terrapin, painted terrapin, roofed turtles, Central American river turtle, and crowned river turtle [*Hardella thurjii*]; Moll, 1980a,b; Obst, 1986; Davenport et al., 1992) (fig. 2.17). River terrapin have two rows of such serrations on the upper jaw and one on the lower jaw which assist the turtles in both severing and maintaining contact with elongated plant material being consumed, while depression of the buccal cavity helps to suck detached pieces into the mouth and esophagus (Davenport et al., 1992). River turtles feed on almost every kind of vegetation they can access including stems and leaves of submerged, floating and emergent vegetation, filamentous algae, inundated terrestrial vegetation during flood periods, terrestrial riparian river bank vegetation which is accessible from the water (black turtles emerge from the water to graze upon shoreline beds of *Ipomoea* and *Dieffenbachia*), floating and sunken leaves of riparian and estuarine (i.e., mangroves) trees and shrubs, and fruits, seeds, and leaves which fall into the water (e.g., Brimley, 1942–1943; Webb, 1961; Legler, 1976; Moll, 1976b, 1989, 1997a,b; Smith, 1979; Georges & Rose, 1993; Fachín-Terán et al., 1995; Moll & Jansen, 1995; Turner, 1995; Kennett & Tory, 1996; Bjorndal et al., 1997; de Lima et al., 1997; Magnusson et al., 1997) (fig. 2.18). A few species of river turtles are known regularly to eat water hyacinths (*Eichhornia* spp.) (Indian tent turtle,

Figure 2.17
An adult Central American river turtle (*Dermatemys mawii*) from the Belize River, Belize. This species is a specialized herbivore throughout its life.

Figure 2.18
The black turtle (*Rinoclemmys funerea*) is a Neotropical herbivore which eats fruit and seeds that fall into the water, aquatic vegetation, and riparian vegetation upon which it grazes terrestrially. It may play a role in seed dispersal of riparian vegetation.

Varghese & Tonapi, 1986; river terrapin, Davenport et al., 1992; yellow-spotted Amazon River turtles, Fachín-Terán et al., 1995). *Eichhornia* has also been recorded in the diet of river cooters (Hart, 1979) (fig. 2.19). As this is an introduced pest plant species which clogs subtropical and tropical waterways all over the world, humans may benefit from the turtles' dietary preference for this plant. Our personal observations suggest that a trend toward herbivory is particularly prevalent in tropical river turtle species.

Although numerous riverine turtles are morphologically and behaviorally adapted for carnivory, it is interesting how few species are known which do not at least occasionally ingest vegetable matter. Examples of forms which are believed to be exclusively carnivorous include a number of the softshelled turtles, especially the giant fish-eating forms in the genera *Chitra* and *Pelochelys*, the long-necked Australian side-necked turtles of the genus *Chelodina*, (with the exception of the recently described omnivorous form *C. burrungandjii* [Thomson et al., 2000]) the Fitzroy River turtle (*Rheodytes leukops*), Maximillian's snake-necked turtle (*Hydromedusa maximiliani*), and many of the map turtles (Legler, 1978; Ernst & Barbour, 1989; Legler & Georges, 1993b; Ernst et al., 1994; Souza & Abe, 1995, 1998; Pritchard, 2001). Totally or mainly carnivorous species eat an extremely wide array of animal foods including zooplankton, sponges, neuston, aquatic insect larvae [including mosquito larvae], nymphs and pupae, drowned terrestrial

Figure 2.19
The river terrapin (*Batagur baska*) is known to eat water hyacinth (*Eichhornia crassipes*), an aquatic pest species which clogs waterways in tropical and subtropical waterways around the world.

insects, aquatic annelids, a range of crustaceans from very small planktonic forms and small shrimp through large crayfish and crabs, gastropod and bivalve mollusks, barnacles, essentially any fishes which can be caught and overpowered, amphibians, turtles and other reptiles, young waterfowl, and other birds which may be ambushed in the water or at the water's edge. Even such large mammals as Nilgai antelope may occasionally be taken, as discussed earlier in this chapter (e.g., Penn, 1950; Knight & Gibbons, 1968; Legler, 1976; D. Moll, 1976b, 1980, 1985, 1990; Shealy, 1976, E. Moll, 1980a; Vogt, 1981a; Cochran & McConville, 1983; Pritchard, 1989; Das, 1991, 1995; White & Moll, 1992; Davenport et al., 1992; Legler & Georges, 1993b; Ernst et al., 1994; A. Tucker et al., 1995; Kennett & Tory, 1996; de Lima et al., 1997; Souza & Abe, 1998; Boycott & Bourquin, 2000; Singh, 2000; Souza & Abe, 2000; Akani et al., 2001; Pritchard, 2001).

Among all the animal prey listed above, a perusal of the literature suggests that small invertebrates—especially aquatic insect larvae and pupae, small shrimp-like crustaceans, snails and clams, and fish (caught live by piscivorous specialists or as carrion as discussed below)—are the most important prey staples for the greatest number of carnivorous (and omnivorous) river dwellers around the world.

Considerable morphological specialization is required for carnivorous species, which specialize on fast-moving, potentially large prey (e.g., fish, large shrimp) or heavily armored prey (e.g., mollusks, large crustaceans). Both require considerable restructuring of the anterior anatomy (i.e., neck, skull, and jaws), the specifics of which depends upon the particular feeding strategy employed (see below) (e.g., Dalrymple, 1977; Obst, 1986; Pritchard, 1984; Lindeman & Sharkey, 2001). The giant softshells and the snake-necked chelid side-necks (*Chelodina* and *Hydromedusa*) for example have very long necks, which can be extended with incredible speed to snatch mobile prey (e.g., Pritchard, 1984, 2001).

Those species (e.g., most softshells, but probably not *Chitra*; Pritchard, 2001), which catch and hold the prey in the jaws (as opposed to "gape and suck" feeders which bring the prey inside the mouth whole) have extremely sharp jaw coverings which move in scissor-like fashion to dispatch their catch (Obst, 1986). The "gape and suck" feeders referred to above (e.g., the mata mata and probably the *Chitra* softshells) do not bite their prey, but use their jaws only to confine prey within the mouth, and therefore have more weakly developed jaws (Pritchard, 1984) (fig. 2.20). Some species like the alligator snapping turtle and big-headed Amazon River turtle have well-developed, raptor-like hooked beaks, which are used to more efficiently secure large, struggling prey (Obst, 1986). Durophagous forms (i.e., consumers of hard-shelled prey) are characterized by very broad alveolar surfaces on the jaws, which are used to crush the shells of their prey. The force necessary for this function is provided by relatively huge jaw muscles which give the heads of durophagous species (and of the large durophagous females in some sexually

Figure 2.20
The mata mata turtle (*Chelus fimbriatus*), a "gape and suck" feeder of Neotropical river swamp habitats. (Photo by Lynn W. Robbins)

(a)

(b)

Figure 2.3
(a) Adult male Central American river turtle (*Dermatemys mawii*) from Belize displaying typical sexual dichromatism. Males in different river systems display dorsal head coloration ranging from tan to day-glo orange while females' are always dark gray.
(b) Adult female Central American river turtle (*Dermatemys mawii*) from Belize displaying typical sexual dichromatism.

(a)

(b)

Figure 2.4
(a) Adult male river terrapin (*Batagur baska*) in breeding coloration. (b) Adult male painted terrapin (*Callagur borneoensis*) in breeding coloration.

(c)

Figure 2.4 (*cont.*)
(c) Adult male red-crowned roofed terrapin (*Kachuga kachuga*) in breeding coloration.

Figure 2.16
Colorful plastron markings of a juvenile mata mata turtle (*Chelus fimbriatus*) suggest a possible aposematic (warning) function for hatchlings and juveniles in this species. (Photo by Lynn W. Robbins.)

dimorphic species) a distinctive, swollen appearance termed "megalocephaly" or "macrocephaly" (e.g., female Barbour's map turtle [*Graptemys barbouri*] and female [*Emydura*] side-necks, especially from western and northern Australian populations [Ernst et al., 1994; Cann, 1998]). Cann and Legler (1994) observed a different form of bivalve feeding behavior in the Mary River turtles, a species which is not morphologically specialized for mollusk feeding. These turtles arrange the bivalve in the jaws and eventually damage the animal within enough by raking along, and within the occlusional edges of the shell to force its "relaxation," and subsequent consumption by the turtle.

Some of the adaptations for effective predation on the focus species (i.e., fish and mollusks) described above probably serve as preadaptations which allow some of the larger, more powerful species to supplement their regular diet with other turtles on occasion. We are aware of several river-dwelling species which may consume other turtles, the alligator snapping turtle (Pritchard, 1984, 1989), the Mexican giant musk turtle (D. Moll, 1990); the big-headed Amazon River turtle (Perez-Emán & Paolillo O, 1997); the Indian softshelled turtle and the Indian flapshell turtle (*Lissemys punctata*) (Das, 1991, 1995); and the Nile softshell (Akani et al., 2001) (fig. 2.21). The extent of cheloniophagy in these species is not well known with the possible exception of the alligator snapping turtle.

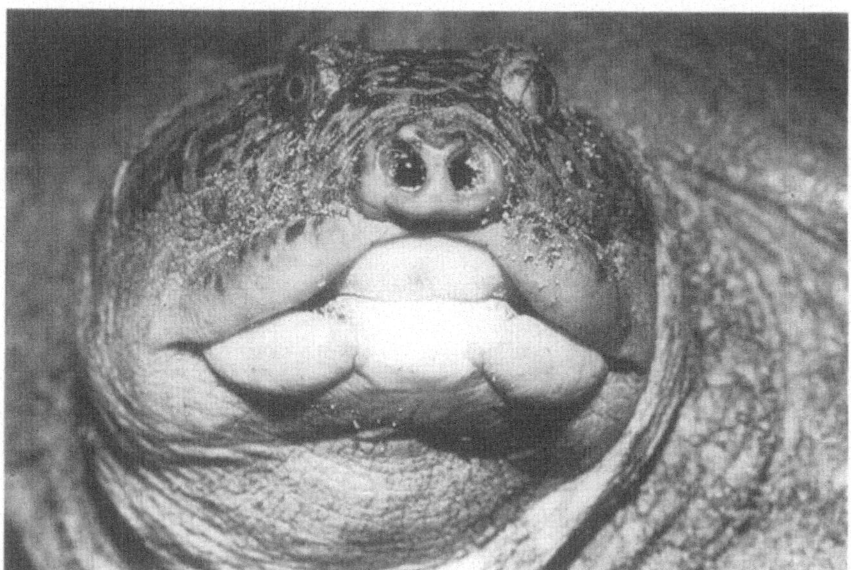

Figure 2.21
The Indian softshelled turtle (*Aspideretes gangeticus*) is an important scavenger and cheloniophage with a widespread distribution across the Indian subcontinent. (Photo by Indraneil Das)

Pritchard (1989) reported the presence of several genera of smaller turtles in the diets of alligator snapping turtles from widespread locations in the southern United States, and he considered the loggerhead musk turtle (*Sternotherus minor*) to be more of a staple food, rather than an incidental dietary item in some locations. The crevice-using behavior and ability of musk turtles to detect alligator snapping turtle's presence from chemosensory cues suggests a long and intense predator–prey relationship between these turtles (Jackson, 1988; 1990).

Completely herbivorous and carnivorous river turtles are relatively few. They represent the extremes of a continuum which grade into the largest dietary category displayed by river turtles, opportunistic omnivory. The actual proportions of animal versus plant foods in the diet shift constantly in most species in response to the many influential factors listed earlier in this section (e.g., ontogeny, shifts in available resources geographically, seasonally, and with changes in water quality, competitive interactions, etc.).

Even some species highly adapted for herbivory as described above may sometimes eat animal foods (e.g., river terrapin will eat mollusks and other invertebrates, fish or carrion as well as plants [E. Moll, 1978a, 1980a], the frugivorous red Amazon side-necked turtle will eat lizards which presumably enter streams to escape terrestrial predators [de Lima et al., 1997]), and morphologically specialized carnivores (e.g., some large softshells, common snapping turtles, alligator snapping turtles, big-headed Amazon River turtles) may eat considerable amounts of plant material (e.g., Lagler, 1943; Loveridge & Williams, 1957; Budhabhatti & Moll, 1990; Sloan et al., 1996; Perez-Emán & Paolillo O, 1997). While accidental ingestion of either category may often occur incidentally during foraging, the large amounts and regularity of occurrence of animal or plant material in the guts of many species specialized for the other indicates voluntary feeding.

Some riverine herbivores and omnivores (e.g., northern Australian snapping turtles, black turtles, river terrapin, giant South American river turtles, yellow-spotted Amazon river turtles, Mexican giant musk turtles, alligator snapping turtles, big-headed Amazon River turtles, pig-nosed turtles, red Amazon side-necked turtle, and Nile softshell) are known to eat fruits and seeds, and are considered as probable seed dispersers for many terrestrial riparian plant species which contribute fruits, seeds, and nuts to their diets (Loveridge & Williams, 1957; E. Moll, 1980a; Vogt & Guzman-Guzman, 1988; Georges & Rose, 1993; Kennett & Russell-Smith, 1993; Kubitzki & Ziburski, 1994; Moll & Jansen, 1995; Kennett & Tory, 1996; Sloan et al., 1996, Magnusson et al., 1997; Perez-Emán & Paolillo O, 1997; Akani et al., 2001; Vogt, in press) (fig. 2.22).

River turtles display a number of interesting behavioral patterns, which allow them effectively to exploit and capture the particular food items they seek. In the most generalized forms, such as some of the short-necked Australian chelid side-necks (*Emydura*), most food is obtained by slowly prowling along the river bottom, often poking the snout into vegetation or under stones where prey may hide. Both visual and olfactory cues are probably used to locate the prey. When food is encountered it is simply swallowed whole if small enough, or torn apart by the jaws and forelimb claws to render the item into pieces small enough to ingest.

Figure 2.22
The Mexican giant musk turtle (*Staurotypus triporcatus*), a Central American omnivore, cheloniophage, and possible seed disperser.

Moderately large food items are ingested by inertial feeding, where the jaws grasp the food item, and then release it quickly, using prey inertia to repeatedly grasp the prey item further along its body until it is swallowed (Gans, 1961). Faster species, such as the more generalized softshells, probably also catch prey by swimming directly toward it, and capturing it with the jaws when within striking distance.

Numerous riverine species are often "sit and wait" or ambush predators (e.g., the long-necked side-necks of Australia and South America, many of the softshells—especially highly developed in the giant Asian blunt-nosed genera (*Chitra* and *Pelochelys*), the alligator snapping turtle, and the mata mata turtle). As described above, these species remain still and inconspicuous, often buried in substrates or blending into the background until they strike the prey with explosive speed (Legler, 1978; Pritchard, 1984, 1989, 2001; Legler & Georges, 1993b). The alligator snapping turtle is the only turtle species known to lure potential (fish) prey closer to facilitate the eventual strike by opening its jaws and wiggling a brightly colored worm-shaped portion of the tongue in angler fish fashion (Drummond & Gordon, 1979; Pritchard, 1989). A study by Harrell and Stringer (1997) predictably demonstrated that predacious fishes (especially centrarchids) were important prey items in the alligator snapper diet. Some of these species may capture the prey in their jaws (e.g., alligator snapping turtle, most softshells), while others (e.g., the long-necked side-necks, the mata mata, and probably the narrow-headed softshells [*Chitra*]) are "gape and suck" or "pharyngeal" feeders. In the latter, a vacuum is created by opening the jaws rapidly, which literally sucks the prey (sometimes a whole school of small fish) into the mouth (Pritchard, 1984; Moll & Moll, 2000). A recent study of the mechanics of feeding of the Australian long-necked turtle demonstrated the importance of the rapid inrush of water through the open mouth caused by the expansion of the throat. The inrushing water not only sucks in the prey, but it jerks the turtle's head and neck forward to meet and more swiftly engulf the prey as well (Summers, 2003).

Some bottom feeders practice a behavior called "benthic bulldozing" where large amounts of benthic substrates and benthic prey items such as insect larvae are ingested indiscriminately. This feeding behavior has been observed in some map turtles and in Geoffroy's side-necked turtles where benthic insect larvae and pupae (mainly Chironomidae) are extremely dense, and perhaps where highly organic substrates may provide some food value also (Quinn & Christiansen, 1972; D. Moll, 1976a,b, 1980, 1985; Souza & Abe, 2000). Similarly, Moll and Legler (1971) found sliders in a Panamanian pond eating large amounts of organic mud—perhaps in order to harvest a thin layer of blue-green algae growing on the substrate surface.

Another remarkable feeding behavior which has been observed in painted turtles, yellow-spotted Amazon River turtles, savanna side-necked turtles (*Podocnemis vogli*), and occasionally in juvenile red-headed side-necked turtles (*P. erythrocephala*) and giant South American river turtles is termed "neustophagia" (Belkin & Gans, 1968; Rhodin et al., 1981). Cann and Legler (1994) also observed possible neustophagia in captive Mary River turtles. This refers to an inertial, sucking action, which is powered by rapid and maximal hyoidal pharyngeal expansion. The turtle opens its jaws slightly at or just below the level of the water's surface film and sucks in particulate matter. It then flushes excess water out through nearly closed jaws by pharyngeal contraction, while retaining the particulate matter, which presumably has food value. Rhodin et al. (1981) considered this behavior as an adaptation for feeding in those podocnemid species and/or life history stages which regularly inhabit lagoons, oxbow lakes, and other less turbulent environments. It is interesting that the behavior has apparently evolved independently in such evolutionarily divergent lineages as the emydid painted turtles and the podocnemid side-necks.

A foraging behavior which has been observed in many different lineages and locations around the world is "windfall feeding," where many turtles will cluster beneath overhanging riparian vegetation which is (usually) dropping fruits or nuts into the water. Aquatic windfall feeding has been observed and described in Antillean sliders (Barbour & Carr, 1940); Panamanian sliders (Moll & Legler, 1971); Nile softshells (Loveridge & Williams, 1957); pig-nosed turtles (Schodde et al., 1973, Georges & Rose, 1983; Georges, 1987); river terrapin and painted terrapin (E. Moll, 1980a,b); black turtles (Moll & Jansen, 1995); and Central American river turtles (D. Moll, 1989). E. Moll's (1980a,b) accounts of long-tailed macaques and leaf monkeys throwing berembang fruits to clustered river terrapin and painted terrapin respectively in the water below is an interesting twist on the standard scenario, and suggests that a loose commensalism between the primates and turtles may have been established. While windfall feeding is usually associated with falling fruit or nuts, Obst (1986) stated that *Podocnemis* turtles (species unspecified) may also gather underneath the branches of trees serving as bird nesting rookeries in order to eat droppings (and presumably nestlings and eggs) that fall regularly into the water.

One of the most unusual behaviors displayed by turtles which live in rivers and streams is the "worm stomping" practiced by wood turtles (*Clemmys insculpta*) in some populations (Harding & Bloomer, 1979; Kaufmann, 1986; Kaufmann et al., 1989). Wood turtles stomp the ground with their forefeet (about once per second) with increasing force on moist ground near streams. This activity attracts earthworms to the surface, and these are then consumed by the turtle. It is thought that the stomping may mimic the vibrations of falling raindrops to which earthworms respond by vertical movement and emergence.

Moll and Legler (1971) described a possible example of intraspecific mutualism in which juvenile Panamanian sliders took turns feeding on epizoic algae from each other's carapaces. Similar feeding behavior has been observed in Florida red-bellied turtles (Meshaka, 1988) and in an observation of a captive Indian roofed turtle (*Kachuga tecta*) eating algae from the carapace of a three-striped roofed turtle (*Kachuga dhongoka*) (Lotze, 1963 in Moll & Legler, 1971). Branch (1988) mentioned a presumably opportunistic, interspecific mutualism in which East African serrated mud turtles (*Pelusios sinuatus*) ate engorged ticks from the legs of drinking cape buffalo and other game. A captive giant South American river turtle has also been observed feeding on loose, dead skin from the neck of a Brazilian radiolated swamp turtle (*Acanthochelys radiolata*) (Lotze, 1963; Moll & Legler, 1971). Krawchuck et al. (1997) observed the regular cropping of leeches attached to snapping turtles by painted turtles, and Vogt (1979) observed grackles feeding on leeches attached to basking Ouachita map turtles. Kaufmann (1991) observed small fishes "cleaning" algae or ectoparasites from the extended limbs and neck of wood turtles in a Pennsylvania creek. He believed the turtles' unusual posture represented a cleaning solicitation recognized by the fish.

Another extremely important role probably played by almost all river turtles to some extent in all turtle-inhabited rivers is scavenging. Even many highly herbivorous species will eat carrion when it is available. Examples include the river terrapin (E. Moll, 1980a; Davenport et al., 1992); the red-headed side-necked turtle (Mittermeier & Wilson, 1974); and the black turtle (Jansen, 1993). The latter two species are attracted to fish-baited lines, and the black turtle readily enters fish carrion-baited traps, although both of their diets are overwhelmingly composed of plant matter in nature. As most carnivorous and omnivorous river turtles are not well-adapted for capturing live fish or other mobile or large prey, carrion is probably an important source of protein for many.

Thompson (1993) estimated that the three Murray River (Australia) side-necked turtles referred to above might have diets composed of at least 20% carrion (based upon data from Chessman, 1986, 1988). Assuming the carrion is predominately protein (and has an energy equivalent of 1030 kJ/kg), Thompson estimated the total consumption of protein in the form of carrion may reach 130 metric tonnes a day, or if the carrion is 30% protein and 70% water, a mass of carrion up to 430 metric tonnes may be consumed daily. He further projected conservatively that if only 100 metric tonnes of carrion is consumed daily for six months of the year (the approximate activity period of the resident turtles), then the river turtles may conservatively scavenge at least 180,000 metric tonnes per year. This crude estimate underscores the magnitude of the clean-up role that river turtles play in their home rivers, and suggests that they are a very important safeguard against excessive eutrophication (Thompson, 1993). Thompson suggested this role might be compromised in the Murray River due to heavy turtle egg predation by an introduced predator, the red fox (*Vulpes fulva*).

The prevalence of scavenging and its importance in rivers has also been noted in several softshell species. For example, the inclination to feed on carrion including a horse carcass by the Euphrates softshelled turtle (*Rafetus euphraticus*) was observed by Taskavak (1995) and Taskavak and Atatür (1998), and Akani et al. (2001) reported the Nile softshelled turtle feeding upon carcasses of bush-pigs and goats along the banks of Nigeria's Sombreiro River. Akani et al. (2001) thought that the Nile softshell could be one of the principal scavengers and therefore a keystone species in the fragile freshwater ecosystems of southern Nigeria. The habitat generalist Indian flapshell is an important scavenger in Indian rivers, as is the larger Indian softshell (Das, 1991, 1995). According to Das (1995), the latter species, along with the Indian peacock softshell (*Aspideretes hurum*) and some other freshwater forms, were formerly raised and recruited to help clear half-burnt corpses from the Ganges River.

Turtles are also important in riverine food webs as prey species which feed an impressive array of invertebrates and vertebrates in their lotic ecosystems (e.g., Moll & Legler, 1971; Pritchard & Trebbau, 1984; and Soini, 1995 for reviews of Neotropical river turtle predators; Carr, 1952; Vogt, 1979, 1980a; Seigel, 1980; Janzen et al., 1992; Ernst et al., 1994; Standing et al., 2000; and Maier et al., 2002 for North American turtle predators; Martín & López, 1990 for southwestern Europe; E. Moll, 1980a and Das, 1991, 1995 for tropical Asian predators; Loveridge, 1941; Loveridge & Williams, 1957; Branch, 1988; Boycott & Bourquin, 2000 for African predators; and Cann, 1998 for Australian predators). Eggs and hatchlings are by far the most predator-vulnerable life history stages of river turtles, with up to 100% mortality in clutches often caused by efficient predators such as red foxes, otters, and raccoons and even dipteran maggots in North America (Vogt, 1980a, 1981b; Mitchell & Klemens, 2000). Most turtles have adapted to cope with high attrition rates of eggs and offspring through the evolution of high lifetime reproductive output and longevity of adults, however (Congdon et al., 1993, 1994). Turtles become less vulnerable as they grow older and their shells harden, but juveniles and adults are still preyed upon by numerous ingenious or powerful predators including other turtles (as discussed above), piranhas (Souza & Abe, 2000), skunks, coyotes, coatimundis, foxes, and raccoons (Minckley, 1966; Thompson, 1983, 1993; Cann, 1993; Pritchard, 1993a; Moll & Moll, 2000), muskrats (Parmalee, 1989), jaguars (André, 1902; Emmons, 1989), several species of river otters (e.g., Branch, 1988; Martuscelli, 1995; E. Moll, pers. obs.), bald eagles, ospreys and other birds of prey (Wiley & Lohrer, 1973; Clark, 1982), and crocodilians (Medem, 1962; Roze, 1964; Delany & Abercrombie, 1986; Jackson, 1986; Rhodin et al., 1993; Ernst et al., 1994; Cann, 1998).

River turtles cope well with high levels of predation under natural circumstances, but many species are seriously threatened by the additional combined predation pressures from "subsidized predators" (i.e., predators which have increased dramatically in population size as the result of human activities and habitat alteration; e.g., raccoons and ravens), introduced predator species (such as the red fox in Australia's Murray River region as described by Thompson, 1983), and human predation (see Mitchell & Klemens, 2000; Thorbjarnarson et al., 2000; and chapters 4 and 5 in this volume for further information concerning these subjects).

Morphology

The remaining morphological traits listed in table 2.2 are thought to be physical adaptations, which enhance the efficiency of one or (usually) more riverine forms in their nektonic niches. Supporting evidence for their value is provided by the independent evolution of a number of these adaptations in distantly related lineages filling similar niches through convergent evolution.

Elongated, tubular snouts with distal nostrils are a good example of convergent evolution, having evolved at least superficial similarity in such divergent forms as the softshelled (most species) and pig-nosed turtles, the river terrapin, and the mata mata. These structures have presumably evolved to allow these species to breathe with minimal exposure above the water's surface—a useful adaptation for hiding from potential predators or prey. Equipped with tubular snouts and long necks, the long-necked side-necks and softshells can presumably remain buried in the substrates of the shallows where they can ambush prey without regularly swimming to the surface for air. Pritchard (1984) suggested that tubular-shaped rostral structures probably evolved to minimize water disturbance during respiration—a key feature for species like the mata mata which must detect mobile prey that approach the head closely, while not disturbing the prey before the strike is accomplished.

Species which are capable of deep, extended dives such as the river terrapin, painted terrapin, Malaysian giant turtles, crowned river turtles, and the larger roofed turtles have bony lung chambers encased by the lateral walls of the carapace, and sometimes reinforced by large carapacial buttresses (e.g., in painted terrapin and river terrapin) (fig. 2.23). Obst (1986) considered these adaptations to help the lungs withstand the extreme pressures which may be experienced during deep dives. E. Moll (1980a,b) thought the heavy buttresses probably also aided these estuarine terrapin in deterring predation by crocodilians, such as the saltwater crocodile (*Crocodylus porosus*), since analogous carapacial reinforcement has evolved in response to alligator predation in North American species such as *Pseudemys nelsoni* and others (Jackson, 1986; Pritchard & Greenhood, 1986).

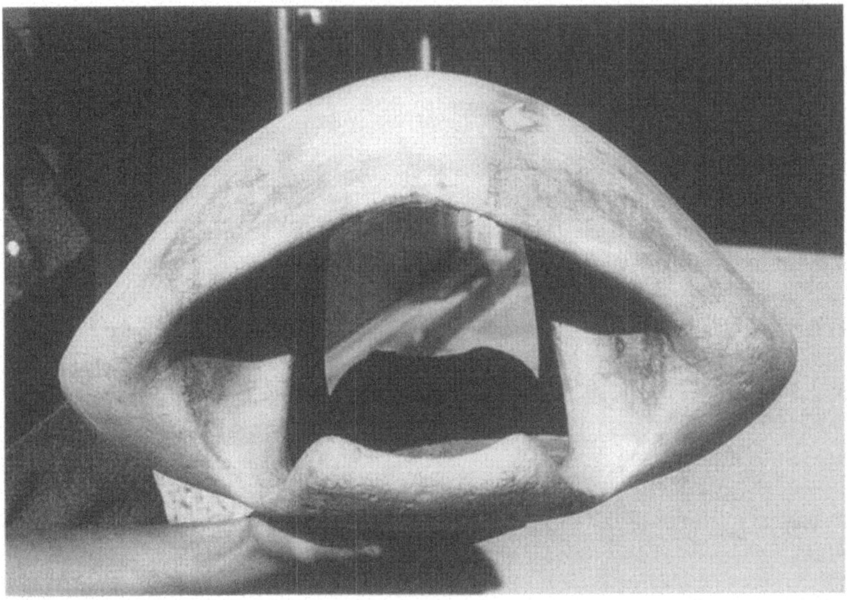

Figure 2.23
River terrapin (*Batagur baska*) shell displaying carapacial buttressing which may help support the lungs during deep dives, and/or help protect these turtles from crocodilian predation.

Another important adaptation for some highly aquatic river turtles is the capacity for underwater respiration made possible through the presence of highly vascularized skin, folds, and villi (i.e., gill-like structures which may be cutaneously, buccophyaryngeally, or cloacally located). Several softshells are known to be capable of both buccopharyngeal and cutaneous respiration (e.g., smooth softshell, Jackson et al., 1976; spiny softshell, Dunson, 1960; Nile softshell, Girgis, 1961; Asiatic softshell [*Amyda cartilaginea*], Davenport & Wong, 1992). The pig-nosed turtle and the Central American river turtle also display buccal-esophageal papillae and are considered to be buccopharyngeal gill breathers (Legler, 1993). All of the chelid side-necked turtles are probably capable of cloacal respiration (Legler, 1993; Legler & Georges, 1993b), and cloacal respiration capability is particularly well developed in Australia's Fitzroy River turtle, northern Australian snapping turtle, and the Mary River turtle (Cann, 1998; Priest & Franklin, 2002; Legler & Cann, in press). Legler and Georges (1993b) consider the Fitzroy River turtle to have achieved the greatest degree of emancipation from surface air breathing of any aquatic turtle, and that submerged and diving individuals can obtain all the oxygen they need through their cloacal bursae. The general advantages of submerged respiration include allowing turtles to remain submerged for much longer periods than would otherwise be possible, thereby avoiding surface predators, and allowing uninterrupted foraging, resting, and other submergent activities. It also allows turtles to stay underwater in winter when their metabolism slows and they become torpid, without the need for rising to the surface for oxygen (Cann, 1998).

Highly nektonic river turtles typically have highly modified limbs, and streamlined body shapes which allow them to move swiftly and powerfully through the water. The legs and feet of most of these species have reduced scalation and claws, are decidedly flattened or palmate in shape, and are heavily webbed (especially the hind limbs) (Obst, 1986). In the pig-nosed turtle the forelimbs have been modified into paddle-like structures analogous to those of sea turtles, and its swimming motion is accomplished similarly also through simultaneous movements of the enlarged pectoral flippers (Walker, 1979, Schultze-Westrum, 1995) (fig. 2.24). All other freshwater turtles swim by paddling with all the limbs. The body is supported by the water, and propulsion is accomplished by alternately pushing the limbs (especially the hind limbs) against the water (Zug, 1971; Walker, 1979). The specialized palmate-shaped limbs and fully extended and webbed hind feet provide the riverine species with the enhanced propulsive force they require to move swiftly through the water. Riverine softshells have perfected the skill of swift swimming to the greatest degree of all the paddlers through loss of the peripheral bones and rigid external scutes, and by reducing the shell lip substantially, and/or increasing its flexibility (i.e., the skin of the neck attaches near the leading edge of the carapace, as most extremely displayed in the *Chitra* and *Pelochelys* softshells) (fig. 2.25a,b). These adaptations reduce drag, facilitate the extension of the limbs

Figure 2.24
Pig-nosed turtle (*Carettochelys insculpta*) displaying a sea turtle-like front flipper—a unique adaptation facilitating fast swimming in this river turtle. (Photo by Richard S. Funk)

(a)

(b)

Figure 2.25
(a) Asian giant softshell (*Pelochelys cantorii*), and (b) Indian narrow-headed softshell (*Chitra indica*) displaying the attachment of the skin of the neck to the leading edge of the carapace—an adaptation which probably enhances both hydrodynamic and sand-burrowing efficiency.

during paddling, and eliminate the energetically costly side-to-side rocking motion (i.e., shell rolling) which occurs in most vigorously swimming hard-shelled species (Moll & Moll, 2000; Pritchard, 2001). These modifications, in conjunction with the flattened shape, also allow the softshells rapidly to bury themselves in bottom substrates where they hide and practice ambush feeding (Graham & Graham, 1991; Pritchard, 2001) (fig. 2.26). The Malayan softshell (*Dogania subplana*) is also characterized by a unique morphological adaptation termed "pankinesis" by Pritchard (1993b). Pankinesis, the elimination of rigid suture zones between carapacial bones, provides great carapacial flexibility which allows the turtle to wedge itself into cracks beneath or between boulders in the streams it inhabits. In this context it may be considered an aquatic ecological equivalent to the terrestrial, flexible, crevice-inhabiting pancake tortoise (*Malacochersus tornieri*) of East Africa (Moll & Klemens, 1996). Both have sacrificed some of the protective rigidity of the archetypical chelonian shell for the even greater security of life within or under boulders (Loehr, 2002).

Zug (1971) noted that nektonic river turtles are able to sink quickly due to having body densities greater than water, while bottom-walking forms and land tortoises have a tendency to float. This characteristic would seem an obvious advantage to the former group which would require energetically efficient movement while diving to search for food or to escape predators, and when choosing to remain below the surface or on the bottom substrates without expending effort. Zug (1971) reasoned that the superior swimming ability of aquatic species would more easily compensate for their relative "heaviness" than could the poor-swim-

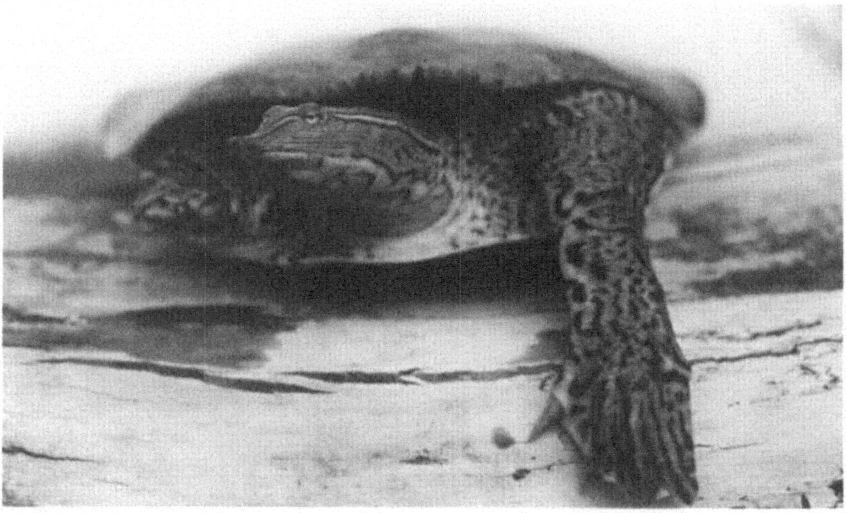

Figure 2.26
A spiny softshell (*Apalone spinifera*) showing flattened form and extensive foot webbing characteristic of the Trionychidae.

ming, bottom-walking forms which could save energy when surfacing to breathe by being generally more buoyant.

Finally, there is an unusual degree of convergence in the structure of the plica media of the penis in two very divergent river turtle lineages—the softshells and the riverine batagurids (Zug, 1966; Moll, 1986). River-dwelling species from both groups tend to display very elaborate flaps formed from the penis lateral fold. We speculate, without evidence, that the flaps could help to hold the penis in place and minimize sperm loss during copulation in flowing water.

There is so much more to learn about the roles which river turtles play in their lotic habitats. The data which have been accumulated to date suggest that the known roles are varied and vital in their contribution to the proper functioning of river ecosystems, though there are undoubtedly others which have not yet even been considered. Further research designed to refine and extend current knowledge is clearly desirable for its own sake, as well as its value in preparing a compelling rationale for the conservation of river turtle stocks. Humans, who have an important stake in properly functioning rivers themselves, must be convinced that turtles contribute significantly in this regard, and are worth more to them alive in the river than they are as just table fare or as a commodity.

Communities and Habitats

3

Profile 3. *The lower Wabash River is still a pretty good place for river turtles to live. To be sure, the waters lack the level of habitat quality that the French found when they first used the river to settle Illinois. Today, the Wabash is somewhat tainted with industrial, municipal, and agricultural pollution and often brown with suspended silt. Nevertheless, the cities along its banks tend to be small and widely spaced, and anthropogenic effects have not been great. So far, only certain pollution-sensitive species such as mussels have seriously declined. The more resilient turtles still prosper in the region.*

The Wabash River from New Harmony, Indiana to where it enters the Ohio River is a wide slow-moving river with sand, gravel, or mud bottom with pools up to 3 meters in depth. The flood plain in this region and along the Ohio River as far as Old Shawneetown, Illinois is dotted with a wealth of lakes, ponds, and backwaters. Turtles are common in both the flowing and quiet waters of this system, but the species composition of these divergent habitats differs markedly. Smooth softshells and Ouachita map turtles dominate the mainstem of the river while red-eared sliders and river cooters are the principals of the flood plain ponds. A scattering of snappers, spiny softshells, painteds, false maps and stinkpots also saunter through these habitats. Actually none of the turtles is completely restricted to either lentic or lotic habitats, but the softshells and map turtles spend most of their time in flowing water while the others are more numerous in the slow-moving or still waters.

On a typical sunny day in late spring, a profusion of softshells haul out onto an exposed sandbank and assume their somewhat comical pose in which legs and neck are extended and positioned to reap maximum exposure to the sun. Some distance from the sandbank, in deeper water, a throng of Ouachita maps queue up along the limb of a stranded tree, lazily basking away the morning. A large, female, smooth softshell swims gracefully through the deeper pools searching for mayflies while her male counterparts patrol the shallow waters just off the sandbanks devouring an array of terrestrial and aquatic insects. Several of the omnivorous Ouachitas forage for caddis fly larvae, filamentous algae and macrophytes along a gravelly stretch of bottom. In a floodplain lake, just a few hundred meters of nettle and poison ivy away, another chelonian community carries on its daily business. Here, basking logs are also blanketed with a myriad of turtles, but in contrast most are red-eared sliders, and river cooters with only an occasional map and painted turtle. In the water, a red-eared slider grazes on a large colony of bryozoans encrusted on the undersurface of one of the basking logs as others swim slowly through a maze of vascular plants, cropping off bits of leaves and stems as they go. Nearby, a river cooter voraciously attacks a snarl of filamentous algae using both jaws and forefeet to pull apart the long tangled strands. The scene changes little throughout the day with the sliders and cooters alternating between basking and feeding. However, by late afternoon, emydid activity begins to wane and other players usurp the stage. A large 15-kg snapping turtle lumbers across the mucky bottom, pushing through stalks of spatterdock, zeroing in on the scent of a dead buffalo sucker partially concealed in a mass of coontail. A musk turtle crawls out of the muck where it has spent most of the day, and with head and neck extended begins slowly prowling the bottom. Periodically, the turtle bulldozes its head into burrows or small mounds of vegetation to grab a snail or aquatic insect. With the onset of darkness, the sliders and cooters settle to the bottom or utilize brush or vegetation for support during their nightly inactivity. The snapper and musk turtle continue to prowl the bottom into the evening hours, then finally they too hunker down for the night.

The species comprising these communities overlap considerably in behavior and diet, but each exhibits specific, differences as well. Both smooth softshells and Ouatchita maps are basking species, but the former typically basks on sand banks in shallow water while the latter prefers basking sites on brush and other debris in deeper water. Both relish a fine repast of aquatic insects, but while smoothies favor mayflies and rarely eat plants, Ouachitas opt for caddis flies and eat moderate amounts of vegetation. Red-ears and cooters are diurnal swimming species that spend considerable time basking out of water on logs and rocks, while snappers are bottom walkers that tend to be more crepuscular in their activities and prefer to elevate their body temperatures by loitering in shallow waters of optimum temperature. Both red-ears and cooters eat large quantities of plants but the latter tend toward algivory (algae-eating) while the former feed more on higher plants spiced with modest quantities of animal food. Snappers at the site eat plant material but invertebrates and fish comprise three-quarters of their diet. The significance of such similarities and differences is the subject of this chapter.

E. O. Moll

Rivers are dynamic and diverse ecosystems composed of a variety of habitats including the main channel, side channel, flood plain and backwater lakes as discussed briefly in chapter 2. Each habitat typically contains a characteristic assemblage of turtles (herein considered synonymous with a chelonian community). Although riverine chelonians may appear anywhere in the river, most species specialize in one or more habitats where they occur in maximum numbers and biomass. Thus, the various habitats of the river may have similar species composition but the rank abundance for each species will differ. Basic divisions of typical rivers are lotic (flowing water) habitats and the lentic (still water) habitats. Beyond this, habitat divisions involve such physical features as gradient (headwaters versus lower reaches), substrate (sand, gravel, and mud), size (feeder streams versus the river proper), shoreline heterogeneity (wooded, marsh, swamp, etc.) and even temperature. One can expect differences in species composition associated with any of these habitat features. The composition of river turtle communities is affected by the biotic as well as the physical environment of the stream. The presence or absence of lower and higher plants, competing species, or predators can all affect community composition. This chapter is chiefly concerned with assemblages of species rather than the individual species comprising these groups. It will examine the composition of river turtle communities, their ecology, and their evolution.

A Brief Review of the Study of Chelonian Communities

The concept that organisms occur in assemblages of species interacting to produce distinctive community characteristics was neglected for many years in the herpetological literature. Prior to the mid-1960s, most ecological research on reptiles was autecological (Scott, 1982). Nevertheless, a scattering of early papers described chelonian assemblages associated with particular habitats. Evermann and Clark's biological survey of Lake Maxinkuckee in 1920 included an overview of the turtle community. In 1942, Fred Cagle delineated the species composition and relative abundance of turtles inhabiting six lentic habitats in southern Illinois. In 1950 he published a similar report with A. H. Chaney on ten lentic and two lotic habitats in Louisiana. Tinkle (1959) compared the species composition and relative abundance of species above and below the "fall line" in five United States rivers emptying into the Gulf of Mexico. Wade and Gifford (1965) described the turtle community in an Indiana lake.

Resource partitioning among closely related turtles was insinuated in I. Y. Mahmoud's (1968a,b) papers comparing the diets and feeding behavior of four sympatric kinosternids in Oklahoma. However, Jim Berry (1975) was first to actually demonstrate behavioral and morphological effects resulting from an interaction between two species of musk turtles inhabiting a Florida Spring. Bruce Bury in a 1979 review of freshwater turtle ecology, diagramed turtle associations and generalized niche separation among turtles in three regions of the United States while stressing the need for new research concerning species interactions, biomass and densities in multispecies communities. Disappointingly, there has not been an overwhelming response to Bury's plea as squamates have proven to be more popular subjects for such studies. Nevertheless, since the 1980s, a scattering of publications dealing with specific communities and species interactions, including Vogt (1981a), Vogt and Guzman-Guzman (1988), Obst (1986), D. Moll (1990), Meylan et al. (1992), Legler and Georges (1993b), Fuselier and Edds (1994),

Fachín-Terán et al. (1995), Congdon and Gibbons (1996), Bodie and Semlitsch (2000a,b), Bodie et al. (2000), and Lindeman (2001c) have helped to sustain interest in chelonian community ecology. As another approach, several authors have compared species richness and biomass over a wide or global scale, among the more recent being Iverson (1982, 1992b,c), Congdon et al. (1986), and Congdon and Gibbons (1989). As the emphasis of conservation programs has expanded from endangered species toward endangered ecosystems, understanding community ecology (species interactions, disturbance regimes, invading species, etc.) has become of particular importance to conservation biology. Nevertheless, community studies still represent but a tiny proportion of the world literature on chelonians.

A Global Overview of River Turtle Communities

Nearctic

Emydids dominate North American river turtle assemblages. Map, cooter, slider, and painted turtles are particularly important components of most communities. Softshells (*Apalone*), snapping turtles (*Chelyda* and *Macroclemys*), and mud turtles (*Kinosternon* and *Sternotherus*) are other regulars. Nearctic turtles reach their northern limits around 50° north latitude. Two well-known generalists, the common snapping turtle and the painted turtle are the northernmost representatives in Canadian rivers. Common map turtles, common musk turtles and spiny softshells, though not as widespread, also reach their northern limits in the St. Lawrence and Ottawa rivers in eastern Canada. Wood turtles range as far north as Nova Scotia. Blanding's turtles, which typically do not inhabit lotic habitats, also range into southeastern Canada and have a relict population in Nova Scotia.

The Pacific coastal drainages of the United States are especially depauperate in river turtles. Rivers west of the Rocky Mountains usually have but one or occasionally two chelonians, the Pacific pond turtle and/or the western painted turtle. The eastern seaboard is somewhat richer. Diamondback terrapins inhabit the estuaries and coastal marshes from Massachusetts all the way to Texas. Most northeastern rivers again contain the conspicuous and near-ubiquitous common snapping turtles and painted turtles. The less noticeable common musk turtle also inhabits many of the slow-moving, mud-bottom habitats. Map turtles and spiny softshells occur sporadically in the northeast. The red-bellied turtle has relict populations in Massachusetts, but the main range begins in southern New Jersey and extends to northeastern North Carolina. It occurs up the Potomac River as far as West Virginia. From Virginia southward, common cooters, river cooters, and sliders join the assemblage. In the central Atlantic drainages, common cooters dwell on the coastal plain while river cooters inhabit the Piedmont (Seidel & Palmer, 1991). River cooters that do not range below the great Okeefenokee Swamp and sliders that barely reach Gainesville, Florida relinquish their niches along the Atlantic drainage of Florida to the common cooter and Florida red-belly.

Similarly, the spiny softshell reappears along the coastal drainage of southern North Carolina only to drop out at the Okeefenokee leaving the Florida coastal drainage to the Florida softshell.

The wealth of the Nearctic turtle diversity lies within the rivers flowing into the Gulf of Mexico. The greatest species density of chelonians in the Nearctic occurs in the lower Mobile Bay Region with 16 species (Iverson, 1992c). A logical starting point in providing an overview of river turtle diversity is to begin with the Mississippi, easily the most extensive river system flowing into the gulf (tables 3.1a,b and 3.2). The "Father of Waters" begins its 3780-km journey to the Gulf far to the north at Lake Itasca at 1463 feet altitude and 47° 15′ N latitude in Minnesota. Here, the river environs are still appreciably wild and long stretches of the river above Minneapolis and St. Paul have been only moderately altered by humans. However, the Mississippi River from St. Paul south would hardly be recognizable to the likes of Hernando DeSoto or Sieur de La Salle. Much of this portion of the river has been greatly modified for flood control and river transportation. The upper Mississippi is now a patchwork of lentic and lotic stretches created by the building of twenty-six locks and dams, while the lower course has been altered by myriads of artificial. levees. While these human alterations do not seem to have eliminated any of the original species in the river, it is virtually certain that the relative densities of particular species have drastically changed. The "pools" mentioned along this descriptive odyssey are impounded portions of the river associated with the aforementioned locks and dams.

As the river flows out of Itasca, painted turtles and snapping turtles are its lone chelonian inhabitants. By the time the river reaches Lake City, Minnesota at 44° 27′ N, three map turtles and two softshells have joined the snappers and painteds in the river proper. Semi-aquatic wood turtles regularly utilize the clean, gravelly, fast- flowing tributary streams while Blanding's turtles inhabit cattail marshes associated with the river. Chelonian communities at this latitude are comprised of only omnivores and carnivores. The latter include molluskivorous common map turtles, male false and Ouachita maps, and the smooth and spiny softshells. The remainder of the assemblage is omnivorous.

Other than loss of the wood turtle near the Wisconsin border, species composition changes little between Lake City and Pool 13 at Bellevue, Iowa (42° 16′ N). From Pool 13 southward to the boot heel of Missouri, however, a number of new turtles join the assemblage including the common slider, river cooter, chicken turtle, two mud turtles, the common musk turtle, and alligator snapping turtle. For the first time herbivores (e.g., river cooters and adult sliders) become part of the community. Two particularly significant changes occur approximately where the river finishes flowing out around the western-most protuberance of the state of Illinois and angles southeastward to rendezvous with the Illinois River. First, the painted turtle which has been the dominant species in lentic habitats since the river's origin is ousted from its exalted position by the common slider. The painteds continue to occupy lentic communities along the river all the way to the Gulf, but now in more moderate densities. Secondly, in this region, the common map turtle largely vacates the main river,

Table 3.1
Species composition, richness, Shannon Diversity Indices (H′) and evenness (J′) for eight lotic (a) and eleven lentic turtle communities (b) in the Mississippi River drainage.

	(a) Lotic communities—Mississippi River Drainage							
	PL4	PL8	PL13	PL26	CPG	TPT	HVI	WAB
Species	(203)	(100)	(176)	(55)	(1032)	(228)	(90)	(73)
CHSE	9.0	8.0	0.5	9.0	1.5	+	3.0	–
CHPI	7.0	13.0	9.0	–	+	–	16.0	6.5
GRGE	3.5	9.0	61.5	–	–	–	2.0	–
GROU	1.0	8.0	0.5	3.5	–	2.0	2.0	18.5
GRPS	48.0	18.0	16.5	11.0	74.0	55.0	3.0	–
PSCO	–	–	–	–	–	–	–	–
TRSC	–	–	–	65.5	17.5	4.0	17.0	7.0
STOD	–	–	–	5.5	+	–	11.0	
APMU	3.5	2.0	3.0	2.0	4.5	34.0	2.0	68.5
APSP	28.0	42.0	9.0	3.5	2.5	5.0	44.0	5.5
Richness	7	7	7	7	7	6	9	5
H′ value	2.01	2.36	1.71	1.74	1.21	1.55	2.37	1.39
J′ (Evenness)	0.715	0.84	0.61	0.62	0.43	0.60	0.84	0.60

	(b) Lentic communities—Mississippi River Drainage										
Species	PL4 (436)	PL8 (1102)	PL13 (636)	PL26 (578)	DRP (218)	MRL (350)	SPL (285)	CHL (322)	HVI (287)	WAB (62)	RDP (376)
CHSE	12.5	6.5	2.0	4.0	8.0	2.0	3.0	–	4.0	5.0	5.0
CHPI	31.0	69.5	76.5	16.0	19.0	20.0	62.0	30.0	5.0	19.5	0.5
GRGE	10.0	4.5	11.0	+	–	5.0	–	1.0	3.5	–	–
GROU	0.5	1.5	–	1.0	–	–	–	–	0.5	1.5	4.5
GRPS	21.0	8.5	7.5	0.5	1.0	20.0	–	–	4.0	–	0.5
PSCO	–	–	–	–	–	–	–	–	–	–	18.0
TRSC	–	–	–	74.0	33.5	35.0	28.0	50.0	63.0	74.0	66.5
KISU	–	–	–	–	15.5	–	–	–	–	–	–
STOD	–	–	–	1.5	15.0	13.0	5.0	12.0	3.5	–	3.0
APMU	3.0	1.0	1.0	0.5	–	–	–	–	0.5	–	0.5
APSP	22.0	8.5	2.0	2.5	8.0	5.0	2.0	7.0	16	–	1.5
Richness	7	7	6	9	7	7	5	5	9	4	9
H′ value	2.37	1.39	1.22	1.31	2.45	2.00	1.42	1.72	1.68	1.08	1.62
J′(Evenness)	0.845	0.50	0.47	0.41	0.87	0.71	0.61	0.74	0.53	0.54	0.51

Lotic habitats are divided into Mississippi River proper and tributaries. N is given in parentheses and values in cells indicate the percentage of each species in the community. Data sources are Dreslik and E. Moll, unpublished data (RDP); Gritters and Mauldin, 1994 (PL13); Gritters and Tucker, unpublished data from the Mississippi River Long Term Monitoring Project (PL4, PL8, PL13, PL26, CPG, HVI); D. Moll, 1977 (CHL, MRL, SPL); E. Moll, unpublished data (DRP, TPT); Pierce, 1992 (WAB). A plus sign (+) indicates species that were seen, but not caught in sampling.

Abbreviations and localities

Species
CHSE – *Chelydra serpentina*, Common snapping turtle
CHPI – *Chrysemys picta*, Painted turtle

(continued)

GRGE – *Graptemys geographica*, Common map turtle
GROU – *Graptemys ouachitensis*, Ouachita map turtle
GRPS – *Graptemys pseudogeographica*, False map turtle
PSCO – *Pseudemys concinna*, River cooter
TRSC – *Trachemys scripta*, Common slider
KISU – *Kinosternon subrubrum*, Common mud turtle
STOD – *Sternotherus odoratus*, Common musk turtle
APMU – *Apalone mutica*, Smooth softshell turtle
APSP – *Apalone spinifera*, Spiny softshell turtle

Localities

Mississippi River:

PL4 – Pool 4, Lake City, Minnesota. 44.27N, 92.18W
PL8 – Pool 8, La Crosse, Wisconsin. 43.48N, 91.15W
PL13 – Pool 13, Bellevue, Iowa. 42.16N, 90.27W
PL26 – Pool 26, Alton, Illinois. 38.55N, 90.10W
CPG – Open river at Cape Girardeau, Missouri. 37.19N, 89.31W
TPT – Open river at Tiptonville, Tennessee. 36.21N, 89.30W
DRP – Borrow pit on Mississippi River floodplain at Deer Park. 31.25N, 91.34W

Mississippi Tributaries:

SPL – Spring Lake, floodplain of Illinois River. 40.28N, 89.49W
CHL – Chautauqua Lake, floodplain of Illinois River. 40.22N, 89.59W
HVI – Illinois River at Havana, Illinois. 40.17N, 90.04W
MRL – Meredosia Lake, floodplain of Illinois River. 39.49N, 90.32W
WAB – Wabash River near Rochester, Illinois. 38.20N, 87.49W
RDP – Round Pond, floodplain Ohio River, Illinois 37.45N, 88.05W

Table 3.2
Species of turtles inhabiting the Mississippi River including distribution on the river and adult feeding guild.

Species	Distribution	Guild
Macroclemys temminckii	S 1/3 Illinois to Mouth	OMNIVORE/c
Chelydra serpentina	Entire River	OMNIVORE/g
Sternotherus odoratus	N Illinois to Mouth	OMNIVORE/c
Sternotherus carinatus	SE Arkansas to S Louisiana	OMNIVORE/c
*Kinosternon subrubrum**	S Illinois to Mouth	OMNIVORE/c
Kinosternon flavescens	N Illinois to NE Missouri	OMNIVORE/c
Chrysemys picta	Entire River	OMNIVORE/g
Clemmys insculpta	Minnesota/Wisconsin Border	OMNIVORE/g
*Emydoidea blandingi**	Central Minnesota to N 1/3 IL	OMNIVORE/c
*Deirochelys reticularia**	Missouri Boot-heel to Mouth	OMNIVORE/c
Graptemys geographica	Minnesota to S 1/3 Illinois	CARNIVORE/m
Graptemys ouachitensis	Minnesota to Mouth	OMNIVORE/g
Graptemys pseudogeographica	Minnesota to Mouth	OMNIVORE/g
Pseudemys concinna	S 1/3 Illinois to Mouth	HERBIVORE/a
Trachemys scripta	N Illinois to Mouth	OMNIVORE/g
Apalone mutica	Minnesota to Mouth	CARNIVORE/i
Apalone spinifera	Minnesota to Mouth	CARNIVORE/g

Guilds are presented as a general category (in caps) and a specialization or preference (lower case). HERBIVORE and CARNIVORE indicate that the turtles' diet comprises >90% of the volume as plants or animals respectively; otherwise they are designated as OMNIVORE. Under guilds, lower-case letters stand for preferences or specialities (>70% of food vol.):
a – algivorous, c – carnivorous, g – general, h – herbivorous, i – insectivorous,
m – molluscivorous (see this chapter for detailed discussion of guilds). Species marked with an asterisk have a strong preference for lentic habitats, inhabiting marshes, swamps etc. near rivers but rarely the river itself. Data drawn from Conant and Collins (1991) and Ernst et al. (1994).

leaving the false and Ouachita map turtles as the chief *Graptemys* in riverine habitats.

Last to join the river's turtle assemblage, at approximately the level of southern Arkansas, is the razor-backed musk turtle, a turtle of the streams and great river swamps. Seventeen chelonians are associated with the river along its course. Some, such as the Blandings, chicken, and mud turtles are not truly river species but are often associated with the still waters of the riverine environment (i.e., marshes, sloughs, backwaters, swamps, etc.). Although not closely related, the Blanding's and chicken turtles are nearly ecological equivalents, sharing a propensity for lentic habitats and an inclination to stalk and catch animal prey with rapid strikes using their long necks. The Blanding's tends to be omnivorous with a carnivorous preference while the chicken turtle is chiefly carnivorous (Rowe, 1992; Jackson, 1996).

To briefly summarize the turtle communities of the Mississippi River from the origin to delta, seventeen aquatic species can be found in the river and its associated habitats. Table 3.1a,b characterizes sample lentic and lotic communities studied along the river. Changes in species composition are largely additive from north to south. In other words, the majority of species, once joining the assemblage, remain a part of the fauna. Exceptions include wood turtles, common map turtles and Blanding's turtles. Individual assemblages associated with specific habitats comprise four to nine species. Turtles of the genera, *Graptemys* and *Apalone* typically dominate the river's lotic habitats, whereas *Trachemys* or *Chrysemys* prevail in backwaters. Relative to the guild structure of these communities, it appears that most Mississippi River chelonians lean toward carnivory, with the river cooter being closest to a true herbivore. Carnivores and omnivores that eat over 70% animal foods comprise 59%, omnivore generalists, 35%, and herbivores, 6% of the species on the river (see table 3.2).

Moving eastward and westward from the Mississippi River are a number of rivers of lesser stature but which in many ways are more interesting due to long periods of isolation from one another. Table 3.3 lists the species composition of seven rivers located between the Mississippi River and peninsular Florida—the Atchafalaya, Pearl, Pascagoula, Tombigbee, Alabama, Apalachicola and Rainbow Run. Table 3.4 lists similarity indices for the chelonians in these five rivers where a value of 0 indicates no overlap and a value of 1 indicates complete overlap.

When comparing the Atchafalaya River with each of the other rivers, we see that similarity is inversely proportional to the distance moved eastward. Nevertheless, even the Apalachicola has a similarity index close to 0.5 when compared with the Atchafalaya. However, as one descends into peninsular Florida (Rainbow Run), the similarity indices drop precipitously, suggesting a quite different history for this region. Despite the periods of isolation, alligator snapping turtles, common snapping turtles, river cooters, common sliders, common musk turtles, and spiny soft shells inhabit all six of the featured rivers above peninsular Florida. However, other lineages have responded to isolation by speciating. Map turtles have been particularly prone to speciation, having endemic species in the majority of the major river systems from the Guadalupe/San Antonio system in Texas to the Apalachicola in Florida, while displaying interesting inter- and intraspecific diversity of feeding guilds (table 3.5). Musk turtles demonstrate species replacement on a smaller scale with the razor-backed musk turtle occupying rivers west of the Tombigbee and the loggerhead musk turtle from the Tombigbee eastward. The flattened musk turtle also inhabits the Black Warrior River above the fall line in Alabama, whereas the loggerhead musk is only found below (Tinkle, 1959). Although smooth and spiny softshells regularly inhabit the same rivers there is little overlap between these two and the Florida softshell, which is the only trionychid in peninsular Florida. Where the spiny and Florida overlap in the Apalachicola, the Florida is most common in the blackwater situations away from the river's mainstem (M. A. Ewert, personal communication). Ewert also reports a parallel situation with river and common cooters where the former occupy the more turbid,

Table 3.3
Species composition of chelonians in seven southern U.S. rivers flowing into the Gulf of Mexico.

	Gulf River Systems						
Species	Atch	Perl	Pasc	Tomb	Alab.	Apal	Rain
C. serpentina	×	×	×	×	×	×	×
M. temminckii	×	×	×	×	×	×	
Ch. picta	×						
G. barbouri						×	
G. flavimaculata			×				
G. geographica		×	×	×			
G. gibbonsi		×	×				
G. nigrinoda				×	×		
G. oculifera		×					
G. ouachitensis	×						
G. pulchra				×	×		
G. pseudogeographica	×						
*P. concinna**	×	×	×	×	×	×	×
*P. floridana**						×	×
P. nelsoni						×	×
T. scripta	×	×	×	×	×	×	
S. odoratus	×	×	×	×	×	×	×
S. carinatus	×	×	×				
S. depressus				×			
S. minor				×	×	×	×
A. ferox						×	×
A. mutica	×	×	×	×	×		
A. spinifera	×	×	×	×	×	×	
Total species	11	11	11	12	10	11	7

Deirochelys not included. Genera abbreviations are: C – *Chelydra*, M – *Macroclemys*, Ch – *Chrysemys*, G – *Graptemys*, P – *Pseudemys*, T – *Trachemys*, S – *Sternotherus*, A – *Apalone*. Rivers: Atchafalaya (LA), Pearl and Pascagoula (MI), Tombigbee and Alabama (AL), Apalachicola and Rainbow Run (FL).

*Taxonomy of the cooters (*Pseudemys concinna* and *floridana*) has been in a state of flux for several decades. The most recent attempt to deal with this complex group (Seidel, 1994) would recognize two subspecies of *Pseudemys concinna* (*concinna* and *floridana*) while elevating *P. concinna suwanniensis* and *P. floridana peninsularis* to full species (*P. suwanniensis* and *P. peninsularis*). Jackson (1995) has argued against these changes, and Seidel (1995) has responded to Jackson. Until this controversy is resolved by future usage of the names, we will follow Ernst et al. (1994).

mainstem of the river and the latter the smaller, clear, blackwater tributaries of the Apalachicola.

As previously stated, turtle assemblages of the Florida peninsula are quite different from the rest of the Gulf Coast. Based on information from our colleague Dale Jackson, the Gulf-draining rivers of the northern peninsula region (e.g., the Wakulla, Wacissa/St. Marks, and Santa Fe/Suwannee) are dominated by river

Table 3.4
Chelonian community similarities of seven rivers flowing into the Gulf of Mexico as indicated by Sorensen Coefficients (Brower & Zar, 1984).

	Gulf River Systems						
River	Atch	Perl	Pasc	Tomb	Alab	Apal	Rain
Atch	–	0.73	0.73	0.64	0.67	0.55	0.33
Perl		–	0.91	0.70	0.67	0.55	0.33
Pasc			–	0.70	0.67	0.55	0.33
Tomb				–	0.91	0.61	0.42
Alab					–	0.67	0.47
Apal							0.82
Rain							–

A value of 0 indicates no overlap in species whereas 1 indicates complete overlap of species. (See table 3.3 for abbreviations.)

Table 3.5
The majority of map turtles (*Graptemys*) have restricted ranges consisting of one to a few small river systems draining into the Gulf of Mexico. Herein, those species with restricted ranges are listed with their distribution and guild.

Graptemys species	River system(s)	Guild
G. barbouri	Apalachicola River	CARNIVORE/i&m
G. caglei	Guadalupe and San Antonio River	CARNIVORE/g
G. ernsti	Pensacola Bay Drainage	CARNIVORE/i&m
G. gibbonsi	Pascagoula and Pearl Rivers	CARNIVORE/g
G. flavimaculata	Pascagoula River	CARNIVORE/i&?
G. nigrinoda	Alabama, Tombigbee, Bl. Warrior Rivers	OMNIVORE/g
G. oculifera	Pearl River	CARNIVORE/i
G. ouachitensis sabinensis	Sabine River	OMNIVORE/c&h
G. pulchra	Mobile Bay Drainage	CARNIVORE /g
G. versa	Colorado River Drainage	CARNIVORE/m?

Adult guilds, HERBIVORE, CARNIVORE, OMNIVORE, are printed in upper case with specialties written in lower case. Letters g, i, m following CARNIVORE indicate general, insectivore, or molluscivore. The letters g, c, or h following OMNIVORE indicate generalist, carnivorous, or herbivorous preference, respectively. If two lower-case letters are given, this means males (first letter) specialize in a different food type from females. (See table 3.2 and this chapter for discussion of guild types.) Data from Ernst et al. (1994).

cooters among the emydids. River cooters/Suhwannee river cooters drop out at the St. Johns River, leaving the common cooter and Florida red-bellied turtle as the chief practitioners of the cooter niche through the remainder of the Peninsula including Florida's "river of grass," the Everglades.

Moving westward from the Mississippi River around the Gulf, the Atchafalaya River is the first encountered. The chelonian assemblage is virtually the same as the Mississippi as expected since it is continuous with the Red River, one of its tributaries. Next is the Sabine River Basin where we begin to see some evidence of isolation, an endemic race of the Ouachita map turtle, the Sabine map turtle (*G. ouachitensis sabinensis*). Cagle and Chaney (1950) captured spiny soft shells and common sliders in the Sabine, while Chaney and Smith (1950) recorded alligator snapping turtles, razor-backed musk turtles, river cooters, Sabine map turtles, and false map turtles. In Texas, several rivers have had sufficient isolation to allow the evolution of endemic or nearly endemic chelonians. *Pseudemys concinna* appears to have spawned two such species: the Texas river cooter (*P. texana*) and the Rio Grande cooter (*P. gorzugi*). The former occupies the Brazos, Colorado, and Guadalupe/San Antonio system and the latter the Rio Grande. Map turtles also have endemic representatives in the Colorado (Texas map turtle, *G. versa*, Vogt, 1981c) and the Guadalupe/San Antonio (Cagle's map turtle, *G. caglei*). Common sliders inhabit the lower Rio Grande, but are replaced in the Big Bend by the closely related Big Bend slider (*Trachemys gaigeae*). Legler (1960a) reported that in the Rio Grande at Lajitas, Texas the turtle community comprises only three species, the spiny softshell (*Apalone spinifera*) (most common), the Big Bend slider (uncommon), and Mexican rough-footed mud turtles (*Kinosternon hirtipes*) (rare). The mud turtle and softshells are chiefly carnivorous, while the slider is an omnivore that feeds heavily on plants as an adult.

Moving westward along the Rio Grande, the climate becomes progressively drier, the rivers smaller, and the chelonian diversity diminished. Mud turtles, because of their ability to adapt to unstable aquatic conditions, are the dominant turtles at many localities. In the Rio Conchos which flows through Chihuahua, Mexico to enter the Rio Grande near the Big Bend, the Mexican rough-footed mud turtle, was the most abundant turtle, followed by the Big Bend slider (*T. s. gaigeae*) and Texas spiny softshell (*A. spinifera emoryi*) (Legler, 1960a; Williams et al., 1960; Smith et al., 1963). Still westward in the Bill Williams, Gila, and Colorado River systems the Mexican mud turtle is replaced by the Sonoran mud turtle (*K. sonoriense*). This may be the only chelonian to occur in these river systems naturally. Spiny soft shells are presently in the Gila and Colorado but are thought to be a recent introduction (see Miller, 1946). Baja California is virtually devoid of permanent rivers, but two species of turtles are associated with the intermittent ones: the Pacific pond turtle in the north, and again a race of the common slider in the south. Northern Mexico, like the Gulf Coast, has an abundance of small isolated river systems and like map turtles, the resident mud turtles seem prone to speciation. The common slider, while not giving rise to full species in these systems, has produced a number of endemic or nearly endemic subspecies (see Legler, 1990; Ernst, 1990; Moll et al., in press).

Typical river systems in northern Mexico accommodate a slider and one or more mud turtles (see table 3.6). Moving southward from the Rio Grande, the rivers of the Mexican Plateau region typically contain one or more of three species of mud turtles, the Mexican rough-footed, Mexican (*K. integrum*), and Sonoran.

Table 3.6
Slider and mud turtles from selected rivers of Nearctic Mexico.

River system	Slider *Trachemys scripta*	Mud turtle *Kinosternon*
Rio Conchos (CH)	Big Bend, *T. gaigeae*	Mexican rough-footed, *K. hirtipes*
Rio Nazas (D & C)	Nazas, *T. s. hartwegi*	Mexican rough-footed, *K. hirtipes*
Rio Verde (SLP)	Huastecan, *T. s. cataspila*	Mexican, *K. integrum*
Rio Panuco (T)	Huastecan, *T. s. cataspila*	Herrera's, *K. herrerai* Scorpion, *K. scorpioides*
Rio Sonora (SO)	None	Sonoran, *K. sonoriense*
Rio Yaqui (SO)	Yaqui – *T. s. yaqui*	Mexican, *K. integram* Mexican rough-footed, *K. hirtipes** Sonoran, *K. sonoriense*
Rio Fuerte (SO & SI)	Fuerte, *T. s. hiltoni*	Mexican, *K. integrum*
Rio Mezquital (D&N)	Ornate, *T. s. ornata*	Mexican rough-footed, *K hirtipes** Mexican, *K. integrum*
San Ignacio (B)	Baja California, *T. s. nebulosa*	

Species prefering temporary water such as *K. alamosae* and *K. flavescens* are not included. Data drawn from Berry and Legler (1980); Ernst (1990); Iverson (1981,1992a); Iverson and Berry (1979); Legler (1990); Smith and Smith (1979). Abbreviations in parentheses are Mexican States: Chiapas, CHihuahua, Duango, Nayareet, San Luis Potosi, SInaloa, SOnora, Tamaulipas.
*Mexican plateau portion only.

While these may occur in the same river, usually there is only moderate overlap. For example, on the Rio Yaqui the Sonoran mud turtle occurs on tributaries west of the continental divide, while the Mexican rough-footed mud turtle occurs in the eastern tributaries (Iverson, 1981). Where the Mexican rough-footed mud turtle and the Mexican turtle occur in the same river in Pacific Coastal drainages, the Mexican rough-footed mud turtle tends to occupy higher elevations while the Mexican mud turtle is more common at lower elevations (for example the Rio Mezquital; see Iverson 1981). At low elevations along the Gulf Coast above Vera Cruz, Herrera's mud turtle (*K. herrerai*), and the scorpion mud turtle (*K. scorpioides*) replace the Mexican rough-footed mud turtle and the Mexican mud turtle.

Neotropics

The Neotropical Region begins as a narrow extension along the lowlands of each coast of Mexico in Vera Cruz and Oaxaca. These coastal prongs merge at the Isthmus of Tehuantepec and then continue through southern Mexico, Central America, and South America (Darlington, 1957). Like the northern portion of

Mexico, rivers of the southern portion and Central America are typically inhabited by a slider and at least one mud turtle, usually the white-lipped (*K. leucostomum*) and/or scorpion mud turtle (*K. scorpioides*). The scorpion mud turtle is more common in ephemeral habitats. Regionally endemic mud turtles (e.g., *K. angustipons, K. acutum, K. creaseri, K. dunni, K. oaxacae*) are also found within different sectors. The diversity of chelonians is somewhat greater than that found in Nearctic Mexico. On the Atlantic side, the Central American river turtle (*Dermatemys mawii*; a herbivore), the inexplicably rare Central American snapping turtle (*Chelydra rossignoni*), and the Mexican giant musk turtle (*Staurotypus triporcatus*), both of which are omnivores, augment the riverine assemblage. The chopontil (*Claudius angustatus*) is another addition to the chelonian fauna in the region. Though not part of the river fauna *per se*, it forages in adjacent flooded grasslands whenever rivers overflow. At the level of Honduras, *Claudius, Dermatemys*, and *Staurotypus* drop out and are replaced by the herbivorous semi-aquatic, black turtle (*Rhinoclemmys funerea*). On the South American side of the Isthmus of Panama (both coasts) the black turtle is itself replaced by another semi-aquatic species, the Columbian wood turtle (*R. melanosterna*).

The riverine fauna of the Pacific side of Central America is more depauperate, consisting chiefly of the slider/mud turtle association with the scorpion mud turtle being the typical *kinosternid* on this side. The Chiapas giant musk turtle (*S. salvinii*) also occurs in flooded habitats and backwaters associated with small rivers from southern Mexico to El Salvador. Table 3.7 provides examples of river turtle assemblages in selected rivers through southern Mexico and Central America. Table 3.8 compares habitat and food preferences of selected Central American river turtles.

Northern South America resembles Central America in that the slider/mud turtle association typifies most of the rivers and the white-lipped and scorpion mud turtles remain the dominant kinosternids. The South American snapping turtle (*Chelydra acutirostris*), another species of northern extraction, inhabits northern South America. Interestingly, the slider turtle and scorpion mud turtle chiefly inhabit rivers of the Atlantic drainage, whereas the snapping turtle and white-lipped mud turtle are found chiefly in rivers of the Pacific drainage. Another kinosternid of the Pacific drainage is Dunn's mud turtle, a large species endemic to the Choco Region of Colombia.

Among the New World batagurids, the most riverine member of the tropical wood turtle group—the large-nosed wood turtle (*Rhinoclemmys nasuta*)—inhabits large, deep rivers as far south as Esmeraldas, Ecuador. The Colombian wood turtle occurs in many of the same rivers, but whereas the large-nosed rarely leaves water the Colombian makes regular sojourns on land and often feeds terrestrially. Moving eastward to the vicinity of the Orinoco River, the Colombian is replaced by an ecological equivalent, the spotted-legged turtle (*Rhinoclemmys punctularia*).

The transition of northern river turtles to southern river turtles begins in the Rio Magdalena of northern Colombia in a novel encounter between emydids and podocnemids (table 3.9). The river harbors both the cryptodiran slider (*Trachemys scripta*) and the pleurodiran Magdalena River turtle (*Podocnemis lewyana*). Presumably coexistence is possible here because while the Magdalena River turtle

Table 3.7
Species of turtles inhabiting selected rivers in southern Mexico and Central America.

Species	Rivers of Southern Mexico and Central America					
	Rio Lacantun Chiapas	Rio Bravo Belize (P)	Chan Chen Belize (T)	St. Domingo Nicaraqua	Tortuguero Costa Rica	Rio Chagres Panama
C. rossignoni	×					
C. acutirostris					×	×
D. mawei	×	×				
R. funerea				×	×	×
T. scripta	×	×	×	×	×	×
K. acutum	×	×				
K. leucostomum		×	×	×	×	×
K. scorpioides			×	×		
S. triporcatus	×	×	×			
Total species	5	5	4	4	4	4

Genera abbreviations are *Chelydra* (C), *Dermatemys* (D), *Rhinoclemmys* (R), *Trachemys* (T), *Kinosternon* (K), and *Staurotypus* (S). Sources include D. Moll (1990); Moll and Legler (1971); Polisar (1991); Vogt and Villareal-Benitez (1997); and unpublished data from E. and D. Moll.

Table 3.8
Habitat preference and adult guilds of selected tropical Mexican and Central American Turtles.

Species	Habitat preference	Guild
C. rossignoni/ acutirostris	Generalized: Deep rivers, streams, estuaries, swamps, ditches.	OMNIVORE/ general
D. mawei	Large bodies of permanent water, particularly deep rivers.	HERBIVORE/general
R. funerea	Semi-aquatic: quiet waters of large rivers, streams, swamps in humid forested areas.	HERBIVORE/general
T. scripta	Generalized; optimum: large, permanent, slow-moving bodies of water, with abundant vegetation, basking sites and open areas for nesting.	OMNIVORE/ herbivorous
K. acutum	Semi-aquatic: small, forested, often ephemeral, ponds and streams.	OMNIVORE/ carnivorous ??
K. leucostomum	Slow-moving areas of rivers, permanent streams, ponds, and swamps.	OMNIVORE/general
K. scorpioides	Semi-aquatic: permanent and temporary water of almost any type.	OMNIVORE/general
S. salvinii	Slow-moving, mud-bottomed, well-vegetated habitats in rivers, streams, swamps.	OMNIVORE/ carnivorous
S. triporcatus	Large permanent bodies of water, rivers, lakes, swamps.	OMNIVORE/ general(durophagous feeding on mollusks, hard seeds, turtles, etc.)

Guilds are presented as a general category (in upper case) and a specialization or preference (lower case). See table 3.2 and chapter 3 this text for information on guild categories. See table 3.7 for genera abbreviations. Sources of information are Dean (1980); Feuer (1966); D. Moll (1990); Moll and Legler (1971); Polisar (1990a); Vogt and Guzman-Guzman (1988).

is a river specialist, the slider is merely a facultative river inhabitant, or river turtle "wanabe," The generalized sliders are the dominant river turtles for much of the region from northern Mexico to northern South America because, other than the small region occupied by the Central American river turtle, there are no river turtle specialists competing for the habitat (see Moll & Legler, 1971; Moll & Moll, 1990). With the occurrence of multiple riverine *Podocnemis* species in the Orinoco River and Amazon River drainages, sliders temporarily disappear from the riverine communities only to reappear below the Amazon basin and outside of the range of *Podocnemis* (table 3.9). Two sliders (*Trachemys adiutrix* and *T. dorbigni*) occur in the state of Maranhao, Brazil, with the range of the latter continuing southward to the Paraná and the Uruguay drainages of northern Argentina.

Table 3.9
Chelonian species composition of seven rivers in South America. The Rio Guapore, Rio Trombetes and Rio Tapajos are part of the Amazon drainage.

Species	Magdalena Colombia	San Juan Colombia	Orinoco Venezuela	Guapore Brazil	Trombetas Brazil	Tapajos Brazil	Paraná Argntina
C. acutirostris	×	×					
R. melanosterna	×	×					
R. nasuta		×					
R. punctularia			×				
T. dorbigni							×
T. scripta	×						
K. leucostomum	×	×					
K. scorpioides	×		×				
A. pallidipectoris							×
A. spixii							×
Ch. fimbriatus			×	×			
H. tectifera							×
Ph. gibbus			×				
Ph. geoffroanus			×	×			×
Ph. hilarii							×
Ph. raniceps			×	×			
Ph. vanderhaegei							×
Pe. dumerilianus			×		×		
P. erythrocephala			×			×	
P. expansa			×	×	×	×	
P. lewyana	×						
P. sextuberculata					×	×	
P. unifilis			×	×	×	×	
P. vogli			×				
Total species	6	4	11	5	4	4	7

Genus abbreviations not found in table 3.7 are: *Acanthochelys* (A), *Chelys* (Ch), *Hydromedusa* (H), *Phrynops* (Ph), *Peltocephalus* (Pe), and *Podocnemis* (P). Data taken from Fachin-Teran et al. (1995); Iverson (1992b); Medem (1962); Pritchard and Trebbau (1984); Vogt and Villareal-Benitez (1997).

South America is graced with two of the world's great rivers, the Orinoco with a length of around 2400 km, and the Amazon, the world's largest in terms of volume is close to 6400 km in length. The Amazon crosses the most extensive tract of rain forest in the world. The Orinoco not only traverses broad tracts of rain forest, but also extensive savannahs known as llanos. Table 3.10 summarizes the species and guilds characterizing these great rivers. The Orinoco and Amazon are home to eleven and twelve chelonians, respectively, a diversity equivalent to that of many of the Nearctic rivers draining into the Gulf of Mexico. However, in the former, the diversity is chiefly of pleurodires (side-necks, turtles that protect their head and neck by turning them laterally under the lip of the shell rather than

Table 3.10
Guilds of chelonians occupying the Orinoco (O) and Amazon (A) Drainages of South America.

Species	Distribution	Guild
Rhinoclemmys punctularia	O,A	OMNIVORE/generalist
Kinosternon scorpioides	O,A	OMNIVORE /generalist
Chelus fimbriatus	O,A	CARNIVORE/piscivorous
Phrynops geoffroanus	O,A	OMNIVORE/generalist
Phrynops gibbus	O,A	OMNIVORE/generalist?
Phrynops raniceps	O,A	OMNIVORE/molluscivorous
Phrynops rufipes	A	OMNIVORE/herbivorous
Peltocephalus dumerilianus	O,A	OMNIVORE/carnivorous
Podocnemis expansa	O,A	HERBIVORE
Podocnemis erythrocephala	O,A	HERBIVORE
Podocnemis sextuberculata	A	OMNIVORE/herbivorous
Podocnemis unifilis	O,A	HERBIVORE
Podocnemis vogli	O	OMNIVORE/generalist

Criteria for guilds provided in table 3.1. A question mark signifies that information found in the literature was inconclusive. Two additional turtles, *Phrynops nasutus* and *Platemys platycephala* from the region have not been included because of their strong preference for lentic habitats. Information obtained from de Lima et al. (1997); Fachin-Teran et al. (1995); Ernst and Barbour (1989); Iverson (1992b); Pritchard and Trebbau (1984).

withdrawing it straight back as do the hidden necks (i.e., cryptodires). Two side-neck families, the Podocnemidae and the Chelidae inhabit South America. Cryptodires are in the minority within these drainages, being represented by two semi-aquatic species, the spot-legged wood turtle (*Rhinoclemmys punctularia*) and scorpion mud turtle.

The podocnemid South American river turtles (*Podocnemis*) dominate the main stem and major tributaries of the Orinoco and Amazon. In fact, only one of the six species in existence (*P. lewyana*) does not occur in either river. New World podocnemids tend to be visualized as large, powerful swimming chelonians with herbivorous preferences, but there is considerable variation. The giant South American river turtle (*P. expansa*) best fits this description. This huge turtle (>100 cm shell length) typically is found in deep portions of the white water rivers and the flooded forests where adults feed chiefly on plants and their fruits. Conversely two species, the red-headed Amazon river turtle and the six-tuberculed Amazon River turtle (*P. erythrocephala* and *P. sextuberculata*) barely exceed 30 cm in shell length, while a third, the savanna side-necked turtle (*P. vogli*)—an inhabitant of the llanos—appears to be the most omnivorous and least riverine of the group, often occupying temporary water (Pritchard & Trebbau, 1984). None of the *Podocnemis* are obligate herbivores and all will take at least small amounts of animal food. The big-headed Amazon river turtle (*Peltocephalus dumeriliana*) is the only other South American pelomedusid. The habits of this large (68 cm), dome-shelled inhabitant of the Orinoco and Amazon are poorly known, but it is most common in black-water habitats. A recent diet study (Pérez-Emán & Paolillo O, 1997) found fruit, seeds, fish and aquatic plants to be the most common items in its diet. The proportion of animal matter found (52% of the volume consumed) is the highest reported for any Neotropical podocnemid to date.

The chelids are not so much river turtles as turtles found in and around rivers. For the most part they are small to moderate-sized, bottom-walkers inhabiting tributaries and backwaters of the big rivers. Two genera, *Chelus* and *Phrynops*, inhabit lotic habitats of the Orinoco and Amazon drainage. The twist-necked turtle (*Platemys platycephala*)—another turtle occurring over much of the area occupied by the drainage of these rivers—is not included here because of its weak swimming ability and preference for lentic habitats.

The mata mata (*Chelus fimbriatus*), a bottom-walker with perhaps the most bizarre appearance of any turtle, successfully stalks fish in a variety of lotic and lentic habitats associated with these major rivers. Reaching 45 cm in length, *Chelus* is the largest of the New World chelids. The toad-headed turtles (*Phrynops*) comprise a diverse omnivorous to carnivorous group occupying habitats ranging from ponds to rivers.

Table 3.10 summarizes guild information and reveals the great similarity in chelonian species between the two great rivers. Ten of the species are shared, and only three species are restricted to one of the two rivers. The similarity index comparing the turtle fauna is 0.87. Obviously these two rivers have a similar history and even today the two achieve some connection through the Casiquiare and the Negro rivers.

Perhaps the greatest ecological contrast one can make between the temperate and tropical faunas thus far reviewed is the trend toward carnivory in the temperate region and the increased importance of herbivory in the tropical region. From a taxonomic standpoint, the dominance of emydids in the Nearctic system shifts to a dominance of podocnemids and chelids in the rivers near the equator. The absence of *Trachemys*-type sliders in the Orinoco and Amazon Basin is particularly striking.

Southward of the Amazon drainage, podocnemids disappear from the assemblage. Other than the emydid genus *Trachemys*, which rejoins the assemblage after the podocnemids disappear, chelids of the genera *Achantochelys* (a genus closely related to *Platemys*), *Hydromedusa*, and *Phrynops* comprise the southern assemblages. Carnivory again replaces herbivory in importance. Based on the Rio Paraná (see table 3.9) all the species represented are either carnivorous or omnivorous. The South American snake-necked turtle (*Hydromedusa tectifera*) is chiefly carnivorous. Limited information on diets of South American swamp turtles (*Acanthochelys*) and the toad-headed turtles (*Phrynops*) suggests carnivory or omnivory with a carnivorous bias. *Acanthochelys macrocephala* feeds heavily on snails. *A. radiolata*, *pallidpectoris*, and *spixii* also appear to be mainly carnivorous. *Phrynops hilarii* is chiefly carnivorous, and *P. geoffroanus* is omnivorous. Presumably *Trachemys dorbigni*, like most other sliders, is an opportunistic omnivore feeding on whatever is available. South of the Rio Paraná little information exists concerning river turtles. *Phrynops hilarii* ranges in the vicinity of the Rio Colorado in Argentina but no other species are reported.

Palearctic

Unlike the Nearctic, river turtles are poorly represented in the Palearctic. In the west many of the rivers are inhabited by the emydid European pond turtle (*Emys obicularis*) and/or one of two species of the batagurid stripe-necked turtles, genus *Mauremys* (*caspica* or *leprosa*). These generalized, chiefly carnivorous, pond turtles possess few specific adaptations to riverine environs, but have opportunistically moved into slow current habitats to utilize this largely uncontested niche. The European pond turtle ranges from northern Africa and Europe eastward to the Aral Sea. The Mediterranean turtle (*Mauremys leprosa*) occupies the western Mediterranean region (northwestern Africa and southwestern Europe) and is replaced by the Caspian turtle (*M. caspica*) in the northeastern Mediterranean region eastward to the Caspian Sea. The Caspian terrapin is joined in the Mid-East by two species of softshells—the Nile softshell (*Trionyx triunguis*) that reaches the northern limits of its range around the eastern end of the Mediterranean Sea and the Euphrates softshell (*Rafetus euphraticus*) inhabiting the Tigris/Euphrates river basin.

Rivers of the eastern Palearctic are similarly poor in numbers of turtle species. The "hardshells" are again pond turtle types, including stripe-necked turtles (*Mauremys*), Chinese pond turtles (*Chinemys*) and Chinese stripe-necked turtles (*Ocadia*). All of these turtles will utilize shallow riverine habitats with slow current. The most common and widespread trionychid of the region, the Chinese softshell (*Pelodiscus sinenis*) ranges from northern Vietnam through China, Korea, Japan all the way to southern Siberia.

The Chang Jiang or, as it is better known to Westerners the Yangtze, is the longest river in Asia, and is selected here to exemplify an eastern Palearctic River. Beginning at some 4900 m (16,000 feet) in the Tanglha Mountains on the eastern border of Tibet, the Yangtze flows across some 5500 km (3430 miles) of China to reach the East China Sea near Shanghai. Two other major rivers of the region—the Salween and the Mekong—originate nearby and parallel the Yangtze for some distance, but where they turn southward toward the tropics, the Yangtze is blocked by a high plateau and continues its course towards the East.

In the mountainous portions of the course, the river makes a precipitous descent through canyons, ravines and gorges which are largely unsuitable habitat for turtles. However, the last 1600 km are wide, deep and passable by ocean-going vessels, and smaller craft can find navigable waters for up to 2400 km (1500 miles) inland.

Despite its size and diversity of habitat, the Yangtze is home to few chelonians. The Chinese softshell is the most common and also the most economically important turtle. Riverine peoples of the region relish the meat and cartilaginous portions of the shell as food, and nearly all parts of the turtle are used for medicinal preparations. This chiefly carnivorous softshell rarely basks and forages for fish and aquatic invertebrates in areas of slow current (Pope, 1935). The enigmatic Shanghai softshell (*Rafetus swinhoei*) could also occur in the drainage, but neither its distribution nor its ecology has been adequately studied. In contrast to *Pelodiscus*, the two "hardshells" known from the river are inveterate basking species and thus are as conspicuous as the softshell is cryptic. The omnivorous Reeve's turtle (*Chinemys reevesii*) also occupies areas of slow-moving current where it feeds on aquatic plants, fruits, and aquatic invertebrates. The yellow pond turtle (*Mauremys mutica*) reaches the northern limit of its range at the Yangtze. It, too, is omnivorous, but seems to lean toward carnivory.

Ethiopian

Despite its large size and having a number of great rivers, the Ethiopian zoogeographic region is peculiarly poor in numbers with regard to river turtles. Five trionychids are the only cryptodires inhabiting the rivers from the great Sahara Desert southward. The Nile softshell (*Trionyx triunguis*) ranges widely over the Nile River and central Africa. Two genera of flapshells, *Cyclanorbis* and *Cycloderma* with two species each also occur in the central region. The remainder of the riverine turtle fauna of Africa is comprised of pleurodires of the chiefly carnivorous genera *Pelusios* (the African hinged terrapins) and *Pelomedusa* (the helmeted terrapin). *Pelomedusa* is best adapted for lentic habitats (temporary pans and seasonal swamps) and becomes a river turtle only outside of the range of *Pelusios* in South Africa (Hewitt, 1935 in Broadley, 1981) and perhaps in southwestern Arabia where it is also the only pleurodire freshwater turtle (Gasperetti et al., 1993). An obvious distinguishing feature separating *Pelomedusa* and *Pelusios* is the hinged plastron of the latter. The African hinged terrapins are a speciose group, with each species having a relatively confined range compared to the eurytopic helmeted terrapin in which a single species (*subrufa*) occurs over most of Africa and Madagascar. Madagascar has an additional side-neck genus, *Erymnochelys*, a large (50 cm carapace length) riverine turtle reminiscent of Neotropical, *Peltocephalus*. In this section three African rivers, the Nile, Zaire (Congo), and Zambezi and one Madagascan river, the Betsiboka, are examined as representative of the Ethiopian Region (table 3.11).

The Nile is not only Africa's longest river, it is the longest anywhere on the planet. No other river has such a great range in latitude beginning in the equatorial lake plateau of East Africa, flowing through the tropical plains of south Sudan eventually reaching the deserts of north Sudan and Egypt, and finally emptying into the Mediterranean Sea.

The Nile softshell (*Trionyx triunguis*) is the most impressive turtle in this expansive river. This huge, aggressive chelonian, approaches 1 m in length and 40 kg in weight. Reports on its diet are contradictory, varying from complete herbivory (Ruppel, 1835 in Loveridge & Williams 1957) to complete carnivory (Anderson, 1898; Worthington, 1929). Villiers (1962) reports that the turtle is an ambush feeder that attacks any prey-sized animal coming within reach (see also

Table 3.11
Species composition and richness of turtles in four rivers of the Ethiopian Region.

Species	Nile	Zaire	Zambezi	Betsiboka
Cyclanorbis elegans	×			
C. senegalensis	×			
Cycloderma aubryi		×		
C. frenatum			×	
Trionyx triunguis	×	×		
Erymnochelys madagascariensis				×
Pelusios adansonii	×			
P. bechuanicus			×	
P. carinatus		×		
P. castaneus		×		
P. castanoides				×
P. chapini		×		
P. gabonensis		×		
P. nanus		×	×	
P. rhodesianus		×	×	
P. sinuatus			×	
P. upembae		×		

Species characteristic of ephemeral and lentic habitats (e.g., *Pelomedusa subrufa* and *Pelusios broadleyi*) are not listed.

chapter 2, this volume). Nevertheless, they also favor ripe fruits which fall into the river. The Nile softshell we consider as a carnivore generalist.

In the upper Nile, sub-saharan flapshells, of the genus *Cyclanorbis* replace *Trionyx* as the common trionychid (Worthington, 1929; Loveridge & Williams, 1957). The name, flapshell, comes from a pair of fleshy flaps at the posterior of the plastron that conceal the retracted hind legs. Two species inhabit the Nile, the Nubian flapshell (*C. elegans*) and the Senegal flapshell (*C. senegalensis*). The natural history of these enigmatic softshells is little known.

Two pleurodire genera, *Pelusios* and *Pelomedusa*, are associated with the Nile. The first is Adanson's hinged terrapin (*Pelusios adansonii*); this is a carnivorous river turtle of savanna regions which ranges to the edge of the Palearctic region, making it the most northerly member of its genus (Loveridge, 1941). Another wide-ranging species, the helmeted terrapin (*Pelomedusa subrufa*), is an ecological generalist preferring lentic and ephemeral bodies of water, but occasionally occurring in habitats associated with the Nile. *Pelomedusa* occupies a particularly broad and interesting feeding niche, eating a wide variety of aquatic animals along with some plants (Villiers, 1958; Boycott & Bourquin, 2000; Branch, 1988). In addition to the typical foraging technique of walking the bottom in search of prey, they occasionally emulate miniature crocodiles, stalking and seizing unwary birds drinking at the water's edge (Branch, 1988; Boycott & Bourquin, 2000) or act as cleaners gleaning ectoparasites from the hides of the large mammals that wallow in their habitat (Rochat et al., 1962) as discussed previously in chapter 2, this volume. D. G. Broadley (unpubl.) has a report from Hwange National Park of a *Pelomedusa* feeding on elephant dung (or more likely beetles working within it).

The Zaire (Congo) is the second longest river in Africa and carries the second greatest volume of any of the world's rivers. Beginning in the Katanga Plateau at an altitude of 1420 meters, the Zaire River flows through Zaire, and the Republic of Congo, crossing the equator twice before finally emptying into the Atlantic Ocean. Providing long navigable stretches exceeding 16 km (10 miles) wide in places, the river also is known for its extensive cataracts (Stanley Falls) and a huge whirlpool (Devil's Cauldron). Rain forest lines the banks for much of its course.

Denizens of the Zaire River System include two softshells (the widespread Nile softshell and Aubry's flapshell (*Cycloderma aubryi*) and a pleurodiran contingent comprising at least seven *Pelusios* (*carinatus*, *castaneus*, *chapini*, *gabonensis*, *nanus*, *rhodesianus*, and *upembae*).

The African forest hinged terrapin (*Pelusios gabonensis*) inhabits the drainage in the rain forests of the Zaire system, whereas the African dwarf hinged terrapin (*P. nanus*) occupies the moist savannas and forest/savanna mosaic along the watershed between the Zaire and Zambezi systems (Broadley, 1981). The carnivorous African keeled hinged terrapin (*P. carinatus*), an ecological equivalent of *P. sinuatus* of South Africa (Broadley, 1981), occupies both lentic and lotic habitats along the river. The West African hinged terrapin (*P. castaneus*) and the variable hinged terrapin (*P. rhodesianus*) are common inhabitants of the many swamps and weed-choked backwaters associated with the river and its tributaries (Broadley, 1981). The latter is reportedly omnivorous but strongly biased toward carnivory. The diet includes invertebrates, amphibians, fish, and aquatic plants (Boycott & Bourquin, 2000; Branch, 1988). Little is known concerning the natural history of the Upemba hinged terrapin (*P. upembae*) or Chapin's hinged terrapin (*P. chapini*)—a species formerly synonomized with *P. castaneus*.

The Zambezi is the principal river of southern Africa. Some 3200 km in length, it is the continent's fourth longest river. The Zambezi begins on a high plateau in northwestern Zambia and eventually enters the Indian Ocean through a myriad of mouths. At Victoria Falls, one of the most magnificent natural spectacles in all of Africa, the Zambezi plunges over 120 m (400 feet) into a winding 1.5-km long gorge. The Zambezi Basin includes the whole of Malawi and parts of Zambia, Angola, Zimbabwe, and Mozambique. The Zambezi flap-shell (*Cycloderma frenatum*), the lone cryptodire, occurs only in the lower Zambezi system below the Caborabassa Gorge and is probably most abundant in shallow waters of southern Lake Malawi (Broadley, unpubl.). This is a carnivorous species that feeds on large pulmonate snails and large bivalves, which are dug from the substrate with the powerful clawed front feet (Mitchell, 1946). Fish, amphibians, and insects are also eaten (Branch, 1988).

Among the pleurodires in the Zambezi watershed, several are lentic adapted species. The nearly ubiquitous *Pelomedusa subrufa* and *Pelusios subniger* are chiefly restricted to ephemeral habitats (temporary pans). *P. rhodesianus* favors weed-choked impoundments such as the nutrient-rich Lake Chivero near Harare, while *P. castanoides* inhabits coastal lagoons and occurs in shallow Lake Chilwa in southern Malawi (Mitchell, 1946).

The lotic habitats of the Zambezi system appear to be partitioned among four species of African hinged terrapins. The dwarf hinged terrapin (*P. nanus*) dwells in small clear tributaries close to the Zaire/Zambezi watershed in northwestern Zambia and eastern Angola. The Okivango hinged terrapin (*P. bechuanicus*) inhabits the deep clear waters of the Zambezi above Victoria Falls, the upper Kafue River and the Kafue Flats. The East African serrated hinged terrapin (*P. sinuatus*), the largest member of the genus (46.5 cm carapace length), is the typical river turtle of the middle and lower Zambezi, whereas the variable hinged terrapin (*P. rhodesianus*) is usually found in marshes and dams in the upper reaches of the tributaries.

Pelusios bechuanicus is carnivorous, feeding on fish and invertebrates. *Pelusios sinuatus* is somewhat omnivorous, but leans toward carnivory, particularly favoring large pulmonate snails (Mitchell, 1946; Boycott & Bourquin, 2000). Like the helmeted terrapin, *P. sinuatus* will both scavenge dead carcasses and pluck ticks from the hides of wallowing buffalo and other large mammals (Boycott & Bourquin, 2000, Branch, 1988). It occupies both permanent lentic and lotic habitats, but where it occurs sympatrically with *P. bechuanicus* above Victoria Falls, *P. sinuatus* is restricted to lagoons and muddy backwaters leaving the river stem to *bechuanicus* (Broadley, 1981).

Madagascar is the only other land mass in the Ethiopian Region large enough to accommodate river turtles. As mentioned previously, only one true river turtle occurs here, the Madagascan big-headed turtle, *Erymnochelys madagascariensis*. *Erymnochelys* lives in large slow-moving rivers, backwaters and lakes in western Madagascar (Kuchling, 1988). It moves between the river stem and backwaters and swamps at various times of the year (Kuchling & Mittermeier, 1993). Two *Pelusios* (*castanoides*, and *subniger*) and the ever-present *Pelomedusa subrufa* have also reached the island. Kuchling found *Erymnochelys* and *P. castanoides* together in a backwater and swamp of the Betsiboka River.

In summary, the Ethiopian river turtle fauna is unusual in its lack of variety and in the fact that carnivory rather than herbivory predominates the feeding guilds of the tropical regions.

Oriental Region

Spectacular rivers with fascinating histories abound in this region—the Indus, Ganges, Brahmaputra, Irrawaddy, Chao Phraya, Mekong, and Xi Jiang to mention a few. Oriental river turtles are as impressive as the rivers they inhabit, comprising huge trionychids with leathery shells exceeding 1 m in length along with massive batagurids weighing 50 kg or more. Softshells come in three varieties:, small flapshells (*Lissemys*), whose closest relatives are *Cyclanorbis* and *Cycloderma* of Africa; typical long-snouted softshells (e.g., *Aspideretes, Amyda, Dogania, Pelodiscus*) spanning the gamut from small to large, and the snub-nosed giants (*Chitra, Pelochelys*) which reportedly can exceed 1.5 m in shell length. The hard-shelled batagurid river turtles include three genera of river specialists known collectively as terrapins (e.g., *Batagur*, *Callagur*, *Kachuga*). A massive, high-vaulted, heavily buttressed shell (see fig. 2.23) is the terrapin's most obvious distinguishing feature. While some have speculated that the large buttresses protect and support the lungs during deep dives, they also likely serve for defense against the crocodiles that formerly teemed in the rivers of the region. Members of less-specialized genera also inhabit rivers and their associated habitats (e.g., *Cuora*, *Heosemys*, and *Orlitia*). Tables 3.12 and 3.13 list the species occurring in eight representative rivers of the region.

In 1876, Alfred Russell Wallace, the celebrated peregrinator and "Father of Zoogeography," recognized four sub-regions of the Oriental Region (the Indian, Ceylonese, Indo-Chinese, and Indo-Malayan) based on the similarity of species occupying each. Other than the Indian, however, these subdivisions do not fit the chelonian fauna very well. Looking at turtle communities inhabiting the larger rivers, we visualize three distinct regions based on chelonian species held in common (table 3.14).

The Indus, Ganges and probably the poorly known Brahmaputra represent a center of similarity which corresponds well with Wallace's "Indian sub-region". Herein this will merely be called division #1. Using Sorensen Coefficients, also known as the quotient of similarity (Brower & Zar, 1984) to show relationship, the Indus and Ganges have a coefficient of 0.64 (where 0 indicates no overlap of species and 1 complete overlap). The *Aspideretes* softshells and the *Kachuga* (excepting *K. trivittata*) characterize this sub region.

With eastern India and Bangladesh as a transition zone, the next distinctive sub-region includes rivers from the Irrawaddy to the Mekong, including the Malay Peninsula, the Oriental portion of Indonesia and the Philippines. Coefficients from comparisons of five rivers examined in this division ranged from a low of 0.33 to a high of 0.84. By contrast, comparing any of these five with rivers from the other two divisions produce coefficients ranging from 0 to a high of 0.16 (table 3.14). *Amyda* softshells with *Cyclemys*, *Heosemys*, and *Cuora amboinensis* are indicative taxa for the region. Wallace viewed this grouping as belonging to two regions, the Indo-Chinese and the Indo-Malayan; herein this will be considered as a single sub-region, #2.

The Xi Jiang (included in the Indo-Chinese sub-region by Wallace) shares only one species, the wide-ranging *Pelochelys cantori*, with the other two regions. Since this region has had a much different biogeographical history than the rivers included in the previous sub-regions, we here refer to southern China as division #3. *Palea* softshells along with *Ocadia*, *Sacalia*, and *Mauremys mutica* are characteristic taxa.

Division #1

The Indus, now chiefly a river of Tibet and Pakistan, is the source of the name India. It is the longest Himalayan river, flowing a distance of some 3330 km from its source in the glaciers of the Kailas range to its outlet in the Arabian Sea. Once outside of the mountains the Indus becomes a multi-channeled, braided river that cuts a relatively straight course to the sea. In places, the river spreads out over 30–40 km in several channels cutting through dazzling sandy deposits. Floods may peak at sixty times the low water level in this semi-desert region. Seven chelonians inhabit the Indus. and its associated habitats—three softshells and four batagurid hard shells. The main stem of the river is the domain of the large softshells. The extensive sandy stretches of the Indus River are particularly favorable for the Indian narrow-headed softshell turtle (*Chitra indica*). With eyes situated distally near its nubbin of a snout, a long muscular neck and a flattened shell, this riverine specialist is admirably suited for ambush feeding. The flat, chapati-like profile of the turtle allows for easy concealment by interment within the sandy substrate. Here, a *Chitra* will lay with its long neck retracted, leaving only the tiny eyes exposed and awaiting some complacent fish, crustacean, or other aquatic morsel to swim near. As discussed previously in chapter 2, the long neck—guided by the distally positioned eyes—is launched towards the intended target to the accompaniment of a veritable explosion of sand. Synchronously, the mouth gapes and throat expands creating an inrush of water that literally sucks the prey into the turtle's maw. The Indian softshell, *Aspideretes gangeticus*, more a generalist than *Chitra*, inhabits either sand or mud habitats where it forages on a variety of plant and animal foods. They are particularly important in their role as "waterlogged vultures," scavenging on the wealth of dead organisms continually being deposited in the river. The Indian flap-shell (*Lissemys punctata*) is another softshell generalist and also inhabits the main stem of the river, but is particularly abundant in ponds, muddy ditches, lakes, and marshes associated with the river. Like the Indian softshell, they are good scavengers and take plant and animal food alike.

Seemingly the most riverine of the Indus' batagurids, the brown-roofed turtle, *Kachuga smithi*, is most common in flowing water, but it also occurs in lakes and ponds communicating with the river (Minton, 1966). It is an omnivore, but relative proportions of plant and animal foods taken in the wild are unknown. Conversely, the Indian roofed turtle (*Kachuga tecta*) is more of a generalist, occurring regularly in lentic and lotic habitats, albeit favoring the former (Minton, 1966). The turtle is chiefly herbivorous, but E. Moll (1987a) captured a male in a hoop trap baited with chicken entrails. The herbivorous crowned river turtle (*Hardella thurjii*) prefers lentic habitats. Minton (1966) found several in vegetation-choked lakes and ponds, and in canals and quiet inlets of the Indus. The spotted pond turtle, *Geoclemys hamiltoni*, also prefers abundantly vegetated lentic habitats (e.g., oxbows, sloughs), but unlike the crowned river turtle, has a strong carnivorous bias with an inclination toward molluskivory.

The Ganges is the most sacred of India's rivers and originates from the junction of two head streams in the mountainous portion of Uttar Pradesh before flowing some 2500 km to Bangladesh and the Bay of Bengal. Major northern tributaries include the Yamuna (Jumna), Sarda, Gomati, Ghaghara, Gandak, and Kosi. Lesser contributions from the south are mainly the Chambal, Betwa, and Son. A great river delta begins some 350 km from the sea. Running through this delta, a formerly important channel—the Bhagirathy/Hooghly—runs southward past Calcutta to the Bay of Bengal. However, the main channel of the river which carries

Table 3.12
Species composition of batagurids in eight representative rivers of the Oriental Region.

Species	Representative rivers of the Oriental Region							
	Indus	Ganges	Irrawaddy	Chao Phraya	Perak	Kapuas	Mekong	Xi Jiang
B. baska		×	×		×			
C. borneoensis					×	×		
Ch. nigricans								×
Ch. reevesii								×
Cu. amboinensis			×	×	×	×	×	
Cu. flavomarginata								×
Cu. trifasciata								×
Cyclemys sp.*			×	×	×	×	×	
G. hamiltoni	×	×						
H. thurjii	×	×						
He. grandis			×	×	×		×	
Hi. annandalii				×			×	
K. dhongoka		×						
K. kachuga		×						
K. smithi	×	×						

K. tecta	×	×						
K. tentoria		×						
K. trivittata			×					
M. subtrijuga				×			×	
Ma. mutica								×
Me. trijuga		×	×					
Mo. ocellata			×					
Mo. petersi		×						
No. platynota				×	×	×	?	
O. sinensis								×
Or. borneensis					×	×		
S. quadriocellata								×
Si. crassicollis				×	×	×	×	
Total batagurids	4	10	7	7	8	6	6	7

Generic abbreviations are: *B* – *Batagur*, *C* – *Callagur*, *Cu* – *Cuora*, *G* – *Geoclemys*, *H* – *Hardella*, *He* – *Heosemys*, *Hi* – *Hieremys*, *K* – *Kachuga*, *M* – *Malayemys*, *Ma* – *Mauremys*, *Me* – *Melanochelys*, *Mo* – *Morenia*, *No* – *Notochelys*, *O* – *Ocadia*, *Or* – *Orlitia Pelochelys*, *S* – *Siebenrockiella*.

*The genus *Cyclemys* is currently in a state of taxonomic confusion. At least two (and probably several) species exist, and because all are very similar in appearance reports of their distribution are unreliable. Hence for the time being we will simply refer to the group as *Cyclemys* sp.

Table 3.13
Species composition of trionychids and the big-headed turtle, Platysternidae in eight representative rivers of the Oriental Region with total species (including batagurids) provided.

	Representative rivers of the Oriental Region							
Species	Indus	Ganges	Irrawaddy	Chao Phraya	Perak	Kapuas	Mekong	Xi Jiang
A. cartilaginea			×	×	×	×	×	
As. hurum		×						
As. gangeticus	×	×						
C. indica	×	×						
D. subplana				×	×	×		
L. punctata	×	×						
L. scutata			×					
N. formosa			×					
P. steindachneri								×
Pe. cantori		×	?	×	×	?	?	×
Pd. sinensis								×
Pl. megacephalum				×			×	×
Total Platysternidae				1			1	1
Total Trionychidae	3	5	3	3	3	2	1	3
Total Bataguridae (from table 3.12)	4	10	7	7	8	6	6	7
Total Species	7	15	10	11	11	8	8	11

Abbreviations are: A – *Amyda*, As – *Aspideretes*, C – *Chitra*, D – *Dogania*, L – *Lissemys*, N – *Nilssonia*, P – *Palea*, Pe – *Pelochelys*, Pd – *Pelodiscus*, Pl – *Platysternon*.

Table 3.14

Chelonian community similarities of eight representative rivers in the Oriental Region as indicated by Sorensen coefficients (Brower & Zar, 1984). A value of 0 indicates no overlap in species, whereas 1 indicates complete overlap of species.

River	Indus	Ganges	Irrawaddy	Chao Phraya	Perak	Kapuas	Mekong	Xi Jiang
Indus	–	0.64	0	0	0	0	0	0
Ganges		–	0.16	0.08	0.08	0	0	0.08
Irrawaddy			–	0.38	0.48	0.33	0.44	0
Chao Phraya				–	0.73	0.63	0.74	0.18
Perak					–	0.84	0.53	0.09
Kapuas						–	0.50	0
Mekong							–	0.10
Xi Jiang								–

the bulk of the water continues southeastwards, where it merges with the Brahmaputra channel known as the Jamuna. The channel continues as the Padma, which in turn joins the Meghna and eventually empties into the sea by way of four mouths. Between the mouth of the Hooghly and the mouths of the Meghna, a myriad of former channels and mouths of the Ganges form one of the world's most extensive mangrove forests known as the Sundarbans. This region of confluence between the Ganges and Brahmaputra has one of the richest assemblages of freshwater turtles of anywhere in the world. As many as seventeen species may inhabit the region (Iverson, 1992b,c) and at least fifteen are associated with the Ganges.

All seven species found in the Indus occur in the Ganges. This is apparently because the Yamuna River, long a part of the Indus system, was captured by (became part of) the Ganges system in relatively recent time. The Ganges, however, has more than twice the species richness of the Indus, indicating that the exchange was unidirectional. In addition to the three softshell species in the Indus, the Ganges supports the peacock softshell (*Aspideretes hurum*) and the giant Asian softshell (*Pelochelys cantori*). The former is a striking species with bright yellow head markings and elaborate ocelli on the shell (at least of the young). It is most common in lotic habitats of the northern tributaries of the Ganges. Limited observations suggest that it is a carnivore with a predilection for mollusks. The giant Asian softshell is another lotic species which frequents estuaries but will move long distances upstream and well out into the ocean, as discussed in chapter 2. *Pelochelys*, like *Chitra*, appears well designed for ambush feeding, but little is known about its habits in the wild. According to Nutaphand (1979) the diet includes fish, crustaceans, mollusks, and aquatic plants. Captives we have observed fed well on fish but shunned fruits.

Regarding hard shells, the Ganges is home to four, chiefly herbivorous, riverine specialists that are not found in the Indus. The smallest of the group is the Indian tent turtle (*Kachuga tentoria*) at 27 cm shell length (fig. 3.1). This species—along with its larger cousins the red-crowned roofed turtle (*Kachuga kachuga*) at 56 cm shell length and the three-striped roofed turtle, *K. dhongoka*, at 48 cm shell length—inhabit lotic habitats above the tidal influence (fig. 3.2). The river terrapin, at 60 cm shell length is chiefly an estuarine species inhabiting the Sunderbans and tidal areas of the Hooghly. Two additional species, the Indian black turtle (*Melanochelys trijuga*) and the Indian eyed turtle (*Morenia petersi*) seem most at home in lentic habitats.

The Brahmaputra has been poorly explored. Inasmuch as it merges with the Ganges in Bangladesh, it would be expected to share most of the same species. Nevertheless, there are no definite records of the large *Kachugas*, *Chitra*, or even *Aspideretes gangeticus*, from the Brahmaputra above its confluence with the Ganges. A recent survey by Bhupathy et al. (1994) reported finding Indian softshells reputed to be from the Brahmaputra in Assam markets, and shells tentatively identified as this turtle in several wildlife sanctuaries. In addition, *Cuora amboinensis* and *Cyclemys* sp.—two division #2 species not found in the Ganges drainage—do appear to be associated with the Brahmaputra.

Figure 3.1
A juvenile Indian tent turtle (*Kachuga tentoria*), a characteristic species of the Ganges River system of India.

Figure 3.2
An adult three-striped roofed terrapin (*Kachuga dhongoka*), one of the larger roofed terrapin of the Ganges River system of India.

In summary, turtle assemblages of division #1 species comprise two families, the Bataguridae and the Trionychidae. Some 37% of the assemblage leans toward herbivory; 30% are generalized omnivores, and 27% are chiefly carnivorous (table 3.15).

Division #2

The Irrawaddy (Ayeyarwady) River of Myanmar (Burma) is the westernmost large river of this division. The chelonian composition of the Irrawaddy shows little resemblance to the Ganges drainage, and in fact only two species are shared (quotient of similarity 0.16). The headwaters of the Irrawaddy arise amid hilly evergreen forest near the border of Myanmar with China. Along its 2100-km course to the Andaman Sea, the river flows chiefly through hilly lowlands with the exception of the wide terraced plains around Mandalay and the expansive delta land southwest of Rangoon.

Table 3.15
Habitat preference and adult guilds of selected river turtles from subregion #1 of the Oriental Region.

Species	Habitat preference	Guild
Batagur baska	Lo; estuaries, large rivers	OMNIVORE/herbivorous
Geoclemys hamiltoni	Le; backwaters, ox bows	CARNIVORE/molluscivorous?
Hardella thurjii	Le; backwaters, ox bows, lakes	HERBIVORE/generalist
Kachuga dhongoka	Lo; large to moderate rivers	F HERBIVORE/generalist M OMNIVORE/ generalist
Kachuga kachuga	Lo; large to moderate rivers	HERBIVORE/generalist
Kachuga smithi	Lo; rivers, backwaters	OMNIVORE/generalist
Kachuga tecta	Lo; Le; ponds, rivers, streams	HERBIVORE/generalist?
Kachuga tentoria	Lo; large rivers to small streams	HERBIVORE/generalist
Melanochelys trijuga	Le; ponds, backwaters, streams	OMNIVORE/generalist
Morenia petersi	Le; backwaters, lakes, ponds	OMNIVORE/herbivorous?
Lissemys punctata	Le, Lo; temporary and permanent aquatic habitats	OMNIVORE/generalist
Aspideretes gangeticus	Lo; rivers, lakes and ponds	OMNIVORE/generalist
Aspideretes hurum	Lo; rivers, lakes and ponds	CARNIVORE/molluscivorous?
Chitra indica	Lo; sandy stretches of rivers	CARNIVORE/generalist
Pelochelys cantorii	Lo; estuaries, rivers	CARNIVORE/piscivorous?

Guilds are presented as a general category (in upper case) and a specialization or preference (lower case). See table 3.2 and chapter 3 this text for an explanation of guild categories. Sources of information are Das (1991); Minton (1966); E. Moll (1980a, 1986, 1987a,b); Nutaphand (1979); and field observations of the authors. Abbreviations: Lo – lotic or flowing waters; Le – lentic, standing water or slow current habitats indicate the most common habitat. A question mark (?) indicates that available information is either insufficient or questionable.

Maxwell (1911) reported that various sand banks along the Irrawaddy delta around the turn of the century were nesting grounds to droves of river terrapin and Burmese roofed turtles (*Kachuga trivittata*). Little nesting can be found here today (Salter, 1983; E. Moll, 1985c), although some Burmese roofed turtles may still nest at upriver sites (van Dijk, 1994). As is typical of most of the rivers of division #2, southeast Asian box turtles (*Cuora amboinensis*), Asian leaf turtles (*Cyclemys* sp.) and giant Asian giant pond turtles (*Heosemys grandis*) are present (fig. 3.3). *Melanochelys trijuga* is widespread, but nowhere abundant (van Dijk, 1997); *Morenia ocellata*, though locally abundant, appears restricted to the delta region (van Dijk, 1997).

Moving eastward, the Chao Phraya is the principal river of Thailand. Originating as four streams in the hills of northern Thailand, it flows past Bangkok to the Gulf of Siam. The headwaters in northern Thailand first flow through a series of narrow precipitous gorges, then expand as they flow southward through broader valleys with extensive alluvial areas that form clearings in the dense forest. The four headwater streams converge at Nakhon Sawan forming the Chao Phraya which then flows leisurely some 250 km across the great plain of central Thailand, one of the preeminent rice-producing areas of the world. The Irrawaddy supports ten and the Chao Phraya, eleven species of turtles, but several

Figure 3.3
An adult Asian giant pond turtle (*Heosemys grandis*), a widespread generalist species which is often a member of the river turtle fauna in southeast Asia.

key differences in species composition exist (see table 3.13). Flapshells and roofed turtles (turtles characteristic of division #1) do not cross the border from Myanmar to Thailand. Conversely, ranges of six division #2 species, the big-headed turtle (*Platysternon megacephalum*), Malayan softshell (*Dogania subplana*), yellow-headed temple turtle (*Hieremys annandalii*), Malayan snail-eating turtle (*Malayemys subtrijuga*), Malayan flat-shelled turtle (*Notochelys platynota*), and black marsh turtle (*Siebenrockiella crassicollis*), do not extend eastward to the Irrawaddy. River terrapins and giant Asian softshells may occur or once have occurred in both rivers, but possibly due to extirpation or inadequate collecting the former is not known from the Chao Phraya nor the latter from the Irrawaddy.

At the eastern limits of this division, the Mekong originates in China's Tanglha Mountains (near the origins of the Salween, Yangtze, and Hwang Ho Rivers) and then flows some 4000 km through China, Tibet, Laos, Cambodia, and Vietnam on the way to its delta in the China Sea south of Saigon. A rushing torrent in the mountainous regions, the river offers suitable habitat only for the big-headed turtle until it enters Indo-China. Here the river is slower, flanked by foot-hills with banks of rich tropical vegetation. From June to October it is swollen with run-off from melting snow in the mountains and rains accompanying the southwest monsoon. From October to May the weather is much drier and the river fills only a portion of its bed.

The species composition of the Mekong is essentially the same as Chao Phraya, except that *Pelochelys* and *Notochelys* have not been reported there (even though the Mekong falls within their range). River terrapins occur just west of the drainage but also have not been recorded for the Mekong.

To the south of Thailand, the appendix-like Malay Peninsula is suspended below the underbelly of Asia. The peninsula serves as a transition zone between turtles that have originated on the mainland and others which originated in the islands of the Indo-Malayan Archipelago. The peninsula together with Sumatra, Java, and Kalimantan (Borneo) have been called "Sundaland." The seas separating these islands from the mainland are relatively shallow. However, if the water level is lowered by a few hundred feet—as happened with each major advance of glacial ice—these islands merge with the Malay Peninsula and connect to mainland Asia by a broad corridor of terra firma. In this fashion, animals evolving in this region were periodically isolated and then reconnected to the mainland. Unsurprisingly, the region today harbors a blend of mainland and island-evolved turtles. River terrapin and *Heosemys grandis* exemplify mainland types that have invaded Sundaland, while *Callagur borneoensis* and *Orlitia borneensis* (which probably evolved in the islands) today range northward only to the Malay Peninsula. Due to their recent common history, rivers within Sundaland also share many species of turtles (see table 3.13). The Perak River of Malaysia and the Kapuas of Kalimantan (Borneo) have a quotient of similarity of 0.84. The Perak differs chiefly in having two mainland forms (*Batagur* and *Heosemys grandis*) which have not reached Kalimantan, and also by the fact that the wide-ranging *Pelochelys* has not yet turned up in the Kapuas. Similarly, the Perak differs from the Chao Phraya by having river terrapin and two of the island forms (*Callagur* and *Orlitia*) which do not range north of the Malay Peninsula. Conversely, the Chao Phraya has three mainland forms—*Platysternon*, *Hieremys*, and *Malayemys*—that do not occur as far south as the Perak River.

Other than the big-headed turtle (*Platysternon megacephalum*), the assemblage again is comprised of batagurids and trionychids. Chiefly herbivorous forms are in the majority at 67%, followed by chiefly carnivorous forms, 26% and generalized omnivores, 7% (table 3.16).

Table 3.16
Habitat preference and adult guilds of selected river turtles from division #2 (as defined in text) of the Oriental Region.

Species	Habitat preference	Guild
Batagur baska	Lo; estuaries, large rivers	OMNIVORE/herbivorous
Callagur borneoensis	Lo; estuaries, large to medium rivers	HERBIVORE/ generalist
Cuora amboinensis	Le; semi-aquatic, ponds, swamps, rice paddies, backwaters, ditches, streams	OMNIVORE/herbivorous
Cyclemys sp.	Lo; streams, tributaries of rivers	HERBIVORE/generalist
Heosemys grandis	Lo, Le; streams, rivers, ponds, swamps	HERBIVORE/generalist
Hieremys annandalii	Le; ponds, canals, swamps, lakes, rivers	HERBIVORE/generalist
Kachuga trivittata	Lo; large rivers	HERBIVORE?
Malayemys subtrijuga	Le; ponds, marshes, streams, rivers	CARNIVORE/mollusciv.?
Notochelys platynota	Lo; streams, upper reaches of rivers	HERBIVORE/generalist
Orlitia borneensis	Lo, Le; rivers, swamps, ponds	HERBIVORE/frugivorous?
Siebenrockiella crassicollis	Lo, Le; rivers, streams, backwaters, swamps, ditches, rice fields, ponds	OMNIVORE/herbivorous
Platysternon megacephalum	Lo; headwaters of rivers, small, rocky mountain streams	CARNIVORE/generalist
Amyda cartilaginea	Le, Lo; ponds, ditches, swamps, streams, rivers	OMNIVORE/generalist
Dogania subplana	Lo; streams, upper reaches of rivers	CARNIVORE?
Pelochelys cantori	Lo; estuaries, large rivers	CARNIVORE/generalist

See legend of table 3.2 and chapter 3, this text for explanation of guilds and abbreviations. Sources of information include: Moll (1980a); Smith (1931); Thirakupt and van Dijk (1995 "1994"); Nutaphand (1979); and field observations of the authors.

Division #3

The final major turnover of species types occurs northeastward of the Mekong in southern China. The typical batagurids associated with rivers here are chiefly small to moderate-sized generalists such as *Chinemys*, *Cuora* (other than *amboinensis*), *Mauremys*, *Ocadia*, and *Sacalia*. Among the softshells, *Amyda* and *Dogania* give way to *Palea* and *Pelodiscus*. The only shared species with division #2 are the big-headed turtle and the Asian giant softshell turtles.

The Xi Jing (Si Kiang), the longest river in southern China, exemplifies the rivers of the region. Arising in the Yunnan/Kweichow tableland, it flows some 2000 km to the South China Sea. Above Wuchow the tributaries comprising the Xi Jiang flow through mountainous and hilly country, habitat which is particularly favorable for big-headed turtles (*Platysternon megacephalum*) and the waddle-necked softshell (*Palea steindachneri*). Below Wuchow, the river spreads out and the current slows as it approaches the broad, low-lying deltaic region near Canton. In the slower-moving areas above the estuary, Reeves turtles (*Chinemys reevesii*), yellow pond turtles (*Mauremys mutica*) and Chinese softshells (*Pelodiscus sinensis*) are typical species.

As in the previous two divisions, rivers in division #3 are comprised chiefly of batagurids and trionychids (table 3.17). However, there is not only a major turn-

Table 3.17
Habitat preference and adult guilds of selected river turtles from the Xi Jiang River sub region #3 of the Oriental Region.

Species	Habitat preference	Guild
Chinemys nigricans	Lo; streams, rivers	OMNIVORE/carnivorous
Chinemys reevesii	Le;Lo; rice fields, ponds, lakes, swamps, streams and rivers	OMNIVORE/generalist
Cuora flavomarginata	Le; semi-aquatic, ponds and streams	OMNIVORE/generalist
Cuora trifasciata	Lo? semi-aquatic, streams	CARNIVORE/generalist
Mauremys mutica	Le; ponds, marshes, swamps, streams	CARNIVORE?
Ocadia sinensis	Lo,Le; rivers, ponds, lakes	OMNIVORE/generalist
Sacalia quadriocellata	Lo; streams, brooks	OMNIVORE/generalist?
Platysternon megacephalum	Lo; upper reaches of rivers, mountain streams.	CARNIVORE/generalist
Palea steindachneri	Lo,Le; upper reaches of rivers, lakes	CARNIVORE/generalist
Pelochelys cantori	Lo; estuaries, large rivers	CARNIVORE/generalist
Pelodiscus sinensis	Lo; rivers, streams, ponds, canals	OMNIVORE/generalist

See legend of table 3.2 and chapter 3, this text for explanation of guilds and abbreviations. Sources of information include: Chen and Lue (1998); Ernst and Barbour (1989); Mao (1971); Pope (1935); Zhou and Zhou (1992).

over in species composition, the guild composition changes markedly from the other divisions where herbivory prevails. In the Xi Jing, carnivores and chiefly carnivorous forms predominate (54%, 6/11 species). Conversely, true herbivores almost disappear. *Ocadia sinensis* females become largely herbivorous as adults, but males retain the carnivorous tendencies of juveniles and immatures (Chen & Lue, 1998).

Australian Region

Australia and New Guinea, the chief turtle-bearing land masses of this region, historically lack land connections with any other region. It is thus not surprising that much of the fauna (including the turtles) is unusual or unique. The resident chelonians are chiefly pleurodires of the family Chelidae which also occurs in the Neotropics. Excellent summaries of the biology and relationships of the Australian snake-necks can be found in Legler and Georges (1993a,b). Australia's chelids comprise five "short-necked" genera (*Elseya, Elusor, Emydura, Pseudemydura*, and *Rheodytes*) and the "long-necked" genus *Chelodina*. At least two additional genera remain to be described. Three pleurodiran genera (*Chelodina, Elseya*, and *Emydura*) occur on New Guinea. The *Chelodina* are chiefly carnivorous, using their extraordinarily long necks for accurate, rapid strikes at prey that are then engulfed using gape and suck means. They either actively search for food or secrete themselves in the substrate to ambush prey approaching too near. "Short-necked" chelids such as *Emydura* actively forage along the bottom [searching] for a variety of plant and animal foods (Legler & Georges, 1993b). Surface feeding has also been described (Legler, 1976). An interesting feeding pattern has been discovered in the genus *Emydura*, a series of largely allopatric species ranging from eastern to northern Australia. Megacephaly (a mollusk-eating adaptation consisting of enlarged head and jaw musculature) is rare in populations of *Emydura* below 26° latitude, but progressively increases to the north and west, culminating in *E. victoriae* in extreme northwestern Australia where all members of the population eat mollusks and nearly all members of both sexes become megacephalic (Legler, 1989; Legler & Georges, 1993b). Lingual and median alveolar ridging (an adaptation allowing for additional shearing and crushing of plant food) occur only in the *Elseya dentata* group among the Australian chelids (Legler & Georges, 1993b). This adaptation parallels that in a variety of herbivorous cryptodires including *Batagur, Callagur, Dermatemys, Kachuga*, and *Pseudemys* (see chapter 2, this volume).

Only two genera of cryptodires occur in the Australian region. The pig-nosed turtle (*Carettochelys insculpta*) occurs in both Australia and New Guinea, while two species of the giant Asian softshell (*Pelochelys*) inhabit only the latter.

Australian chelonians are largely restricted to the external drainage system comprising a series of small rivers around the periphery of the continent. The interior is too dry to support turtles except for a portion of the easternmost part of the internal drainage (Legler & Georges, 1993a). Table 3.18 lists the species reported to occur in five selected rivers. The 1600-mile (2500-km) Murray River is the largest in Australia. Species richness is relatively low, and Legler and Georges

Table 3.18
Chelonian communities from selected Australian rivers.

		River system				
Species	Guild	Alligator (South)	Daly	Fitzroy/ Dawson	Mary	Murray/ Darling
Ca. insculpta	O/h	×	×			
Ch. burrungandjii	O/g	×	×			
Ch. expansa	O/c			×	×	×
Ch. longicollis	C/g			×	×	×
Ch. rugosa	C/g	×	×			
El. dentata group	O/h	×	×	×	×	
El. latisternum	O/c	×*	×*	×	×	
Elu. macrurus	O/g				×	
Em. australis "a"/ Em tannybaraga	O/h	×	×			
Em. krefftii	O/g			×	×	
Em. macquarii	O/g					×
Em. victoriae	O/m		×			
Em. worrelli	O/m		×*			
R. leukops	O/i			×		
Total species	–	6	8	6	6	3

Information compiled from Cann (1998); Cann and Legler (1994); Chessman (1986, 1988); Legler (1980, 1982); Legler and Cann (1980); Legler and Georges (1993a,b); Georges and Kennett (1989); Kennett and Tory (1996); Thomson (1996). Genera abbreviations: Ca. – *Carretochelys*; Ch. – *Chelodina*; El. – *Elseya*; Elu. – *Elusor*; Em. – *Emydura*; R. – *Rheodytes*. Guild abbreviations: Generalized guild, C = CARNIVORE; H = HERBIVORE; O = OMNIVORE. Specialties: c = carnivorous; g = generalized; h = herbivorous; i = insectivorous; m = molluscivorous (see table 3.2). The taxonomy of some Australian chelonian populations in northern Australia is still being debated, and the identifications below are tentative at this writing.
*Located above the fall line.

reported six to be the maximum in any river. However, Thomson (1996) recognized eight species in the Daly River (see table 3.18). The South Alligator, Nourlangie Creek system of the Northern Territory, and the Fitzroy/Dawson and Mary Rivers of Queensland all accommodate six species; however, only two to four species are usually found in microsympatry.

Legler (1980, 1982) provided a detailed description of the chelonian assemblage in the tropical South Alligator/Nourlangie Creek River system. The following (with some taxonomic revision) is based on Legler's observations. The South Alligator and several of its tributaries begin on the Arnhem Plateau. Only two species—the serrated snapping turtle (or sawshell) *Elseya latisternum* and the recently described *Chelodina burrungandjii* which is closely related to *C. rugosa*—occur in the tableland, and both are chiefly carnivorous. A yet undescribed species of Chelodina occurs further to the south and west in the Kimberley region

(Thomson et al., 2000). A complete replacement of species occurs below the fall line. Here, between the falls and the tidal reaches, the yellow-bellied snapping turtle, a member of the *Elseya dentata* group, is most abundant. The pig-nose turtle (*Carettochelys insculpta*) and the northern Australian snake-necked turtle (*Chelodina rugosa*) are common, and the northern yellow-faced turtle (*Emydura tanybaraga*) also occurs within the drainage. Of the four, *C. rugosa* is the most carnivorous. The others are omnivorous tending toward herbivory in the dry season when they feed chiefly on fruits, leaves and flowers that fall into the river. *Elseya*, with its broad-ridged alveolar shelf, seems best equipped to process this diet, and this may partially explain its abundance.

Further west a similar assemblage occurs in the Daly River drainage. Based on accounts by Legler and Georges (1993b), Thomson (1996), and Cann (1998), the following species are reported for this river. *Elseya dentata, Emydura victoriae, E. tannybaraga* (formerly *Emydura australis* "a"), *Chelodina rugosa*, and *Carettochelys insculpta* inhabit the lower catchment below the fall line. The Daly appears to provide optimum habitat for *Carretochelys* which may obtain their peak densities in Australia here. Worrell's short-necked turtle (*Emydura worrelli*) inhabits the drainage above the fall line (Katherine River tributary), but to what extent it extends downstream to the Daly River remains to be determined (Cann, 1998). *Elseya latisternum* and *Chelodina* sp. (the sandstone long-neck) also have been taken above the fall line.

On the East Coast, the Fitzroy/Dawson Drainage flows northwest for a considerable distance paralleling the coastline until it finally turns sharply eastward and southward to empty into the sea. As described by Legler and Cann (1980), the long northwestern flowing section comprises large deep pools alternating with shallow, rapid-flowing riffles. The Fitzroy River turtle (*Rheodytes leukops*) dominates the riffles. These agile turtles swim effortlessly over the stony substrate of the clear fast waters, examining promising rocks for favorite foods, including insect larvae and freshwater sponges, which they scrape off the surface using their projected even-edged tomium (Legler & Cann, 1980). *Rheodytes* rarely surface in this oxygen-rich environment, being able to obtain adequate gaseous exchange using nonpulmonary surfaces, particularly a pair of prominent vascularized sacs in the cloaca known as cloacal bursae (Legler, 1993; Legler & Georges, 1993b; chapter 2, this volume).

In the deep slower-moving pools, Krefft's river turtle (*Emydura krefftii*) predominates, with the northern snapping turtle (*Elseya dentata*) and the giant snake-necked turtle (*Chelodina expansa*) occurring in lesser numbers. The omnivorous Krefft's turtle roams the bottom while periodically grazing on aquatic vegetation and snapping up small invertebrates as they are encountered (Legler & Georges, 1993b). The northern snapping turtle is similarly opportunistic but biased toward herbivory. In the Northern Territory during the dry season fruit, leaf, and flower windfalls from riverside arbors are key components of the diet (Kennett & Tory, 1996). The carnivorous giant snake-necked turtles will actively search for highly motile prey (e.g., decapod crustaceans, aquatic insects and fish), but may also ambush quarry from concealment while buried within the substrate with only the tip of the snout and eyes exposed (Legler & Georges, 1993b). When an unsuspecting animal swims within range of the long neck, the force of the rapid strike causes the turtle virtually to explode from the bottom. The target is engulfed by a gape and suck action of the mouth and throat too rapid for the eye to follow.

Although uncommon in the river proper, two additional opportunistic carnivores, the common long-necked or snake-necked turtle (*Chelodina longicollis*) and the serrated snapping turtle (*Elseya latisternum*) also inhabit the drainage. The common snake-neck prowls the backwaters of the river while feeding on carrion and just about any animal small enough to be subdued (Chessman, 1988). The serrated snapping turtle chiefly inhabits and forages in the tributaries joining the Fitzroy.

The turtle assemblage of the Mary River south of the Fitzroy is similar in species richness (six) but the Fitzroy River turtle is replaced by another short-necked species, the Mary River turtle (*Elusor macrurus*). Perhaps the largest of the short-necked chelids, this unusual turtle is characterized by an exceptionally large, laterally compressed tail bearing haemal arches on the distal vertebrae. *Elusor* is omnivorous including mollusks and filamentous algae in its diet (Cann & Legler, 1994).

Further southward, the Murray/Darling River System winds through New South Wales and Victoria. Despite being the largest of Australia's rivers, only four or five species of turtles occur there. Chessman (1983, 1984, 1986, 1988) has capably reviewed the diet and habitat preferences of three members of this assemblage. The Murray River turtle (*Emydura macquarrii*), an opportunistic omnivore, is the dominant form in both the river proper and backwaters. The common snake-necked turtle (*Chelodina longicollis*), a carnivore with catholic tastes, is the dominant species in oxbows, anabranch ponds, and swamps associated with the river. The carnivorous giant snake-neck (*Chelodina expansa*) is intermediate between the above but tends to lean closer to the habits of *Emydura*. Two additional populations related to *Elseya latisternum* have been recently discovered in tributaries of the Murray/Darling that originate in the Tablelands of New South Wales and southern Queensland. Cann (1998) lists these as Bell's turtle (*Elseya belli*) in the Namoi-Gwydir Rivers and the Bald-rock Creek short-neck, (*Elseya* sp.) from the extreme upper reaches of the Darling River. Whether these populations represent separate species or two conspecific subspecies remains to be determined.

Turtle richness declines to a single species in the rivers of the arid southwest and western portions of Australia. Short-necked turtles are absent, and those rivers which can support chelonians accommodate the narrow-breasted snake-neck (*Chelodina oblonga*) in the extreme southwest or Steindachner's snake-neck (*C. steindachneri*) in the west. Both are generalized carnivores. Steindachner's snake-neck can inhabit temporary streams which are dry for long periods, apparently estivating in the mud until the water flows once more.

New Guinea's fresh-water turtles are unevenly distributed over the island, with most inhabiting only the southern lowlands of the island. A formidable mountain range running from northwest to southeast along the axis of the island provides a barrier preventing many chelonians from reaching the northern lowlands (Rhodin et al., 1993). Nevertheless, populations of the giant Asian softshell (*Pelochelys*) and New Guinea snapping turtle (*Elseya novaeguineae*) have reached the northern region. Whether populations on either side of the divide differ specifically is currently being studied (Rhodin et al., 1993). Webb (1995) recently restricted the name *Pelochelys bibroni* to the southern population while reinstating an older name, *P. cantorii* for the northern population and those throughout the rest of the range.

The chelonian fauna of New Guinea consists of two cryptodire genera (*Carettochelys* and *Pelochelys*) and seven pleurodires, including five Chelodina (*novaeguineae*, *parkeri*, *pritchardi*, *siebenrocki*, and *reimanni*), an *Elseya* (*novaeguineae*), and an *Emydura* (*subglobosa*).

Liem (1983) reported the chelonian species composition of the Purari River as the pig-nosed turtle (*Carettochelys insculpta*), the giant Asian softshell (*Pelochelys bibroni*), Siebenrock's snake-necked turtle (*Chelodina siebenrocki*), the New Guinea snapping turtle (*Elseya novaeguineae*), and the red-bellied short-necked turtle (*Emydura subglobosa*). Far less is known of the habits of the New Guinea turtles than those in Australia. If we assume that these species have similar habits to their Australian counterparts, then this assemblage would comprise two omnivore types that are chiefly herbivorous (*Carettochelys* and *Elseya*), an omnivore generalist (*Emydura*), and two carnivores (*Pelochelys* and *Chelodina*).

Community Composition, Species Richness, and Diversity

The terms species richness and diversity are commonly used when comparing communites or assemblages. The former is simply the number of species present in an area or community, whereas the latter is a composite of species quantity weighted by some measure of their importance, such as abundance, productivity, or size. Herein diversity was quantified by the Shannon–Wiener function (H') and eveness (J') as given in Krebs (1989). H' takes into account both species richness and the relative abundance of each species in the community. Higher H' values indicate greater community diversity. The measure is open-ended, increasing with species richness. Conversely, J' ranges from 0 (when no two species are equally common) to 1 (when all species are equally common). Other important attributes of the community concern the species composition and the abundance or rarity of each species.

Community Characteristics, Species Composition, and Replacement

Species composition, richness and diversity are important aspects of a riverine turtle assemblage, and can be influenced by a variety of physical and biological factors. However, most of the available information concerns individual species and very few published data are available relating community diversity and composition to riverine habitats. Based on a survey of the scant published literature, unpublished reports, and observations of the authors, we managed to assemble detailed information on nineteen assemblages of turtles occupying various sites and habitats in the Mississippi River drainage (see table 3.1a,b). Although these data were collected by a variety of techniques, all represent moderate to large collections (55 to 1102 turtles) which should provide some useful implications concerning species composition and diversity in different riverine habitats. At the least, the interpretations drawn herein may stimulate confirmatory studies.

Turtle communities worldwide tend to be relatively low in species diversity. Only a few species—often only one or two—make up the bulk of the turtle numbers in any one habitat. At least seventeen species inhabit portions of the Mississippi River (see tables 3.1, 3.2, and 3.3), but only four to nine (mean = 7; $n = 19$ communities) occur together in specific habitats. In fourteen of nineteen communities, the two most dominant species comprised greater than 75% of the numbers in the sample. Softshells and map turtles dominated lotic habitats, while common sliders and/or painted turtles dominated the lentic ones, as also observed by Bodie et al. (2000 and discussed in chapter 2). The remaining members of each community were a mix of uncommon residents and transients from other habitats. Habitats with higher species richness tended to have more rare species.

H' values for the nineteen communities ranged from 1.08 to 2.45, and J' values from 0.43 to 0.87. Lotic habitats ($n = 8$) varied from 1.21 to 2.36 (mean = 1.79, $s = 0.40$) and from 0.43 to 0.84 (mean = 0.66, $s = 0.13$) respectively. Lentic habitats ($n = 11$) ranged from 1.08 to 2.45 (mean = 1.66, $s = 0.43$) and from 0.47 to 0.87 (mean = 0.61, $s = 0.15$). Lotic habitats appear to have a greater variety of microhabitats and thus might be expected to have a greater diversity of species compared to lentic. In fact, the H' and J' values of the eight lotic habitats averaged higher than those of the eleven lentic habitats, but the difference was not significant. However, a comparison of the six localities having both lentic and lotic habitats (using a one-tailed paired *t*-test and a Wilcoxon paired-sample test) indicated that the diversity and evenness of these lotic habitats was significantly greater than that of the associated lentic habitats ($P =< 0.04$). The subject deserves more study. Substrate type was unavailable for some of the habitats, making it impossible to observe how community composition changes with substrate. It may be significant, however, that the three lowest H' values recorded for lotic habitats were areas of sand substrate which typically have lower primary productivity.

Both intrinsic and extrinsic factors can influence the composition and turnover of species comprising turtle communities. Intrinsic factors include the adaptability of the component species of a community. "Specialist" and "generalist" are relative terms lacking absolute definition, but for the purposes of this section are loosely defined by the following criteria. Herein, the terms refer to an organism's habitat requirements, food requirements or a combination. Habitat specialists are stenotopic, occupying relatively few habitat types and usually having a small geographic range, whereas generalists are eurytopic species occurring in a variety of habitats and ranging over wide areas. The Fitzroy River turtle of Queensland, Australia exemplifies a consummate habitat specialist, living in riffle areas of a single river system in Eastern Australia (Legler & Cann, 1980; Cann, 1998). In contrast, the common slider is a good example of a generalist occupying lentic and lotic habitats in rivers, streams, lakes, and ponds from northern United States into South America.

Food specialists have a narrow range of diet, while generalists are rather catholic in their tastes. The Escambia map turtle of Alabama and Florida is a food specialist, with females consuming mostly mollusks and males mostly insects. The Murray River turtle from Victoria and New South Wales is a food generalist that consumes carrion, periphyton, aquatic invertebrates, filamentous algae, and macrophytes depending upon what is plentiful in different habitats (Chessman, 1986). Legler (unpubl.) relates an instance of rapid dietary shift in a New South Wales population of this species that had been feeding on filamentous algae and the invertebrates it contained. Following a flood, most of the algae were scoured from the streambed and the population immediately switched to pure carnivory.

Often, but not always, food generalists or specialists are also habitat generalists or specialists. Overall, strict specialists are rare among the chelonians, as one of the key survival features of the group has been their adaptability. The majority today are omnivorous opportunists inhabiting a variety of habitats.

The significance of being a generalist or a specialist to turtle distribution patterns can be seen in the overview of the Nearctic region. The lotic habitats in the Mississippi River drainage and other Gulf Coast rivers appear to favor food and habitat specialists, while the backwaters are home to the generalists. In less favorable areas, however, such as near limits of turtle occurrence or under harsh environmental conditions, generalists tend to predominate in both lotic and lentic situations.

Other factors influencing community composition are extrinsic to the turtles. Rivers comprise a variety of lentic (still water) and lotic (flowing water) habitats. Within each of these general categories a variety of physical and biological factors interact to determine the turtle species that occur there. Among the more obvious are strength of current, depth, type of substrate, the amount, type, and seasonal variation of aquatic vegetation, and the availability of basking and nesting areas (see chapter 2, this volume, for further discussion of this subject).

As a river flows from its source to its mouth, current slows, depth increases and eventually salinity becomes a factor. At any section along a river, current will be greatest in the main stem, reduced in side channels, and absent in backwaters and flood plain lakes. Certain sections of the river will have sand or gravel substrates, others mud bottoms. Some backwaters remain connected to the river except during extreme drought conditions, while others—such as flood plain lakes or oxbows (former meanders now cut off from the river)—may only connect with the river during flood conditions. Some habitats may have abundant aquatic vegetation, whereas others may be barren. All of these extrinsic factors can and do influence the composition of turtle communities along a river. A replacement or turnover of some or all species can be expected to accompany changes in external factors.

Optimum conditions vary for each species. As such, species composition and relative abundance of species differ in different habitats. Examples of both phenomena can be seen in Tinkle's 1959 study of turtle community composition above and below the fall line in Gulf Coast Rivers. The "fall line" in this study divides older Paleozoic geologic strata from strata of Cretaceous age and still more recent coastal plain sediments. Visibly, the landscape is a zone of low hills, while shoals and rapids denote where rivers flow across the fall line. Tinkle observed that some species occurred only above the fall line (e.g., flattened musk turtle) while others occurred chiefly below (black-knobbed map turtle). In species found both above and below the fall line, some were of approximately equal abundance (loggerhead musk turtle and the Alabama map turtle) while others occurred in different proportions (e.g., the common cooter was much more abundant above than below the fall line on the Tombigbee and Apalachicola Rivers).

As another example of species replacement along a habitat gradient, consider a typical large river of peninsular Malaysia such as the Perak or Terengganu. In the fast-flowing upper reaches or headwaters, the Malayan softshell (*Dogania subplana*) and the Malayan flat-shelled turtle (*Notochelys platynota*) are characteristic species. Farther along in the wide middle reaches lined by a gallery forest comprised chiefly of neram (*Dipterocarpus oblongifolius*), the Asiatic softshell turtle (*Amyda cartilaginea*) and the giant Asian pond turtle (*Heosemys grandis*) are common species. Finally, approaching the mouth, where the wide tidally influenced lower reaches of the river are lined by mangrove forest and nipah palm, the river terrapin and painted terrapin are the dominant batagurids and the giant Asian softshell (*Pelochelys cantori*) is the common softshell (see also Pritchard, 2001).

The turnover or replacement in species composition exemplified above could be termed "ecological replacement." Turnover and changes in relative abundance of species can also be expected to take place in communities over time either over geological time or short term. As an example of changes over geological time, Holman (1995) has reviewed changes in herpetological communities during the Pleistocene. More specifically for turtles he reports that, in Michigan, painted turtles, common snapping turtles, and wood turtles were among the primary invaders of the rivers following the retreat of glaciations. The next wave of invaders included two additional riverine species, the spiny softshell and common map turtle. The final wave of invasion included common sliders and stinkpots, completing the riverine assemblages presently found in the state.

Of more concern to many ecologists and conservationists are the changes that are taking place over shorter periods of time. How have the multitude of anthropogenic changes in riverine habitats affected the turtle communities inhabiting these sites (see chapter 6, this volume)? Considering the extensive changes that have occurred and are occurring to the world's rivers, long-term studies of the effects on chelonian communities are crucial to conservation planning. Models for such long-term studies can be found in the exhaustive study programs of turtle communities that have been carried out since 1953 at the E. S. George Reserve in Michigan (Congdon & Gibbons, 1996) and since 1967 at the Savannah River Ecology Laboratory in South Carolina (Gibbons, 1990b). However, these studies have mainly concerned turtle communities in lentic habitats, and presently no comparable studies are available concerning river turtle communities.

Nevertheless, some of the same information can be obtained by other means. Researchers can search the literature and/or museum lists to obtain information on the composition of riverine communities at a site during former periods and then carry out follow-up surveys to see if or how it has changed over time. Three examples of this approach follow.

During the late 1970s, D. Moll (1980) compared findings from his field surveys of turtle communities on the Illinois River with old museum collections of turtles made before the riverine environment was altered by development projects and pollution. He found that populations of a few specialist species had seriously declined or disappeared (e.g., smooth softshell, yellow mud turtle), but that a number of generalist species (e.g., snapping turtles, false map turtles, common sliders, and spiny softshells) were thriving in the altered conditions (fig. 3.4a,b).

Dr. A. J. Acharji, a former head of the Zoological Survey of India, provided some information on the relative abundance of species occupying the Ganges near Rajmahal, Bihar. In a survey of this region in 1955, Acharji (1955) found four chelonians to be most common—the Indian softshell (*Aspideretes gangeticus*), the crowned river turtle (*Hardella thurjii*), the three-striped roofed turtle (*Kachuga dhongoka*), and the red-crowned roofed turtle (*K. kachuga*). Twenty-nine years later, E. Moll (1985c, 1990a) surveyed the same area and found that only the three-striped roofed turtle was still plentiful. The Indian softshell was only occasionally caught by fishermen, and the crowned river turtle and red-crowned roofed terrapins were caught only rarely or not at all.

In 1990, Meylan et al. (1992) surveyed the chelonian community of the Rainbow River, a site that had been previously studied by Marchand (1942). Whereas emydid turtles of the genus *Pseudemys* predominated in the 1940s, kinosternids of the genus *Sternotherus* predominate there today. Whether these changes were due to environmental effects or human exploitation was unclear.

Species replacement of still another kind takes place between different river systems. This type of turnover requires both ecological and evolutionary explanations. The Gulf coastal drainages again provide a good illustration (see table 3.3). An examination of species in six Gulf Coast Rivers from the Atchafalaya (LA) to the Apalachicola (FL) indicates that when isolated in neighboring river systems, some turtles speciated while others did not. The map turtle group (*Graptemys*) has been very prone to speciation and most of Gulf Coast rivers have an endemic or nearly endemic species of map turtle (see table 3.5). The *Sternotherus carinatus* group has diverged into three species over the six rivers. The razor-backed musk turtle occurs in the Atchafalaya, Pearl, and Pascagoula (fig. 3.5), while the flattened musk turtle inhabits the Warrior River above the fall line. The loggerhead musk turtle is the only member of the group in the Tombigbee, Alabama, and the Apalachicola Rivers and is sympatric with the razor-backed musk turtle in the Pearl and Pascagoula and with the flattened musk turtle on portions of the Warrior River.

(a)

(b)

Figure 3.4
(a) The yellow mud turtle (*Kinosternon flavescens*), a species extirpated from the Illinois River due to habitat degradation during the twentieth Century. (Photo by Richard S. Funk). (b) The common snapping turtle (*Chelydra serpentina*), a habitat generalist which continues to thrive in the Illinois River.

Figure 3.5
The razor-back musk turtle (*Sternotherus carinatus*), a common species of the southeastern United States' slow-flowing rivers and associated wetlands. (Photo by Indraneil Das)

Other species are present in all of these rivers (e.g., *Chelydra serpentina, Macroclemys temminckii, Chrysemys picta, Pseudemys concinna, Trachemys scripta, Sternotherus odoratus*, and *Apalone spinifera*). In an evolutionary sense they have been unresponsive or less responsive to the vicariant events that stimulated the *Graptemys* and *S. carinatus* groups to speciate. What is the difference? One hypothesis touted to explain such situations hinges on the premise that specialized species may have higher rates of speciation than generalized species. This hypothesis predicts that generalized clades should have few species that range over broad areas and a variety of habitats. Specialized clades should be more speciose with individual members being restricted in range and habitats occupied.

The phenomenon described in the Gulf Rivers fits this description and may be the turtle equivalent of "William's Pattern" (Eldredge, 1985). Around the turn of the century, H. S. Williams—a paleontologist from Cornell working with brachiopods—recognized two general types of fossil species: (1) Specialized types characterized by little variability, restricted habitat, limited geographic range and a relatively brief existence in the fossil record; and (2) Generalized types characterized by considerable variability, a diversity of habitat types, a wide distribution and extended existence in the fossil record. Particularly good examples in extant groups are evident among the fauna of Africa including cichlid fishes (Fryer & Iles, 1972), the impala and wildebeest/hartebeest clade of antelopes (Vrba, 1980, 1984) and the pelomedusid turtles. In the cichlid fishes, the generalist *Tilapia* has rather few species, unspecialized feeding habits, and broad ecological tolerance. Conversely, some of the largest species flocks known are in the genus *Haplochromis*, a group which is specialized in feeding and habitat requirements. Among the antelopes, the generalized impala group is a widespread, generalized feeding species with but one extant and one or two fossil species. The wildebeest/hartebeest clade differs markedly, being more restricted in food and habitat and having many extant and fossil species. Finally, in the pelomedusid turtles the genus *Pelomedusa* contains but one species, a habitat generalist and omnivore that occurs over most of Africa, Madagascar, and southwestern Arabia. In contrast, its sister genus, *Pelusios*, is a speciose group (sixteen species) each occupying a confined range in sub-Saharan Africa or associated islands.

The explanation that specialists evolve or diversify more rapidly than generalists has been suggested by a variety of authors over time, and it has come to be known as the resource-use hypothesis (Stebbins, 1950; Simpson, 1953; Rensch, 1959; Eldredge, 1979, 1985; Vrba, 1980; Eldredge & Cracraft, 1980). Vrba (1987) has further refined this hypothesis stating that the history of the availability of particular resource patches has been more important than being specialized or generalized in respect to speciation rates within a group. She points out that specialists whose resource patches have persisted over extensive periods would be expected to have as low a speciation rate as a generalist.

Is speciation a specialist's strategy for dealing with a changing environment? According to the punctuated equilbrium theory (Eldredge & Gould, 1972), species tend to maintain their adaptations over long periods of time without change. It contends that most changes in adaptation are associated with speciation events and that once an adaptation is established, little change occurs thereafter. Specialist species are viewed as being tied to particular habitats. Rather than adapting to new habitats as environmental conditions change, these species track (i.e., follow) their preferred habitat. Where new conditions prevail, open niches are filled by speciation events in which former small satellite populations pre-adapted to these new conditions expand to become the resident population (a process that has been called "species selection"; see Eldredge, 1985 for a review). Conversely, generalist species are less affected by changing environments. Because of their broad niches, they can fit into new conditions with the adaptive baggage that they bring with them and thus persist longer as species than specialist types. Assuming that the common slider, *Trachemys scripta*, is a single species, it represents an excellent example of a wide-ranging generalist which has extended its range from the north temperate region through the tropics to northern South America with relatively little speciation or any significant changes in adaptation to the tropics (Moll & Legler, 1971; Moll & Moll, 1990).

Returning to Gulf Coast Rivers, four of the seven species that occur in all six rivers are among Nearctic turtledom's most generalized species. The common snapping turtle, painted turtle, common slider, and stinkpot all inhabit a broad range in habitats, are decidedly catholic in their food tastes and are distributed over most of the eastern half of the United States. Map turtles tend to contrast sharply with the above. A majority is restricted to one or two river systems where they are decidedly more abundant in clear, clean, lotic habitats. Regarding diet, most are chiefly carnivorous (with males and females utilizing somewhat different portions of the available food spectrum. Similarly, the *Sternotherus carinatus* group tends to be more specific in habitat and diet than their wide-ranging cousin the stinkpot (*S. odoratus*). The specialized requirements of these groups (possibly their heavy reliance on aquatic insects and mollusks—organisms which are particularly vulnerable to environmental changes) may have led to more frequent speciation than in the associated generalist species.

The river cooter and alligator snapper appear more specialized than the five generalists listed above, but still did not respond to the speciation events that affected the map and musk turtles. River Cooters are chiefly herbivorous but are flexible in habitat requirements, being found in both lotic and lentic situations. Alligator snappers, though highly varied in diet, are considerably more restricted in habitat than their close relative, the common snapper. Perhaps these exemplify specialized species whose resource patches have persisted through long periods of time as suggested in Vrba (1987). The mollusk and aquatic insect foods of map and musk turtles—may be more vulnerable to environmental perturbations than the aquatic vegetation eaten by the river cooter or the variety of fish, turtles, and acorns eaten by the alligator snapper.

Before leaving the subject, it is of interest to draw one additional parallel between William's brachiopods and map turtles—anatomical peculiarity. Williams noted that his specialists tended to be highly distinctive in appearance such as having elaborate shell ornamentation not seen in the generalized species. The map turtles' high-peaked shell decked with knobs or spines is a particularly distinctive feature. Only the Pangshura group of the Asiatic batagurid genus *Kachuga* have similar shell characteristics. Both groups are largely riverine species, but the *Kachugas* differ as a group from the map turtles by being chiefly herbivorous (juveniles and males take some animal food) rather than carnivorous (E. Moll, 1987b). At present, too little is known of the habits of the pangshuras to draw other parallels with the *Graptemys*. Vrba (1984) explained the trend toward morphological specialization in the wildebeest group of antelopes as another result of the high rate of speciation characterizing narrow-niched species. Whether this applies equally well to map turtles and pangshuras remains to be seen.

Guild Composition of Chelonian Communities

Usually there is a great deal of overlap in the foods consumed by different species within a chelonian community. As mentioned above, most turtles are opportunistic omnivores and only a few turtles are purely carnivorous or herbivorous (see also chapter 2). Even those whose diets in the wild seem to be purely herbivorous are likely to take meat in captivity or enter a hoop trap baited with meat. Adaptations and general appearance of species can be misleading. Common snapping turtles with their long necks and powerful jaws might be assumed strictly carnivorous, but studies such as those by Lagler (1943) have demonstrated that plants comprise over a third of the diet in Michigan. Individual snappers may be chiefly herbivorous at some times. A large male observed by Budhabatti and Moll (1990) in northern Illinois fed almost exclusively upon duckweed from 15 June to 10 August. Hammer (1972) reported that plants made up the majority of the turtle's diet in a North Dakota marsh. D. Moll (1997a) found that fallen leaves from riparian vegetation are an important dietary component of Costa Rican snapping turtles during the dry season. The closely related alligator snapping turtles are also surprisingly catholic in their diets. Sloan et al. (1996) found that, while they ate a wide variety of animal foods, acorns were by weight and volume the most abundant item in the diets of 65 individuals examined from Louisiana.

Softshelled turtles are another group which appears well-adapted for carnivory, but here again particular species within the group are quite omnivorous. The Asiatic softshell (*Amyda cartilaginea*) is such an opportunist, regularly feeding on clams, snails, prawns, snakes and fish but it is not unusual to find fruits and seeds comprising 70–90% of an individual's digestive tract contents (Moll & Khan bin Momin Khan, 1990). The Indian softshell turtle (*Aspideretes gangeticus*) and the Indian flapshell (*Lissemys punctata*) are other omnivorous softshells (Minton, 1966).

Nevertheless, sufficient levels of specialization exist in natural communities for distinct guilds to be recognized. A guild is a group of species with similar ecological requirements and foraging strategies that therefore play similar roles in a chelonian community. Fachín-Terán et al. (1995) and Vogt and Villareal-Benitez (1997) have attempted to categorize trophic equivalents in several New World river turtle communities.

Unfortunately, a majority of the information available in the literature concerns only the composition of the diet and relatively little about how the food is obtained. Where feeding strategies are mentioned, often the information is incomplete or inadequate (based on a few observations, or on only one sex or age group, etc.). For example, the smooth softshell (*Apalone mutica*) has been studied better than most species, but its feeding behavior is still not thoroughly understood. Ernst et al. (1994) describe their insectivorous feeding behavior as consisting of capture from ambush, by active pursuit, or by searching along the bottom on submerged debris and over exposed sandbars. Whether one or more of these techniques is more important than the others is unknown. Williams and Christiansen (1981) reported that smooth softshells feed more often on insects in the water column, while sympatric spiny softshells were more benthic feeders. However, in this study, food information from both sexes was analyzed together and so it is unclear whether feeding in the water column applies more to one sex than the other. This appears significant in that a study in Kansas by Plummer and Farrar (1981) found that while females foraged in deep water and chiefly took aquatic insects, males foraged at the shallower interface between land and water and took a majority of terrestrial insects.

One of the best techniques for obtaining information on the feeding behavior is to observe the turtle in its own environment. Using the technique of goggling or snorkeling to study turtles was pioneered by Lewis Marchand (1945). Some very detailed information has been gained by this technique. John Legler and John Cann for example have used this technique very successfully in obtaining observations on the unusual Australian chelids. Unfortunately, turbid waters prevent this technique from being used in many habitats. Observations on captives can also be instructive, but as stated previously food habits are often altered in captivity. To date, accurate information on feeding behavior is available for only a small percentage of the world's river turtles (see also chapter 2).

Due to the general deficiency of information on feeding behavior, for the purposes of this chapter we will consider a guild in a simplified sense, turtles with the same diet preferences regardless of how the food was obtained. Herein we have compiled general information on dietary preferences within turtle communities on the local and global scale (see tables 3.2, 3.5, 3.8, 3.10, 3.16, 3.17, 3.18, and 3.19). General diet preferences are presented in caps and specializations in lower case (e.g., CARNIVORE/piscivorous). For species where quantitative data are available on the diet, HERBIVORE and CARNIVORE indicate that >90% by volume or mass of the turtles diet is plant or animal respectively, otherwise they are designated as OMNIVORE. Preferences or specialties imply that >70% of the diet is a specific food type. Specialty categories range from fairly general to fairly specific and include the following: algivorous (algae eaters), carnivorous (meat eaters), durophagous (feeding on hard-bodied organisms including seeds, mollusks and turtles), frugivorous (fruit eaters), herbivorous (plant eaters), insectivorous (insect eaters), and molluskivorous (snail and clam eaters). For example, the diet of the smooth softshell (*Apalone mutica*) studied in Indiana's Wabash River comprised 96% animal food and approximately 80% of the diet was insects (Pierce, 1992). Herein, we categorize it as CARNIVORE/insectivorous. The diet of the river terrapin on the Perak River of Malaysia is 45.2% leaves and stems of riverine plants, 25.2% fruits, and 29.6% mollusks (E. Moll, 1980a). It would be categorized as OMNIVORE/herbivorous.

Unfortunately, much of the information on diets of the world's turtles lacks quantitative analysis, and in these cases we have relied on published reports in the literature as to how these species should be classified. While far from a precise method, this does allow us to look for geographic trends concerning the distribution of guilds.

In looking over the relative distribution of riverine guilds on a global scale, one trend stands out: tropical regions tend to have more herbivore guilds and temperate regions more carnivore guilds than would be expected by chance (table 3.19). Wilbur and Morin (1988) found no correlation between herbivory and latitude, but their data set included few batagurids and podocnemids which are mostly herbivorous. Table 3.19 depicts the composition of carnivores, omnivores, and herbivores in selected rivers from each biogeographic region. In particular, the Neotropical Region and the Oriental Region—two chiefly tropical regions—have especially high percentages of herbivorous species (37% and 52%, respectively). This might seem to indicate that herbivory is an adaptation to the tropical environment. However, if this is true it is difficult to explain why there are no herbi-

Table 3.19
The percentage of carnivores, omnivores and herbivores among aquatic turtles occurring in each biogeographic region and in selected rivers within each region.

River system	Zone	Carnivores	Omnivores	Herbivores
Australian (20)	Trop/Temp	50% / 15%	20% / 5%	10% / 0
S. Alligator (6)	Tropical	50%	33%	17%
Fitzroy/Dawson (6)	Tropical	67%	17%	17%
Murray/Darling (3)	Temperate	67%	33%	–
Ethiopian (13)	Trop/Temp	85% / 0	15% / 0	0 / 0
Zaire (9)	Tropical	78%	22%	–
Zambezi (7)	Tropical	100%	–	–
Nearctic (40)	Trop/Temp	10% / 50%	0 / 27.5%	2.5% / 10%
Mississippi (17)	Temperate	53%	41%	6%
Alabama (12)	Temperate	50%	42%	8%
Rainbow Run (7)	Subtropical	43%	14%	43%
Neotropical (35)	Trop/Temp	26% / 3%	28% / 6%	37% / 0
Orinoco (11)	Tropical	18%	45.5%	36.5%
Amazon (12)	Tropical	17%	41.5%	41.5%
Parana (5)	Subtropical/ Tropical	60%	20%	20%
Oriental (31)	Trop/Temp	32% / 0	16 / 0	52% / 0
Chambal (7)	Subtropical	21.5%	28.5%	50%
Perak (11)	Tropical	18%	9%	73%
Xi Jiang (11)	Tropical	54.5%	36.5%	9%
Palearctic (7)	Trop/Temp	0 / 72%	14% / 14%	–
Yangtze (3)	Temperate	67%	33%	–

In this classification Temperate, Subtropical and Tropical refer to rivers located >30° latitude, between 23.5 and 30° latitude, and 0 to 23.5° latitude, respectively. For each biogeographic region, the abbreviation Trop combines tropical and subtropical categories. Food preferences were obtained from designations in the literature or where adequate data were available; carnivore and herbivore indicate species in which >70% of the diet is meat or plants, respectively, and omnivores include all others. The numbers in parentheses indicate the species for which diet information was available.

vorous riverine chelonians in the tropical Ethiopian Region and very few in the tropics of the Australian Region.

The answer to this question likely relates in part to phylogeny and zoogeography. Lineages evolving from an herbivorous or carnivorous ancestry demonstrate what could be called phylogenetic inertia. In other words, at the genus, and even the family level, there is a tendency for a lineage to lean toward a particular food preference. For this reason, using related organisms as independent data points in comparative studies can often bias the results (see Harvey & Pagel, 1991). For example, the chelonian fauna of the Oriental region is rich in batagurids; in turn, the majority of the Bataguridae are herbivorous. Is this because the tropical environment has selected for herbivores, or is it simply that the ancestral batagurids were herbivores and the majority of those around today inherited this tendency? Similarly, chelonian communities of the Neotropics have a high percentage of herbivores because they also include batagurids (genus *Rhinoclemmys*) and river turtles of the family Podocnemidae that are chiefly herbivorous. If we remove batagurids and podocnemides from the global analysis, then the trend for herbivory within tropical communities vanishes.

Table 3.20 demonstrates that carnivorous and herbivorous tendencies do tend to follow phylogenetic lines. The Chelidae, Pelomedusinae, Kinosternidae, and Trionychidae have strong carnivorous tendencies, while the Bataguridae, Dermatemydidae, and Podocnemidae are predominately herbivorous. However, despite the important role of history in determining food preference, other evidence still suggests that the global distribution of herbivores among freshwater chelonian families goes beyond mere phylogenetic similarity. For example:

- All predominately herbivorous families (or subfamilies) are subtropical or tropical in distribution.
- Only four of the 37 herbivores distributed over six chelonian families have extratropical ranges (>30° latitude) and all of these are emydids.

Table 3.20
The numbers of carnivores, omnivores, and herbivores occurring in tropical and temperate latitudes for each family of chelonians.

	Carnivore		Omnivore		Herbivore	
Family	TROP	TEMP	TROP	TEMP	TROP	TEMP
Chelidae (31)	15/48%	6/19%	8/26%	1/3%	1/3%	–
Carettochelyidae (1)	–	–	–	–	1/100%	–
Pelomedusinae (11)	10/91%	–	1/9%	–	–	–
Podocneminae (6)	–	–	1/17%	–	5/83%	–
Bataguridae (33)	4/12%	3/9%	6/18%	1/3%	19/58%	–
Chelydridae (4)	–	–	2/50%	2/50%	–	–
Dermatemydidae (1)	–	–	–	–	1/100%	–
Emydidae (32)	2/6%	11/34%	–	9/28%	6/19%	4/13%
Kinosternidae (15)	4/27%	7/47%	3/20%	1/6%	–	–
Platysternidae (1)	1/100%	–	–	–	–	–
Trionychidae (15)	8/53%	3/20%	4/27%	–	–	–
Total (150)	44/29%	30/20%	25/17%	14/9%	33/22%	4/3%

The number in parentheses following the family indicates the species for which adequate information on food habits was available. TROP species include those in which the mean latitude of the range falls is 0 to 30°. TEMP species include those in which the mean latitude of the range is >30°. In each box the actual number of species is followed by percentage of the total number examined.

- *Trachemys*, the only emydid genus to make a major inroad into the Neotropics, is chiefly herbivorous in tropical areas.
- *Chelydra* is chiefly herbivorous in the tropical parts of its range.
- The few extratropical batagurids are either carnivores or omnivores.

It seems likely that either herbivory is advantageous in the tropics or it is disadvantageous at higher latitudes, or a combination of both. Available data suggest that herbivory is likely disadvantageous for ectotherms living at high latitudes or cool climates (Schoener, 1977; Zimmerman & Tracy, 1989). Most reptiles are primarily carnivorous, and herbivory and omnivory occur only in some turtles and lizards. Schoener (1977) argues that herbivory is not more common among reptiles because of the following disadvantages associated with it: (a) slow growth to maturity; (b) a low assimilation frequency; and (c) a need for high body temperature to process enough basking and consequent exposure to predation. He concluded that these factors, combined with competition from endotherms, disfavored herbivory except in warm, seasonally constant environments. It can also be argued that in warm seasonally constant environments, where the disadvantages of herbivory are ameliorated, herbivory is advantageous in that food is abundant year-round. Competition for animal food may also be greater in complex tropical ecosystems (compared to temperate ones), thus favoring herbivory, but this remains to be determined.

Correlates of Turtle Richness

Because of its simplicity, species richness has typically been used in comparing diversity over broad geographic areas. A few researchers have looked at correlates of turtle species richness over a broad scale and discovered that they occasionally deviate from trends observed in vertebrates as a whole.

Schall and Pianka (1977, 1978) found that species richness of reptiles correlated positively with mean July temperature in the Iberian Peninsula, and with mean annual temperature and annual rainfall in the United States and Australia. J. S. Rogers (1976) and Owen and Dixon (1989) found turtle species richness to be positively correlated with annual precipitation in Texas.

Iverson (1992c), the only worker to examine correlates of species richness on the global scale, used findings from other vertebrate groups and predicted that chelonian species richness should be greatest in the tropics; in larger drainage basins; in regions with warmer winter or summer temperatures or with less annual range in temperatures; and in areas of higher rainfall or low seasonal variation in rainfall. However, the only highly significant correlate which he found was annual rainfall. Maximum species richness was observed in the lower Ganges and Brahmaputra river basin (seventeen aquatic species) and in the lower Mobile river basin in Alabama (sixteen species). This does not mean that factors such as temperature, river size, and latitude are unimportant in determining species richness. It does mean that on the global scale other complicating factors involving the ecology, evolution, and historical biogeography of the groups being considered may be more important. For example, one need only examine the range maps in Iverson (1992b,c) to observe how temperature affects turtle species richness at high latitudes. Freshwater turtles reach 50°N latitude in the Nearctic and almost 60°N in the Palearctic, but in both cases only a single emydid reaches these extremes (*Chrysemys* and *Emys*, respectively). Similarly, the southern limits for freshwater turtles are approximately 38°S in South America (one chelid), 40°S in Australia (one chelid), and 35°S in Africa (one pelomedusid). Actually, turtles simply run out of terra firma (and consequently rivers) in Africa, but in South America rivers are still available to 52°S and in Australia turtles reach King and Flinders Islands but do not occur on Tasmania at 41°S. Moving toward lower latitudes from these extremes in either hemisphere, species richness increases at least initially. The trend does not last; at some point before reaching the tropics, the trend for greater species richness reverses. For example, consider the Nearctic region.

At its source (47°N), the Mississippi River has only two species, yet at its confluence with the Red River (31°N) in Louisiana the species richness peaks at fourteen. The aforementioned lower Mobile River Basin (with its sixteen aquatic species) is also at this latitude, just east of the Mississippi River. However, if we shift westward to Texas in order to again move southward by a land route, the trend for greater species richness at lower latitudes quickly reverses. Instead of communities of ten or more turtles in the rivers flowing into the Gulf, the rivers of south Texas and northern Mexico typically harbor one to four species, usually including one or two mud turtles, an emydid and/or a softshell. Species richness picks up again in the tropical fingers that extend northward through Central America and along the Mexican coast to Vera Cruz, but no rivers approach a species richness of ten until the Orinoco of Venezuela is reached. Explanations of this reversal in the latitudinal trend involve ecology, evolution, and geological history.

At high latitudes, low temperature is an important limiting factor, but not at the latitude of south Texas where reduced rainfall becomes the most significant limiting factor (Owen & Dixon, 1989). History has been important here as well because the emydids which predominate in the Nearctic turtle communities appear to have evolved in temperate latitudes and with the exception of *Trachemys*, have been unable to cross the northern Mexico barrier to reach the Tropics. (The kinosternids which have their greatest diversity in tropical America were also able to cross this barrier, but in the other direction.)

Similarly, most of the diversity of South America is provided by side-necked turtles of the families Podocnemidae and Chelidae that have evolved at southern latitudes but have no established populations in Central America and Mexico. Local and regional species richness tends to be positively correlated. Hence, without a diversity of emydids and/or pleurodires to draw upon, Mexico and Central America would not be expected to have a high diversity of species. Geological history is significant here. The relatively recent joining of North and South America (ca. 6 million years) by way of Central America likely has not provided adequate time for extensive invasion of this transition zone between the two continents by these turtle groups.

Even in regions where turtles are abundant and diverse, turtle communities tend to be relatively restricted in species diversity and only a few species (often only one or two) occur together in any particular habitat.

Islands and Peninsulas

Islands tend to have less than their share of species diversity compared to equivalent areas of mainland. MacArthur and Wilson in their seminal 1967 book, *The Theory of Island Biogeography*, identified three factors that influenced the species richness of islands—remoteness, age, and area of the island, with the latter being most important. The species–area relationship predicting that the number of species in a region correlates positively with the area is not a new concept, but has been variously stated at least since the mid-nineteenth century. Among the more important papers on the subject are included Arrhenius (1921), Gleason (1922), Darlington (1948, 1957) and Preston (1960, 1962a,b, 1968). However, like so many other important concepts involving island faunas, MacArthur and Wilson brought the importance of island size into focus. One of their examples supporting the species richness/area correlation involved the herpetofauna of the West Indies. Based on information derived from Darlington (1957), they demonstrated a linear relationship between numbers of amphibians and reptiles and island size in the West Indies when species numbers and island size are plotted on a logarithmic scale. However, as in the case of a number of other truisms mentioned in this book, turtles are not necessarily crowd followers.

West Indies

Turtle numbers and island size in the West Indies show little relationship. With the exception of the scorpion mud turtle, *Kinosternon scorpioides*, that occurs on Isla de San Andrés, a small Caribbean island near Nicaragua, the sliders, *Trachemys*, are the only freshwater turtles that have managed to naturally colonize the oceanic Caribbean Islands. Several sliders are endemic to the West Indies. Of the larger islands, Hispaniola (75,600 km^2) has two species, whereas Cuba, Great Inagua, Jamaica, and Puerto Rico (ranging from 14,500 to 108,000 km^2) have only one. Obviously, size has not been an especially important determinant of turtle species richness in the West Indies. The presence and amount of permanent freshwater habitat as well as opportunities for colonization were possibly other significant factors determining which islands were colonized by the *Trachemys*.

Another question concerns why only the emydid genus *Trachemys* has successfully colonized the West Indies when several other families are available in the source areas. As sliders appear to have evolved after the separation of the proto-Antillean plate from North America, their arrival in the West Indies is better explained by dispersal rather than vicariance (Hedges, 1996; Seidel,

1996). As one might expect, sliders have significant salinity tolerance. In Georgia and Carolina, *Trachemys scripta* occur in and around salt marshes on the mainland and coastal islands (Gibbons & Coker 1978; Gibbons et al., 1979; Gibbons & Harrison, 1981). Mesoamerican sliders have been observed in mangrove swamps and estuaries (Moll & Legler, 1971; Legler, 1990) and in Costa Rica they enter the ocean in order to nest on sea beaches (D. Moll, 1994). The rate of electrolyte transfer in sliders is relatively low (Dunson, 1967). Sliders can survive 100% sea water for at least a week, and one of the West Indies group of sliders, *Trachemys decussata*, appears to be intermediate in salinity tolerance between truly freshwater forms and the highly specialized diamond back terrapin, *Malaclemys terrapin* (Dunson & Seidel, 1986). However, other families available in the potential source area of Central America (Chelydridae, Dermatemyidae, and Kinosternidae) also appear to have salt-tolerant members. At least they have been found in brackish habitats (Neill, 1958; D. Moll, 1986). Consider the family Kinosternidae. Neill (1958) provides a selection of anecdotal reports of kinosternids inhabiting brackish water habitat. However, the actual salinity of the habitats mentioned in these examples is never cited. The most rigorous studies of salinity tolerance in a mud turtle are those of Dunson (1979b, 1981). He reports that the striped mud turtle, *Kinosternon baurii* is the only non-specialized estuarine species occurring in aquatic habitats of the Lower Florida Keys (Dunson, 1981). Dunson found that while the island form of the species showed greater physiological adaptation to saline habitats than four other kinosternids tested (including the mainland form), it was not capable of long-term survival in salinities above 15% sea water. He concluded that the striped mud turtle had been able to invade the Florida Keys because of its behavior rather than any unusual salinity adaptation. Specifically it appears to be their ability to utilize terrestrial habitats for estivation whenever the aquatic habitats become too saline that is significant. However, such terrestrial adaptation would be of little value in invading the West Indies. Survival times (50%) reported for *Kinosternon baurii*, *K. leucostomum*, *K. subrubrum*, and *Sternotherus odoratus* in 100% sea water were >5, 3.9, 5.8, and 2.9 days, respectively (Dunson, 1981, 1986). If these figures are representative of all kinosternids, then they may simply have insufficient salinity tolerance to cross extensive gaps of salt water. Other possible contributing factors include their tendency to be bottom walkers and their small size. Bottom walkers tend to be poor swimmers (Zug, 1971). Small size may be disadvantageous for active dispersal across expanses of salt water for a variety of reasons including increased predation risk, increased rate of electrolyte flux due to the greater surface to body mass ratio, and reduced swimming power.

Snapping turtles do not have the size problem, but again are bottom walkers and not well-adapted physiologically to a salt water environment. Snapping turtles studied by Dunson (1986) in a tidal creek on Virginia's eastern shore seemed to rely heavily on behavioral osmoregulation (avoiding long-term exposure to high salinities by migrating between more and less saline portions of the creek). Dunson found that the snappers' integument is quite permeable to water but relatively impermeable to sodium. Net water loss in seawater was inversely proportional to body mass, providing larger turtles with a significant advantage in osmoregulating in saline water.

The Central American river turtle, *Dermatemys mawii*, also feeds in estuaries (D. Moll, 1989), but its osmoregulatory capabilities have not been studied. Of all of the species in the Central American source area, this large and powerful swimmer would appear to have the best chance of actively dispersing to the West Indies. Either it does not have the physiological capability to make the crossing or some inadequacy of the habitat available in the West Indies precludes colonization.

Perhaps the trait or adaptation that ultimately allowed sliders to reach the Greater Antilles was behavioral rather than physiological, and their dispersal was passive rather than active. In at least some areas of Mesoamerica, sliders spend the early part of their lives associated with floating grass mats that tend to accumulate in backwaters of the river (Moll & Legler, 1971). During the rainy season, high water and strong currents break up these mats and carry them down river and eventually out to sea. It may have been such rafts which carried sliders to the Antilles. Grant (1932 in Neill, 1958) was told by a resident of Vieques Island that after severe hurricanes, sliders floated over from Puerto Rico (ca. 25 km) on masses of broken branches and limbs. *Rhinoclemmys funerea* has also been observed being carried to sea on mats of water hyacinth in Costa Rica (Jansen, 1993), but to date they have only managed to colonize offshore islands such as Bastimentos Island in Panama (Moll & Legler, 1971).

Legler (1990) raised the alternative possibilities that humans may have introduced sliders into the Antilles, and those parts of Central and South America may have been colonized by slider turtles from the Antilles rather than the reverse. Seidel (1988, 1996) has reviewed the affiliations of the Antillean sliders and the biogeographical theories of their origin, concluding that the populations of the Greater Antilles are likely a para- or polyphyletic compilation whose origin involved early dispersal events followed by intra-Caribbean vicariance and dispersal.

Sundaland (East Indies)

Species richness in the East Indies (Greater Sunda Islands) fits the species area pattern imperfectly but much better than the West Indies as indicated in figure 3.6, a species/area diagram based on nine islands. Java (126,000 km^2), the smallest of the large islands, has the fewest aquatic turtles (eight). However, Borneo (742,000 km^2)—the largest—has nine species, and smaller Sumatra (405,000 km^2) has eleven species. Lovich (1994) published a species/area diagram for six islands of the East Indies and Peninsular Malaysia and found that the relationship between land area and number of chelonian species is significantly positive ($r = 0.76$; $P = 0.05$).

Perhaps the major difference between the East and West Indies in regard to turtle diversity is the fact the Greater Sunda Islands are continental islands having been part of mainland Asia 13,000 to 14,000 years ago. During periods of glaciation, so much of the world's water became tied up as ice that the shallow seas connecting Asia and these islands became dry land. In other words, turtles did not have to move across seawater to reach these islands. No salt tolerance was neces-

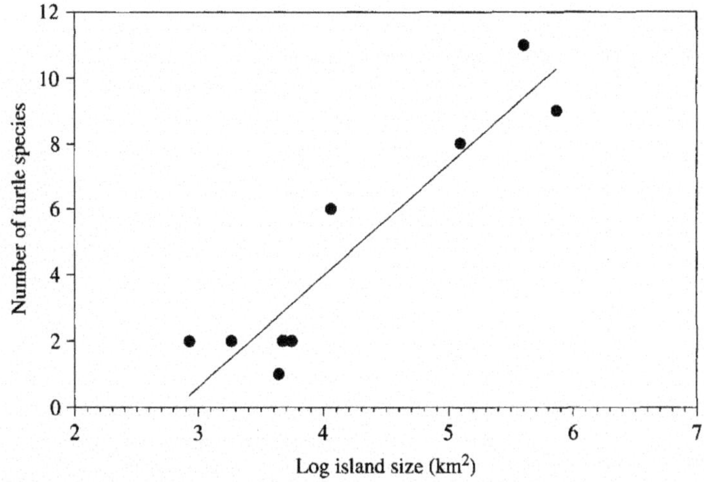

Figure 3.6
Chelonian species per area (log km^2) on nine representative islands of the East Indies islands (Sundaland). Islands and area (km^2) plotted are Borneo 742,000; Sumatra, 405,000; Java, 126,000; Banka, 11,500; Bali, 5560; Nias, 4772; Siberut, 4400; Simeulue, 1844; and Lingga, 850.

sary to become widely distributed over the Sunda Islands. Present-day differences in species composition and richness are likely due to differential extinction and distance from source areas. For example, the greater number of species on Sumatra compared to the larger Borneo is likely due to its closer proximity to the Malaysian source area (which has thirteen species). Considering that the Straits of Malacca separating Sumatra and Malaysia are only 50–75 km wide in places, the dispersal of turtles across this gap would not be unusual. The equilibrium theory of biogeography predicts that smaller islands should have higher extinction rates than larger, but a mitigating factor is distance from the source area (MacArthur & Wilson, 1967). Close proximity to the source area increases the colonization rate which can override the effects of a higher extinction rate. Yet another difference between the East and West Indies is habitat; there are substantially more major rivers available for colonization in the former.

Peninsulas

As with islands, some authors have recognized a reduction of species on peninsulas compared to associated mainland. G. G. Simpson (1964) found that North American mammals show a decreasing density of species from the base to the tip of peninsulas. MacArthur and Wilson (1967) reported that while bird species density tended to increase toward the tropics, the Florida, Baja California and Yucatan peninsulas were impoverished in diversity compared to mainland areas. They explained the impoverishment as resulting from reduced immigration rates due to the absence of land around three sides of a peninsula. Busack and Hedges (1984) questioned the applicability of the peninsular effect to squamates and as a general phenomenon. They examined the squamate faunas of Baja California, Florida, Yucatan, and Iberia and found that only Florida exhibited a decreasing gradient in species density from the base to the tip. As a result, Busack and Hedges concluded that the absence of the effect on the other peninsulas suggests that historical factors were more important in determining species diversity than a peninsula effect. A reduction of suitable habitat on peninsulas compared to the mainland is still another explanation for the depauperate peninsular faunas (Williamson, 1981).

Evidence from aquatic turtles is equivocal. Baja California and the Iberian Peninsula have too few species (two) to be of much comparative value. Both Florida and the Yucatan Peninsula are less diverse in turtle species compared to adjacent areas of mainland. The Florida peninsula has fourteen aquatic species at the base, grading to eight at the tip, while the Yucatan Peninsula has nine species at the base grading to four at the tip. Conversely, the Malay Peninsula shows the reverse of the predicted effect. The base in Thailand has at most twelve aquatic turtles compared to the Malay Peninsula's thirteen. Presumably, the richness of the Malay Peninsula is at least in part attributable to its position between two centers of evolution, the Asian mainland and the Sunda Islands, which both have contributed to its fauna. In addition to its mainland forms, peninsular Malaysia has at least four aquatic species which likely evolved in the Sunda Islands (based on Sundaland distributions which do not (or barely) extend north of the Malay Peninsula). Hence, the Malay Peninsula differs from Florida and Yucatan in that in addition to the mainland source area at the proximal end, there is a secondary evolutionary center and source area, the Sunda Islands, at the distal end.

Resource Partitioning and Species Interactions

Alfred Russell Wallace, a renowned English naturalist of the last century, noted in a paper published in 1855 that: "Every species has come into existence coincident both in space and time with a pre-existing closely allied species." Thirteen years later a German naturalist Moritz Wagner (1868) took Wallace's observations a step further by stating that closely related forms tend to occur in adjacent ranges, usually separated by some barrier. In 1905 the American ichthyologist, David Starr Jordan, reintroduced Wagner's observations to the scientific public, and thus the phenomenon is occasionally referred to in the American literature as Jordan's rule. Another better known precept, the competitive exclusion principle (Gause's rule) has usually been invoked to explain the phenomenon. In its simplest form, Gause's rule states that no two species can occupy the same niche indefinitely, the thinking being that if two species occupy the same niche, then interspecific competition eventually will result in the elimination of the less-adapted form (either by extirpation or by forcing it into another niche).

The Wagner/Jordan Rule is well exemplified by Gulf Coast *Graptemys*. Four sister species of map turtles—Barbour's, the Escambia, the Alabama, and the Pascagoula—occupy adjacent river systems flowing into the Gulf of Mexico (see table 3.5). Females of all four are double the size of males, have markedly enlarged heads with broad alveolar crushing plates on the jaws, and feed heavily on mollusks. Although their occurrence in adjacent river systems may partially reflect a recent separation from a common ancestor (i.e., too little time to invade adjacent river systems), the competitive exclusion principle (Gause's rule) predicts that microsympatry among these closely related mollusk specialists probably cannot occur. Conversely, in cases where sympatric congeners do occur, Gause's rule predicts that these species have been able to partition the available resources so as to allow distinct or somewhat distinct niches for each species, thus limiting interspecific competition.

Again using map turtles as an example, the Pascagoula and Alabama map turtles occur sympatrically with three narrow-headed map turtles, the ringed, yellow-blotched, and black-knobbed map turtles (three species whose ranges, like those of their broad-headed cousins, do not overlap). This situation does not violate the competitive exclusion principle because narrow-headed map turtles are not mollusk specialists. Rather, they tend to be insectivorous (ringed, and yellow-blotched) or omnivorous (black-knobbed) and mollusks are usually not a major part of their diets (see table 3.5). Males of the broad-headed and both sexes of the narrow-headed map turtles do tend toward insectivory. Possibly this overlap presents less problems to coexistence because insects have always been sufficiently abundant in these rivers to support both species (i.e., similar species can coexist provided resources are not limiting). Presuming mollusks to be the more limited resource over time, a reasonable working hypothesis is that partitioning of the female's diets allows narrow-headed and broad-headed map turtles to coexist in these river systems.

Although the above examples emphasize dietary similarities and differences, other dimensions of these animals' niches can also affect coexistence of similar species. Pianka (1994) stated that closely related sympatric species typically have subtle differences of three basic types: (1) they eat different things; (2) they utilize different habitats or microhabitats; or (3) they are active at different times. The separation of sympatric forms by either, food, space, or time partitions these resources, thus reducing interspecific competition and allowing coexistence.

Nevertheless most studies of resource partitioning in turtles have concerned the food habits of related species. Foods of associated species may differ in type or relative importance in the diet for a variety of reasons, including differences due to preference, microhabitat, feeding behavior and/or adaptations, and size of the turtle's feeding apparatus. Mahmoud (1968a,b) compared the food habits of four sympatric kinosternids (two mud and two musk turtles). He concluded that while the four species were generalized omnivores, mud turtles showed a preference for fast-moving prey compared to musk turtles.

Vogt (1981a) studied food partitioning in three map turtles inhabiting the Mississippi River near Stoddard, Wisconsin and reported that the three species used different foods and feeding strategies. Most of the differences observed by Vogt concerned the females. The only broad-headed map turtle, *Graptemys geographica*, was carnivorous, specializing on mollusks. *Graptemys ouachitensis*, the most common species, was omnivorous but specialized in surface feeding, whereas *G. pseudogeographica* was a general omnivore that overlapped the other two species in both food and feeding sites. Another1981 study by Williams and Christiansen examined the niches of two sympatric softshells in Iowa. Relative to food they found that spiny and smooth softshells essentially eat the same organisms, but that the relative composition of food items differed markedly. The authors concluded that a chief cause of the observed differences was that the softshells partitioned the foraging habitat. Smooth softshells usually fed in the water column while spiny softshells were chiefly bottom feeders. Plummer and Farrar (1981) found that male and female smooth softshells ate different foods in part because they foraged in different microhabitats where different prey was available to them.

Several authors have investigated food partitioning in tropical turtle communities (Vogt & Guzman-Guzman, 1988; D. Moll, 1990; Fachín-Terán et al., 1995). The latter two examples concern species in flowing water habitats. D. Moll studied food habits of four turtles inhabiting a tropical freshwater stream turtle community in Belize where an herbivore, *Trachemys scripta*, was the dominant species. Three kinosternids comprised the remainder of the community (in order of abundance), *Kinosternon scorpioides (cruentatum)*, *K. leucostomum*, and *Staurotypus triporcatus*. The *Kinosternon* were chiefly insectivorous, whereas the large-headed *Staurotypus* was durophagous, feeding on large snails and other turtles. Dietary overlap between the mud turtles was highest during the time the populations were at their lowest density in the stream (January) and least at their greatest density (April). When populations were most dense, *K. leucostomum* ate relatively greater amounts of small snails and benthic insects and relatively less of drowned terrestrial insects, while *K. scorpioides* did just the reverse.

Fachín-Terán et al. (1995) studied the food habits of five sympatric pleurodires inhabiting the Rio Guapore in Rondonia, Brazil. The assemblage contained two herbivores (*Podocnemis unifilis* and *P. expansa*), an omnivore (*Phrynops geoffroanus*), a molluskivore (*P. raniceps*), and a piscivore (*Chelus fimbriatus*). Although collections for three of the species were small, it appears that there was no appreciable overlap of diet in this assemblage, with the possible exception of the herbivores. Too few *Podocnemis expansa* were examined, however, to determine if they partitioned available plant foods with *P. unifilis*.

Partitioning of space or habitat is also apparent between turtle communities within a river, and particularly among closely related species. Legler (1978) provided a particularly striking example of spatial partitioning among Australian chelids in the shallow Namoi River in New South Wales. *Chelodina expansa*, *C. longicollis*, and *Emydura macquarii* foraged together in the deeper pools that had formed at bends where uprooted willow trees had lodged during floods. Three levels of vegetational structure were recognized and described by Legler as follows: "the bottom (lowest 0.5 m) was either clean, covered with leaf litter, or contained submergent rooted vegetation. The middle depths consisted of tree limbs, ranging from 5 to 20 cm in diameter, in which a variable amount of vegetational debris had collected. The top layer consisted of a dense mat formed by the willow weeps, their adventitious roots, and vegetational debris which had accumulated in them (chiefly leaves)." *Chelodina longicollis* usually foraged in the lowest level, while *Emydura* utilized the middle layer and *C. expansa* used the upper mat.

In a somewhat different type of habitat partitioning, the river terrapin and the painted terrapin both inhabit tidal regions of the Perak River in Malaysia, but a degree of spatial separation still exists (Dunson & Moll, 1980). In collections made near the mouth, the river terrapin outnumbered the painted terrapin 6:1 (192:31), but in collections taken some 30 km upriver, the painted terrapin predominated 11:1 (155:14).

In southeastern and south central Africa, the hinged turtles, *Pelusios sinuatus* and *Pelusios bechuanicus* are riverine species whose ranges are mutually exclusive except in the upper Zambezi River. Where their ranges overlap above Victoria Falls, *P. bechuanicus* inhabits the deep clear waters of the river while *P. sinuatus* is restricted to the lagoons and muddy backwaters where large game animals drink (Broadley, 1981).

In salt marshes along the Virginia coast, the potentially competitors, diamond back terrapins (*Malaclemys terrapin*) and common snapping turtles (*Chelydra serpentina*) overlap in range and diet (Dunson, 1986), but differ somewhat in habitat. Terrapins inhabit the highly saline channels of the marsh, while the snappers remain in the fresher portions of the marsh and spend much of their time burrowed beneath the substrate.

In their examination of the niches of spiny and smooth softshell turtles in Iowa, Williams and Christiansen (1981) found the smooth softshell to be more of a habitat specialist being restricted to large downstream portions of rivers. The spiny softshell occurred throughout the rivers and a variety of ponds and backwaters. Where the two occurred together, the smooth softshell was taken in more open waters near sandy beaches, while the spiny was found around submerged structures such as brush, trees and other debris.

The most detailed report of habitat partitioning in turtles of which we are aware concerns a study of three sympatric species of map turtles in Kansas by Fuselier and Edds (1994). The authors measured thirty-two environmental variables at each trap site to analyze habitat overlap between the common map turtle, the Ouachita map turtle, and the false map turtle. The common map turtle was restricted to shady streams with rock and gravel substrates. Other factors distinguishing common map turtle aquatic habitat from those of false maps or Ouachitas were a higher percentage of bare shoreline, higher CO_2 content, and greater acidity. *Graptemys pseudogeographica* preferred sites with a mud substrate and ample basking sites, and was never taken on sand substrate. Conversely, *G. ouachitensis* could be found on mud, sand, or rock substrates and co-occurred with *G. pseudogeographica* on some of the mud bottom habitats. Other factors distinguishing Ouachita sites were higher than average water temperatures, deeper/wider runs, higher dissolved O_2, less emergent vegetation, and less shade.

Besides providing an excellent example of how closely related turtle species partition habitats, the Fuselier/ Edds study provides another significant take-home message: conclusions drawn from one part of a species' range do not necessarily apply to the rest of the range. For example, the authors found that their characterization of the map turtle habitat differed significantly from that of researchers working with the turtle farther east. A number of authors (see review in Fuselier & Edds, 1994) have reported that map turtles inhabit lakes, backwaters, and oxbows having abundant macrophytes, good basking sites, and soft clay or mud bottoms, none of which applied to Kansas populations. Fuselier and Edds supported Vogt's (1981a) suggestion that since common map turtles select different habitats within its range, food availability rather than habitat may be the more important limiting factor for distribution in this species. Similarly, Fuselier and Edds (1994) never found false map turtles associated with sand substrate in Kansas. However, we have found them to be the most common *Graptemys* in sand substrate habitat of the Mississippi River at least from Tennessee to northern Illinois. However, in Wisconsin, Vogt (1981a) found Ouachita map turtles to be more abundant than either common map or false map turtles in pool 8 of the Mississippi River, a chiefly lentic habitat liberally punctuated with sand islands. Also, the Ouachita map turtle is the common map turtle in sand substrate habitats of the Wabash and Ohio rivers bordering southern Illinois where false map turtles are rare.

Temporal partitioning has been little studied among chelonians. Perhaps the most obvious type would be coexistence of diurnal and nocturnal forms. Pianka (1977) generalized that most turtles are diurnal. While this may be true of the temperate zone, it does not hold for turtles in the tropics (Kearney, 1972; Moll & Moll, unpubl. obs.). Moll and Legler (1971), Legler (1990), and Moll and Moll (1990) have emphasized the diurnal behavior of *Trachemys scripta* as being one of the traits which could have been significant in its invasion of tropical America. The slider's closest relatives in the middle American tropics are the batagurids, *Rhinoclemmys funerea* and *R. melanosterna*. All three feed chiefly on vegetation, but the niche appears to be partitioned both temporally and spatially. Tropical *Trachemys* forage aquatically throughout the day and are inactive at night, while the *Rhinoclemmys* have been observed foraging in late afternoon, at dusk, and at night, and at least partially on land (Jansen, 1993; Legler, 1990; Medem, 1962; Moll & Jansen, 1995; Moll & Legler, 1971; Moll & Moll, unpubl. obs.). It is our impression that *Trachemys* is the most diurnal of the aquatic chelonians inhabiting tropical Mesoamerica. Potential competitors *Chelydra*, *Dermatemys*, *Kinosternon*, *Staurotypus*, and *Rhinoclemmys* tend to be more crepuscular and/or nocturnal (D. Moll, 1989, Vogt & Guzman-Guzman, 1988; Moll & Moll, unpubl. obs.). Conversely Dr. John Iverson (pers. comm.), an authority on kinosternids and who has collected extensively in temperate and tropical America, believes nocturnal behavior is uncommon in the family and would classify them primarily as crepuscular and/or diurnal.

Therefore, is *Trachemys scripta* occupying a relatively uncontested chelonian niche in regard to its diurnal activity in tropical America? The case remains to be demonstrated, since much of the evidence is anecdotal and few rigorous studies have been made of the activity times of the aforementioned turtles. Carefully controlled 24-h trapping experiments in a variety of communities and habitats throughout tropical America will be needed to answer this question.

At least some species show no preference for day or night feeding. Relative to the latter, D. Moll (1989) found that in Belize, inland *Dermatemys* foraged nocturnally but in a tidally influenced areas they foraged during periods of rising tide at any time of day. This parallels the behavior of the superficially similar river terrapin and painted terrapin on the Perak River that feed on the rising tide at any time of day (E. Moll, 1980a,b, and unpub. obs.).

Species Interactions

Although we have suggested that the underlying cause of the examples of resource partitioning provided above is interspecific competition, competition is extremely difficult to demonstrate in nature. Part of the problem is that the partitioning effects that we presently see could have resulted from past conflicts for limited resources under environmental conditions which no longer exist. Much of the evidence for competition in wild populations is circumstantial (e.g., closely related species occupying different habitats, feeding on different foods, or occupying adjacent, but non-overlapping ranges).

One of the most convincing sorts of circumstantial evidence comes from species interactions. Observable differences in habitat, diet, temporal activity, or morphology in two closely related species where they overlap in range from their exclusive range supports a competition hypothesis. Berry's (1975) study on the musk turtles, *Sternotherus odoratus* and *S. minor* is still one of the best studies of species interaction among chelonians. Although both species occur in the same geographic range, *S. minor* prefers a lotic habitat and *S. odoratus* lentic. Overlap occurs in intermediate habitats with considerable aquatic vegetation and slow current. Berry studied these turtles in several Florida localities both where they were sympatric and allopatric. In their exclusive ranges *S. minor* was a mollusk specialist with hypertrophied head and jaw musculature, whereas *S. odoratus* fed as a generalized carnivore with no obvious feeding specializations. Where sympatric, these congeners overlapped considerably in diet and showed no evidence of spatial or temporal partitioning. Most interesting here is that *S. minor* has reduced mollusk-eating adaptations in sympatry. Berry attributed this loss to the phenomenon of character convergence (MacArthur & Levins, 1964, 1967) which results when resources are limiting and of a generalized nature. Under these conditions, natural selection favors individuals of a specialized species that are more generalized and capable of competing with their generalized relative(s).

Better known is the phenomenon of character divergence (Brown & Wilson, 1956), in which two species tend to be more different in localities where they occur together than where they occur apart. Often these differences concern the size of the species or their feeding adaptations. Either can bring about differential taking of food types and thus reduce interspecific competition in the area of sympatry. Vogt (1981a) provides an example of classical character displacement in map turtles. In Arkansas and Louisiana, large-headed molluskivorous *Graptemys pseudogeographica* occurs sympatrically with narrow-headed *G. ouachitensis*. However, in Wisconsin where *G. pseudogeographica* is sympatric with the broad-headed molluskivorous *G. geographica*, it has a narrow head and feeds as a generalized omnivore.

Another related phenomenon suggesting competition is ecological release, or niche expansion resulting from reduced interspecific competiton. In India, ranges of the closely related Indian roofed turtle (*Kachuga tecta*), Indian tent turtle (*K. tentoria*), and the Brown roofed turtle (*K. smithi*) overlap in the Ganges River system. In the Indus, only *K. tecta* and *K. smithi* are present, while in the eastern drainages south of the Indus only *K. tecta* is present (E. Moll, 1987a; but see Bhaitt, 1989). In the Ganges drainage, *K. tecta* is chiefly confined to lentic habitats while *K. smithi* and *K. tentoria* are most common in lotic situations. In the Indus, *K. tecta* remains an inhabitant of lentic habitats while *K. smithi* is the common lotic form (Minton, 1966). However, in the eastward flowing Narmada River where neither *K. smithi* or *K. tentoria* are known to occur, *K. tecta* was moderately common in lotic habitats (E. Moll, 1987a).

4 Traditional Exploitation Methods, Efficiency, and Consequences for River Turtles

Profile 4. *"For hours together the fisher stands on the shore watching for the appearance of the chelonians. Hardly is the shoal in sight when he bends his bow, places an arrow in it, and waits. At the instant when the floating mass is passing before him, he sights it horizontally, then suddenly raising his bow and arrow he makes the weapon describe a trajectory, the descending line of which has for its point of intersection the shell of a turtle. Sometimes several individuals throw themselves into a canoe, pursue the shoal, assail it with their arrows, which describe parabolic curves, and continue the chase until their canoe is so filled with the booty that it is ready to sink. To judge by the cries, the hurrahs, and the bursts of laughter which accompany this fishing, it might be supposed that it is an amusement rather than a toil for the Conibos."*

St. Cricq, 1874

A wide variety of ingenious methods for collecting river turtles have been developed over time. None requires a particularly high level of technology but many require a great deal of skill, patience, and sometimes physical ability by the collectors, as well as a detailed knowledge of the ecology of the species being sought. Many parallel collecting methods have developed independently in turtle-dependent cultures around the world, leading Nicholls (1977) to state in regard to Bates's (1863) description of an Amazonian turtle hunt, "With some allowance for small differences in technique, his descriptions provide an accurate image of turtle hunting as it was practiced anytime, anywhere, during the past thousands of years." We thought that a summary of these techniques with comment upon their variation in different areas and with different species, their effects on populations when this can be ascertained, and examples of their practitioners would be an appropriate addition to our treatment of river turtle exploitation patterns. We will limit our discussion mainly to techniques employed by subsistence and commercial turtlers for obtaining animals and largely omit reference to the growing body of information concerning the collection of turtles for scientific purposes (many of which are largely modifications of the former techniques). For information concerning the latter category the reader is referred to the excellent summary of equipment and techniques by Plummer (1979) and papers by Carr and Marchand (1942), Chaney and Smith (1950), Legler (1960b), Ream and Ream (1966), Wahlquist (1970), Bider and Hoek (1971), Braid (1974), Robinson and Murphy (1975), MacCulloch and Gordon (1978), Iverson (1979), Petokas and Alexander (1979), Vogt (1980b), Frazer et al. (1990), Kennett (1992), Graham and Georges (1996), Jensen (1998), and Kuchling (2003b).

Capturing Turtles in the Water by Diving

Free diving for turtles is of course a time-honored, effective, and nearly cosmopolitan approach to collecting turtles that requires little or no equipment. While diving mask, fins and sophisticated breathing gear certainly enhance the process, they are not required by skilled divers in order to harvest large numbers of turtles. Smith (1975) wrote of the Paumari of the Purus River and the Mura of the Madeira (both in Brazil) who could swim to a depth of 30 feet (10 m) to catch the giant South American river turtle. Australian aborigines catch pig-nosed turtles, by diving on top of them from the bank when they come near the surface or after waiting quietly in the water while the turtles are herded near by others (Georges & Rose, 1993). Crowned river turtles, were caught by free-diving by the Tontal tribe of India (Anderson, 1878–79). The Tontals tied up large bundles of marsh grass which could be used as floats to support the diver between dives, and to help transport turtles to shore after they were collected.

Alligator snapping turtles in clear southern rivers have also been occasionally collected by swimmers equipped with mask (or goggles) and fins. Ross Allen searched the Suwannee River for specimens beginning in the 1930s and was usually successful in obtaining display animals (Allen & Neill, 1950). Once located, the turtles were hand-caught and hoisted into a canoe traveling behind the swimmer. The most detailed account of the dynamics and effects of turtle diving for any species is provided by Polisar (1995) concerning its use in obtaining Central American River turtle in Belize. Free diving, usually with the aid of mask and fins, is one of three commonly used techniques for collecting the Central American River turtle (with harpooning or "pegging" and netting discussed below). The method here involves repeated deep dives to examine substrates and submerged logs where turtles are resting. The technique is mainly used in the spring dry season when water levels in the rivers inhabited by *Dermatemys* are low and clarity is very high. At such times it is highly effective and large numbers of turtles can be captured for market, especially in habitats where submerged aquatic vegetation (which may hide turtles) is sparse. In

Polisar's main Belize River study area free diving was a relatively well-organized activity. Diving teams consisted of several divers, one or more boats with crew members who watched over collected turtles, and other dories with harpooners ("strikers") on the alert for turtles which surfaced while trying to flee the divers. Formerly, river sections being worked were also blocked off on the downstream end with a trammel net to catch escaping turtles, but conservation concerns by local turtlers have led to a reduction in this practice. Diving typically entailed plunging into and chasing turtles at depths of 6–9 m for six continuous hours—a very physically demanding activity. Polisar found free diving to be the most effective of the three collecting methods during the dry season—resulting in more captures than either pegging or netting (although more people were involved in the other two activities). Polisar studied the system in 1990, a particularly wet year with a short dry season. He thought it was likely that a more typical, longer dry season would probably allow an even greater catch by the diving method. Free diving also yielded a larger set of size classes of collected animals than the other collecting methods. This is because all sizes were accessible to divers, but peggers selected only larger animals and large meshed nets allowed smaller turtles to avoid capture. More of the smaller (250–300 mm carapace length) specimens were collected by diving than in the other methods, but divers rejected even smaller animals available to them as they were of little financial value. All animals collected represented the sub-adult through adult size classes. Of 309 turtles collected by diving, 24% were males, 16% females, and 60% juveniles. Polisar found free-diving capture rates in densely human-populated sections of the Belize River to be less than one-fifth of those in sparsely or unpopulated sections. He also found fewer large adults (especially females) in the more populated sections of the river. This familiar scenario led to the movement of diving teams into more distant locations along the river and into adjacent lagoons where more and larger turtles were still available (and which in turn were quickly decimated by collecting).

Wading or "Muddling"

Where turtles are apt to hide, on or in the substrates of shallower waters near river banks or lake margins, finding them by feel with hands or feet can be an effective technique. Probing undercut banks with hands or a short pole (sometimes called "noodling") where turtles may find refuge is a variation on this theme. The Karens of Myanmar harvest Asian leaf turtles (*Cyclemys dentata*) from the sides of deep pools in hill streams using this latter technique (Theobald, 1868b), and it is used in America to collect snapping turtles (D. Moll, pers. obs.). The Australian aborigines have long been adept at probing for benthic turtles, a process first described by Grey (1841, in Goode, 1967). Aboriginal turtlers would slowly wade along the margins of freshwater habitats until a turtle was encountered. When the turtle dove to the bottom to escape, the hunter would feel for it with his toes and then pin it to the substrate with his foot. It could then be grabbed and brought ashore. Grey recounted that two or three hunters could catch up to fourteen 1–3 lb (0.5–1.5 kg) turtles in the course of a very short time. The turtles caught in this manner were baked in the shell in campfires and relished by the aborigines. They considered the main turtle season (December–January) to be a very important part of their annual foraging pattern, and presumably many turtles were harvested in this manner. A similar account of the aboriginal method of hunting common snake-necked turtle and Murray River turtle in South Australia was recounted by Waite (1929, in Goode, 1967). He described the process as involving wading in water up to waist deep and finding turtles by groping with the hands as well as locating them by sight. Das (1991) also wrote that Indian flapshells were often collected by probing in the bottom mud with the hands and feet. Fugler (1984) reported that experienced turtle collectors in Bangladesh sought concentrations of bottom-dwelling turtles with their feet, then dove to retrieve the animals.

Commercial snapping turtle hunters in the United States, usually working from a boat, have learned to probe for turtles with a long metal rod with a sharpened hook at one end (Clark & Southall, 1920; Bishop & Schoonmaker, 1921; Carpenter, 1955; Vogt, 1980b). The hunters seek out muskrat burrows, old submerged logs, and locations where springs feed into larger bodies of water which serve as hibernacula for large congregations of snappers in winter. After cutting through the ice the hunters probe with the rod until they feel a turtle, then pull it to the surface with the hook. Huge numbers could be harvested from such areas as the Illinois and Mississippi Rivers, and Iowa's Cedar River in the early part of the twentieth century in this manner. Clark and Southall (1920) wrote that as many as 5 tons of turtles had been taken from various muskrat holes near Muscatine, Iowa, in one season, and that as many as twenty-six turtles could be taken from one muskrat burrow. From 500 to 1000 lb (220–440 kg) of turtle were estimated as a normal daily catch with 1420 lb (650 kg) collected on one particular run. At another Mississippi River slough twenty snappers weighing 10–20 lb (5–10 kg) each were collected from under just one log.

Poling is still used in the southern states for catching alligator snapping turtles for market (Pritchard, 1989). Alligator snappers often congregate in winter hibernacula as described for common snappers above. Hunters use hooked poles, which may have detachable sections that allow the pole to be extended in length for use in deep water, to probe the bottoms of bayous in search of these concentrations. Experienced hunters learn to recognize the characteristic sound of a pole striking a turtle's carapace and then can pull the lethargic turtle to the surface with the hooked end of the pole. In Louisiana, between fifteen and twenty alligator snappers have been pulled from a single hibernaculum (often deep alligator holes). Aber (1982 in Pritchard, 1989) listed a variant poling technique in which a turtle is located and pinned to the bottom with a long, slender pole while a diver captures the turtle by hand. Ross Allen suspended bait from a non-hooked bank line at night in the Suwannee River, then used a hooked pole to retrieve alligator snappers which gathered (illuminated by a searchlight) around it (Allen & Neill, 1950).

Archery, Harpooning, and Spearing

The acquisition of turtles with specially designed arrows, harpoons, and spears is an ancient and widespread practice which requires a great deal of skill and knowledge of turtle ecology. The main difference in techniques and the design of the equipment used relates to the need to propel or throw the first two devices respectively with enough force and accuracy to pierce the carapace of the prey and then restrain it from escaping with the projectile. In the latter technique (spearing) the point is simply thrust into the turtle and it is then restrained by the spear itself until the animal can be brought from the water. The design of the arrows and "striking" or "pegging" instruments is universally similar in that formerly a stone, or now usually a sharp metal point fits into the shaft (usually via a wooden peg) in such a way that it will detach when it is propelled into the carapace of the turtle. The embedded point is attached by strong twine or cord to the shaft or to the boat of the hunter which can then be used to haul in the turtle. Bates (1863) described the archery version of the process (and provided a diagram of a turtle-shooting arrow) as practiced by aboriginals to obtain large *Podocnemis* in an isolated pool near Brazil's Solimoẽs River (also see St. Cricq's [1874] narrative of the Peruvian Conibo's skill in this practice which was used as the lead-in to this chapter). Some archers constructed small platforms of wooden poles and cross-pieces secured with lianas to provide firm footing and shot at turtles from the bank, while others shot from launches. The hunters could perceive the presence of submerged turtles before they surfaced to breathe by slight water movements which indicated their presence below. Bates was astonished at the skill with which these submerged turtles could be hit and harvested, even when the arrows were launched from a considerable distance requiring a parabolic trajectory in the flight of the arrow (fig. 4.1). The archers, in fact, preferred these longer shots as the arrow could penetrate the victim's carapace perpendicularly with great force and with no warning. When a turtle was hit it was played gingerly with its attached cord as it tried to escape. When near enough, a second arrow was launched into the beast and the two attached cords now provided sufficient purchase for the turtle to be hauled in. Bates' group harvested about twenty sub-adult turtles in about 2 hours of shooting with this method. Smith (1975) described similar equipment and techniques in his account of the destructive exploitation of giant South American River turtles. He also provided examples of numerous tribes, which used archery or harpooning (as well as other techniques) to harvest the turtle.

Pritchard (1989) reported that alligator snappers are sometimes harvested at night with bow and arrows on the Sabine River on the Louisiana–Texas border. Smith (1979) described and diagrammed a harpoon called a tapua with a typical design for turtling (fig. 4.2). It is composed of a cord-attached detachable point, as described above, fitted to a 2-meter hardwood shaft. Principally used for the yellow-spotted Amazon River turtle, in Amazonia, the tapua may be jabbed or thrown to impale the submerged turtle—usually detected by a trail of bubbles reaching the surface above it. The smaller six-tubercled Amazon River turtle is also sometimes harvested by tapua (Smith, 1979).

CAPTURE OF TURTLES.

Figure 4.1
A plate from St. Cricq (1874) showing Peruvian Conibos harvesting giant South American river turtles (*Podocnemis expansa*) in the Ucayali River via archery. (See text for further description of the techniques.)

In Belize, harpoons or "pegs" used to catch Central American River turtles (Mexican giant musk turtle or slider may also be pegged opportunistically) are similar but 3–4 meters long and tipped with a cord-attached, detachable, barbed metal point made from an old triangular file (Polisar, 1995). Live marketed animals which have been collected with this device are easily identified as they have a small but conspicuous triangular-shaped hole, usually plugged with clay, in the carapace (fig. 4.3). Most turtles are obtained at night as they rise to the surface to breathe or forage, but during periods of rising water (when turtles are stimulated to forage) or during the peak striking months of April and May when the water is low and clear and turtles are concentrated, daytime harpooning is also successful. Polisar (1995) considered the number of strikers to be many times greater than the number of turtle netters and divers in Belize, and that their catch was usually used for family consumption or local trade. D. Moll saw many struck animals for sale in the main Belize market in 1983–84, however. Polisar found that far fewer *Dermatemys* were taken in Belize by striking than by either free diving or netting, and that the animals collected by this method were generally larger. In his study area over a 15-month period, 36 *Dermatemys* were collected by striking, of which 30% were adult males, 17% adult females, and 53% juveniles. In Bangladesh, barbed harpoons with between three and twenty-one points and with a cord attached are thrown from a boat and the impaled turtle is retrieved accordingly (Fugler, 1984).

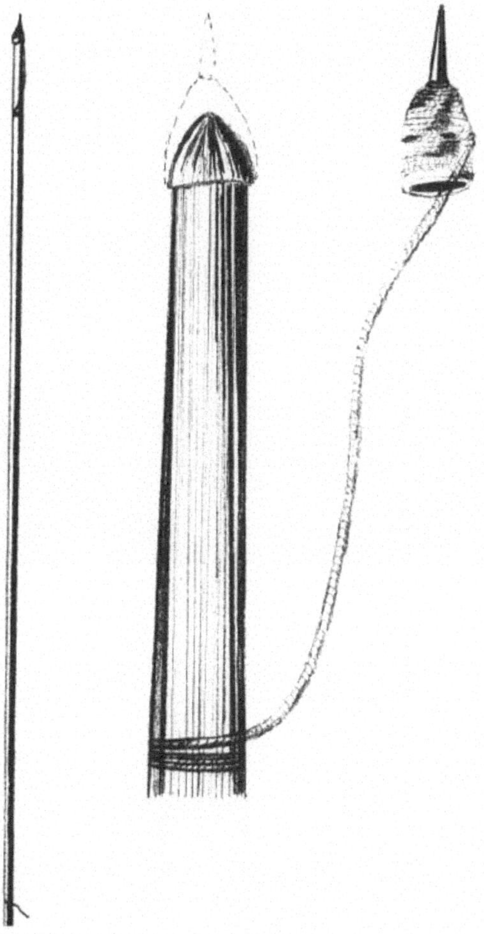

Figure 4.2
A generic "peg" or "tapua" used to harpoon turtles. Key features include the sharpened detachable head with attached line used to strike and retrieve the harpooned turtle, mounted upon a wooden, hand-held shaft. (Adapted from figure in Smith [1975] by Janice Moll)

Spearing is also practiced by India's Sahnsias. They lure Indian softshells near shore with carrion, while waiting hunters surround and impale the animal with spear thrusts. In Bangladesh, a spear with between three and twenty-one points is tapped along the shallows by the wading hunter. When a turtle is felt it is impaled with a sharp thrust (Fugler, 1984). Khan (1987 in Das, 1991) alluded to the use of this technique to catch and kill Indian narrow-headed softshell in Bangladesh. In northern Australia, Aborigines would hide in trees sloping out from the water's edge and spear rising pig-nosed turtles through the front flipper with a pronged spear (Georges, 1987). Fiasson (1945, in Pritchard & Trebbau, 1984) stated that inhabitants of the Venezuelan llanos were skillful at spearing yellow-spotted Amazon River turtle, as they surfaced to breathe. André (1902) reported that spearing was the primary method for taking this species during flooding periods on the Caura River (a tributary of the Orinoco) when the turtles entered the shallows to feed upon submerged vegetation.

Figure 4.3
Central American river turtle carapace (*Dermatemys mawii*) in Belize City market, displaying the characteristic triangular peg tip entry hole that identifies a turtle collected by "pegging."

Netting and Trapping

The use of various types of nets and traps for catching turtles is another widespread and successful technique. Polisar (1995) again provides detailed information concerning the use and effects of traps and nets on Central American River turtle populations in Belize. He found that some cylindrical, hoop-type traps originally designed for commercial harvesting of blue catfish were occasionally used for capturing *Dermatemys*. The traps were secured on the closed end to a tree, baited with fig leaves, and sunk in midstream with the funnel-like trap opening facing downstream. These traps could be effective in capturing smaller animals bottom-walking in rapidly flowing, turbid waters, and Polisar sometimes found large numbers of turtles in the Belize City market during high water periods which had probably been collected with this technique. Polisar thought the use of these devices was probably not widespread in Belize, and that the use of spread nets was much more common.

The large hoop-type commercial nets described by Polisar (also called Fyke or frame nets) are widely used around the world for capturing fish, turtles or both, however. When properly placed (baited or unbaited) in areas traveled by turtles, and often equipped with "wings" and "leads" (which are sections of netting extending outward from the trap that lead turtles toward its funnel-like entrance), these traps can catch many turtles. Pritchard (1989) described one alligator snapper trapper's model as being composed of metal hoops of about 1 meter in diameter, covered with 4-cm (1.5-inch) nylon mesh, a 5-cm (10-inch) diameter release ring on one end (which allowed small turtles to escape) and a funnel-shaped entry on the other. The traps were baited with carp and pegged to the river bottom (usually facing the current) at night. The trapper estimated that he was able to catch about 500 kg (1000 lb) of alligator snappers per day with fourteen of these traps on Georgia's Kinchafoonee River (a tributary of the Flint) during the 1970s. A total catch per day on the Flint by all trappers during this time period was estimated at 3–4 tons per day. Alligator snappers caught in submerged traps may live 4–12 hours before drowning, or when not pegged down, the traps might be forced to the surface by trapped turtles to obtain air. In some areas, such as Florida's Apalachicola and Ocklockonee Rivers, extensive mortality of emydid turtles incidentally caught and drowned in the submerged hoop traps accompanied the alligator snapper harvest (Pritchard, 1989). Various forms of spread, set nets, and seines (pulled nets) are effective turtle catchers under particular conditions.

In Belize, several types of nets were employed for collecting *Dermatemys*, and two types were locally constructed and commonly used within Polisar's (1995) study area on the Belize River. Relatively short eddy nets (made of polypropylene or braided nylon) were set in short calm sections of flowing water and baited with fruits (usually *Ficus* or *Inga edula*) in the Belize River and adjacent streams. They are most effective when low water is rising but nearly useless during flooding periods. They yield relatively few turtles and mainly supply animals for private use and local trade. Lagoon nets are much longer and deeper, polypropylene or braided nylon, square mesh nets used in slow moving waters. These nets are not baited and are designed to entangle moving turtles. They are not particularly effective during flooding periods and are therefore mainly employed during the dry season for both small-scale subsistence and large-scale commercial collecting of *Dermatemys*. Polisar found that these types of nets collected a relatively broad range of sizes of animals intermediate between the size ranges collected by harpooning (largest animals) and free diving (greatest size range). The large mesh size of the lagoon nets (13 cm square) allowed most animals under 300 mm carapace length to escape, and Polisar found no turtles under 210 mm carapace collected by this method. The sex ratios of 141 net-collected animals were composed of 24% adult males and 14% adult females (62% juveniles)—similar to the adult sex ratios collected by harpooning and free diving.

An early and detailed account of the use of a seining technique to capture young *Podocnemis* river turtles in Amazonia was provided by Bates (1863). This collecting approach followed the use of archery in the forest pool near the Solimoēs River described previously. The collectors used a large net nearly 70 meters long, which was spread in an arc at one end of the turtle pool. It was pulled forward by vertical poles attached at each end while others in the party across the pool shouted and beat the water with poles and bundles of vegetation to drive turtles toward the net. After about an hour of driving, the seine was quickly drawn forward to completely encircle the concentrated turtles. The men then easily caught the collected turtles within the enclosure and deposited them in waiting canoes. Bates stated that three boatloads or about 80 turtles were collected in about 20 minutes, and that the collecting continued until the canoes were twice filled.

A variant of the technique of driving or beating the water to catch turtles without use of a net was described by Bates (1863). Yellow-spotted Amazon River turtles which lived in isolated lagoons in the forest near the riverbank were apparently distressed by the beating of the water with rods for several hours during the day. After dark the turtles would try to escape the pond, but sentinels would collect them as they moved overland toward the river.

Several other net types and netting techniques are used in modern times to either purposefully catch turtles or which yield turtles incidentally along with fish. Commercial fishermen's gill nets and trammel nets (composed of two larger-meshed gill nets overlaying a smaller-meshed central section) are often used. For example, Smith (1979) stated that Brazilian six-tubercled Amazon River turtles were usually captured in gillnets set for fish, particularly at night. Klemens and Thorbjarnarson (1995) reported that many individuals of this species are caught using long nets called "malhadeiras" set in bays along rivers. They reported that the catch supplying just one shop in Tefe, Brazil, which roasts the turtles may consist of up to 12,000 animals per year. Fugler (1984) and Das (1991) reported that Indian and peacock softshells and flap-shells are commonly taken with bag-shaped nets, fixed to weighted ropes and dragged along river bottoms against the current in Bangladesh and India. The current sweeps the turtles into the net. Cast nets are also used for these species in Bangladesh (Fugler, 1984). Smaller individual cast nets may be thrown as the collector wades or stands in a boat, while larger versions (called drag nets) are only cast from a boat. These are allowed to descend to the bottom, and then slowly retrieved by the caster (Fugler, 1984). In the Sunderbans of India, river terrapin were caught with a tiger prawn net set between two long poles in a river channel (Das, 1990). Similar nets (called "ambai") in Malaysia are also used (E. Moll, 1980a). Their wide mouths face into the current and their long funnel-shaped bags trail behind. River terrapins often moved upstream and into tidal creeks to feed on a rising tide and would be likely to be caught during this foraging activity. In a slight variant of this technique in Malaysia, "Belat" nets were set parallel to a sloping bank or across a small tidal creek. River terrapin (and presumably painted terrapin) swim over the nets during flood tide to reach foraging areas, but are stopped by the nets as they attempt to return to deep waters on the ebbing tide (E. Moll, 1980a). They typically bury themselves in the mud at the base of the net where they may be felt by probing and captured by hand (E. Moll, 1980a; fig. 4.4a,b). In Bangladesh, grated bamboo barricades with a net in the center span small streams and are used to catch a diversity of turtle species (Fugler, 1984).

(a)

(b)

Figure 4.4
(a) A Malaysian river terrapin (*Batagur baska*) extracted from the mud at the base of a collecting net (b) stretched across a tidal creek it had entered in order to feed.

Hook and Line Techniques

Single baited hook and line, and multiple baited hook arrays are common turtling methods with numerous variations. The single baited hook and line approach is often more suited to subsistence harvesting, and turtles caught may be the objective or incidental to the main goal of catching fish. Nevertheless, turtles which are landed often end up as human food or are killed vindictively and discarded by the fisherman. Many species are easily caught in this manner. An example is provided by the pig-nosed turtle, caught by Australian aborigines with hand lines baited with wallaby or buffalo meat (Georges, 1987). Polisar (1995) stated hook and line fishing with natural bait (fruit, etc.) was used to catch Central American River turtle in Belize, but that it probably yielded few animals. Das (1991) stated that single baited hooks were used to catch crowned river turtles. Loveridge and Williams (1957) recorded several Nile softshell caught on hook and line. Series of individual bank lines may also be purposefully set for turtles. Pritchard (1989) reports an interviewer who described the use of large baited hooks with a reed pole and hanging twine (which would be likely swallowed rather than cut by turtle jaws) for catching alligator snappers in Florida's Ochlockonee River before populations there were depleted. Pritchard (1989) also described a "jug-fishing" approach to catching alligator snappers in Louisiana in which a large hook baited with mullet attached to a 4-meter (12-foot) length of nylon line was wrapped around a floating motor oil can. Caught turtles would gradually unwind the line as they attempted escape, and the completely extended line would indicate to the fisherman that a turtle was hooked.

Another variant of float fishing is used to catch yellow-spotted side-necks in Venezuela's Orinoco River system (Pritchard & Trebbau, 1984). Hooks baited with plantain are attached to both styrofoam floats and sinkers so that the baits are suspended in the water column. Pritchard witnessed three large females caught in this manner.

The multiple hook arrays, variously called trotlines or short lines, are widely employed for turtles and fish and supply both subsistence and commercial interests. These are generally similar in that short lines with sturdy hooks are tied to, and spaced at regular intervals along, a main cord which is secured at each end. These may extend across the water for considerable distances. The hooks may be baited with any available animal bait for carnivorous turtles or with fruits or other vegetable material for herbivores. Fugler (1984) considered trotlining to be the principal technique for commercial harvesting of Indian and peacock softshells, flapshells, and giant narrow-headed softshells in Bangladesh. Alligator snappers are also frequently caught on baited trotlines, especially on very dark nights during the summer (Pritchard, 1989). Caught turtles are kept alive until enough have accumulated to warrant their transportation to major market centers such as New Orleans. Smith (1979) described a trotline technique for capturing yellow-headed side-necks in Brazil. Elevated trotlines of four to six hooks each are set under trees with falling fruit which attract the turtles during high water periods when the forests are flooded. Turtlers may set out up to twenty-four trotlines spaced several hundred meters apart, depending upon the position and availability of the fruiting trees. The hooks are baited at sunrise with the appropriate fruits and suspended so that the fruit baits just touch the water's surface. The diurnally feeding side-necks which take the bait can stay alive at the surface until the line is checked as they can breathe, and are less likely to be attacked by piranhas. Smith (1979) said that one trotliner caught 120 turtles (about equally apportioned among males and females) during a seven-week period in May and June (the peak months), 1977. Smith (1979) also recorded that fish baited trotlines took the smaller species, six-tubercled Amazon River turtle. River terrapin in Bangladesh and Malaysia are also taken using trotlining techniques.

Other Aquatic Methods

A variety of other techniques may occasionally be used to catch turtles in aquatic settings. Schultze-Westrum (1995) described the use of tall, conical, open-ended baskets which are used in New Guinea to capture pig-nosed turtles when they are buried in the mud. At low tide the open end of the basket is placed over depressions in the mud of the river bottom which mark the location of buried turtles. When the turtles surface inside the basket they are captured for eating.

Local residents of the brackish Muthurajawela Marsh area of Sri Lanka build circular brush pile traps near the mouth of the Attanagalu Oya River and in the adjacent Nagombo Lagoon. The traps are designed to attract and capture fish, but they also attract Indian flapshell turtles and saltwater crocodiles which are also consumed (Porej, 2001). Parts of the marsh with dry grasses are also set ablaze here in order to capture escaping or burned terrapins (Indian black turtles [*Melanochelys trijuga*] or Indian flapshells) and water monitors for consumption.

Electric shocking was mentioned for collecting unspecified turtle taxa in Vietnam by Le Dien Duc and Broad (1995) and in Sumatra (Shepherd, 2000). It was also used as a collecting technique for alligator snappers (with a 230 V AC shocker) by Dobie (1971), but whether it is ever used for commercial collecting of this species is unknown.

Collecting River Turtles Ashore

Another common method of collecting river turtles is to ambush them or capture them in pitfalls when they come ashore to bask on mud or sand banks or when they emerge from the water to nest. In the latter technique turtlers have, of course, the bonus of acquiring eggs as well as the nesting turtle.

Sinha (1995) described a technique called "Bhitha" which is utilized by turtlers along the middle Ganges of India for luring and capturing basking turtles (usually large softshells). A hard platform is constructed by raising mud, wet silt or sand at the riverbank to attract a turtle looking for a good basking site, while a bunker camouflaged by grasses is constructed to hide the turtler about 6 meters (20 feet) inland. When a turtle emerges onto the platform a long bamboo stick with a hook on the end is used to secure and drag the turtle toward the blind where it can be captured.

The most unusual technique we encountered for capturing basking turtles was described by Gosse (1851, in Barbour & Carr, 1940) for obtaining basking, Jamaican sliders, *Trachemys terrapen*, in Jamaica. They were caught in a rat trap on the end of a pole which was baited with meat—preferably "a small bird with the feathers singed."

The capture of nesting turtles may be a rather fortuitous process practiced by individual turtlers who snatch up turtles that are observed emerging onto the nesting beach. With the large, colonial species especially, however, their harvest at the nesting beach may be a highly organized and coordinated process. The taking of nesting "terecailles" (presumably yellow-spotted Amazon River turtles) by Indians on sand banks of the Caura River (André, 1902) and the giant South American River turtle on the Orinoco's Pararuma nesting beaches first described by Mosqueira Manso (1945, 1960 in Pritchard & Trebbau, 1984) exemplify the latter. Pritchard and Trebbau (1984) summarized the Pararuma procedure as follows:

> Workers were principally divided into those who turned the turtles on their backs ("volteros"), the animals being immobile once on their backs; and those who carried the turtles ("caleteros") to consumers distributed throughout the populated parts of the Orinoco between Puerto Ayacucho and Ciudad Bolivar, and along its major tributaries. Turning of the turtles commenced between 0030 and 0100 h, and took place under the direction of the beach guardian. The number of men designated for this duty depended upon the number of turtles to be turned, the normal quota being 50 turtles per man per night. The volteros lay in wait very quietly until the word was given by the supervisor, at which point they marched in single file, again without making any sound, to the nesting area. If the turtles were nesting in a reasonably compact group, the leader turned the first turtle, and the workers followed, turning their quotas of 50 as quickly as possible, but avoiding those animals that were actually in the process of laying. The turners were paid five bolivares per night before 1945; subsequently they were not paid at all, but were allowed to keep one or two turtles for their own use.
>
> Prior to March 1946, the caleteros would pick the animals up where they had been turned and then carry them to the holding stockade about 700 m away. However, this apparently proved an impractical arrangement if the remaining turtles were to be left until they had the maximum opportunity to lay. A new scheme was therefore developed whereby the turtles were carried about 200 m to a designated area south of the laying zone, from which they could be picked up at any time by the caleteros without disturbing those females still laying. The entire quota of turtles was thus removed from the nesting zone by 0300 h instead of 0600 to 0700 h as had formerly been the

> case. Each caletero carried a single turtle on a trip, on or above his head, with the carapace down, although a strong caletero could carry two relatively small turtles tied together. In addition, a caletero might encounter another turtle returning to the river while on his way to the stockade, and interrupt his journey to guide this additional turtle to the stockade with him. (Pritchard & Trebbau, 1984)

Hundreds of turtles were typically turned each night during the first week of the nesting season with this efficient system in the 1940s (Mosqueiro Manso, 1960 in Pritchard & Trebbau, 1984). Sometimes the harvest was limited by the shortage of sufficient boats to transport turtles to market. At other times, other activities, such as almond collecting took precedence over turtling, and some nesters were spared (Mosqueiro Manso, 1960 in Pritchard & Trebbau, 1984). The caleteros were paid according to the number of turtles transported, and allowed to keep one turtle each night for personal use (Mosqueiro Manso, 1960 in Pritchard & Trebbau, 1984). Some 26,600 turtles were taken from Pararuma beaches between 1945 and 1948, and many thousands were harvested from nearby beaches as well. These Pararuma turtles were estimated to have yielded 159,594 kg of meat, 39,898 kg of fat products, and 2,431,148 eggs (Mosqueiro Manso, 1960 in Pritchard & Trebbau, 1984). Flower (1933, in Loveridge & Williams, 1957)

Figure 4.5
A plate from St. Cricq (1874) depicting an activity that predates recorded history—the harvesting of river turtle eggs from nesting beaches.

also noted that natives of Sudan have specialized in capturing Nile softshell when they came ashore to lay their eggs. Danforth (1925, in Barbour & Carr, 1940) reported that Puerto Rican commercial turtlers caught female Puerto Rican sliders (*Trachemys stejnegeri*) with the aid of lights when they emerged on April nights to lay their eggs.

Harvesting of the eggs of river turtles rather than, or in addition to, the nesting females is, of course, another important form of exploitation which may be conducted in a highly organized and efficient manner (fig. 4.5). This has traditionally been the case wherever colonial nesting species congregate (e.g., Pezzuti & Vogt, 1999). Characteristic examples of this process will be described in chapter 5 of this volume, concerning exploitation of river turtles for food and other uses.

River Turtle Exploitation: Past and Present

5

Profile 5. "*It is in vain that the forests and the waters offer to the Conibo an abundant and varied commissariat. He has no taste for anything but turtles, and that predilection, according to a mania, has converted him into the most ruthless exterminator of these animals. Essentially a chelonaphagist, he passes long hours at the river-side studying the manners of this dull reptile, from the period when it lays its eggs to that of its migrations. If I should ever undertake to write a special treatise on the Emys, Chelys, Mata mata, or Testudo families, it is to the Conibos I should go for the necessary information.*

Between the 15th of August and the 1st of September, the epoch when the turtles lay their eggs in the Ucayali—not to be confounded with the affluents of this branch of the Amazon, where the period is three weeks or a month later—the snow having ceased to fall on the summit of the Andes, the river becomes less impetuous, its level is lowered, and vast spaces of sand are left bare. The sinking of the waters is the signal for the Conibo turtle-fishery. On a fixed day the natives and their families embark, furnished with all necessary utensils, and travel up or down the river as caprice may dictate or instinct suggest. These voyages extend from thirty miles to sixty, or nearly a hundred.

When the fishers have discovered on any part of the shore those disjointed lines, the claw-marked furrow traced by the turtle when walking, they call a halt, and having built at some two hundred yards from the edge of the water their temporary ajoupas, they patiently await in ambush the arrival of their amphibious prey. The instinct of the fishers is so unerring, that their encamping hardly precedes by more than a day or two the appearance of the turtles.

On a dark night, between midnight and two o'clock, an immense swirl agitates the river, and its waters seem to boil. Thousands of turtles come clumsily tumbling out of the water, and spread themselves over the shore.

Our Conibos, squatted or kneeling under their leafy sheds and keeping profound silence, await the moment for action. The turtles, who separate themselves into detachments on leaving the water, dig rapidly with their fore-feet a trench, often 200 yards long, and always four feet broad by two deep. The zeal with which they apply themselves to this work is such that the sand flies about them and envelops them as in a fog.

When they are satisfied with the capacity of the trench they deposit in it their soft-shelled eggs, forty at least, but perhaps seventy or more, and with their hind-feet quickly fill up the excavation. In this contest of paddling feet, more than one turtle, tumbled over by his companions, rolls into the trench and is buried alive. Half an hour suffices to accomplish this immense work. The turtles then make a disorderly rush for the river; the moment has arrived for which our Conibos have anxiously waited.

At a signal given by one of them, the whole band suddenly rise from their lurking-place, and dash off in pursuit of the amphibia; not to cut off their retreat—they would be overthrown and trampled under foot by the resistless squadrons–but to rush upon their flanks, seize them by their tails, and throw them over on their backs. Before the corps-d'armée *has disappeared, a thousand prisoners often remain in the hands of the* Vireurs. *(Note: From* virer, *to turn,* chavirer *to upset. This name is given by the missionaries of the Ucayali, and the river-side inhabitants of the Upper Amazon, to the individuals who hunt or fish the turtle by running after it, and throwing it on its back.)*

On the first appearance of daylight the massacre commences: under the axes of the indigenes the shells of the amphibia fly to splinters, their smoking intestines are torn out and handed over to the women, who separate from them a fine yellow fat, superior in delicacy to the fat of the goose. The disembowelled corpses are finally abandoned to the vultures, the vulture-harpies, and the fisher-eagles, who flock from all sides to the scene of carnage."

St. Cricq, 1874

Turtles and their eggs have long served as an important source of food for humans—almost certainly since very early in the evolution of the hominid lineage, and surely for at least the last 20,000 years (Nicholls, 1977). Evidence in the form of shells and skeletal material (some showing burn marks as evidence of cooking) in the middens of Paleolithic aboriginal cultures, and from eyewitness accounts of explorer-naturalists in more recent times is available from numerous locations around the world (e.g., Bates, 1863; St. Cricq, 1874; Goode, 1967; Rhodin, 1992, 1995; Pritchard, 1994; Lee, 1996; Stiner et al., 1999). Skeletal evidence of river turtles, in particular from such locations as Mohenjodaro and Harappa in the Indus Valley (e.g., Indian narrow-headed softshells and river terrapins), Mayapan, and many other Mesoamerican Mayan sites (e.g., Central American river turtles), and Naga ed-Der of Upper Ancient Egypt (e.g., Nile softshell) suggest that river turtles have helped to support the rise of the world's great civilizations as well (de Treville, 1975; Nath, 1959 in Groombridge & Wright, 1982; Das, 1991; Lee, 1996).

Their role continues and, in fact, has expanded as human populations have burgeoned and spread throughout the modern world. River turtles have always been too convenient and succulent a source of protein to ignore. Often large, fecund, and easily collected with simple techniques and equipment, especially in communal nesters which may concentrate at nesting sites in helpless thousands (at least formerly), river turtles are ideal prey. Much of the harvesting has been, and continues to be, conducted in relative obscurity in many parts of the world. Occasionally, however, the sheer magnitude of the resource and its slaughter has attracted the attention of literate observers, such as the early explorer-naturalists of the New and Old World tropics. Their accounts have given us some idea of the former truly spectacular abundance of some riverine species, and the equally spectacular levels of consistent exploitation which have brought them to their modern, much-diminished condition. Summaries of the exploitation of the two best documented examples of destruction of formerly abundant riverine species, the Asian river terrapin, and the giant South American river turtle, are provided under their appropriate geographic sections below. They, more than any other species, epitomize a process of destructive, relentless exploitation which continues in many of the Earth's rivers today.

Modern Subsistence Use of River Turtles

In spite of increasingly reduced stocks in many areas resulting from the types of overexploitation described throughout this chapter, river turtles are still an important dietary component of flood plain dwellers and others within foraging distance of rivers. They often serve as primary prey of such groups, harvested by simple but effective techniques which have been handed down from generation to generation, and have changed little over the centuries (see turtling techniques in chapter 4). Turtles may also be an important incidental catch in fin fisheries.

The most intense use of river turtles as subsistence prey in the modern world occurs in the tropics, but the collection of many species for subsistence consumption (as well as commercial trade) has been occurring in temperate North America for many centuries also (Carr, 1952; Moodie & van Devender, 1979; Rhodin, 1992, 1995; Klemens, 1993; Klemens & Thorbjarnarson, 1995). Probably all natives species have been used as food at one time or another, but Carr (1952) listed thirteen aquatic species, several of which are primarily or exclusively river dwellers (or estuarine in the case of the diamondback terrapin), which have served as subsistence prey (table 5.1). Several species were—and still are—esteemed as food items in particular regions. Carr (1952), probably having sampled more species than anyone else before or since, was in a unique position to evaluate the degree of justification for these preferences—proclaiming some imaginary and some based upon subtle variations in flavor and other qualities of turtle meat among different species, and even among populations of wide-ranging forms. Perhaps the first North American non-marine species to be used as food by European colonists was the estuarine diamondback terrapin (*Malaclemys terrapin*). It had long been intensively harvested for private consumption by coastal native Americans, colonists, and later coastal residents before it briefly became a

Table 5.1
United States aquatic turtles which have been utilized for subsistence and commercial food use (summarized from Carr, 1952).

Species	Subsistence use	Commercial use
Alligator snapping turtle, *Macroclemys temminckii*	Yes	Yes
Common snapping turtle, *Clelydra serpentina*	Yes	Yes
Pacific pond turtle, *Clemmys marmorata*		Yes
Blanding's turtle, *Emydoidea blandingi*	Yes	Yes
Diamondback terrapin, *Malaclemys terrapin*	Yes	Yes
Common map turtle, *Graptemys geographica*	Yes	
*False map turtle complex (*Graptemys pseudogeographica* and *Graptemys ouachitensis*)	Yes	Yes
Painted turtle, *Chrysemys picta*	Yes	Yes
Slider, *Trachemys scripta*	Yes	Yes
Red-bellied turtle, *Pseudemys rubriventris*		Yes
Florida red-bellied turtle, *Pseudemys nelsoni*	Yes	Yes
Cooter, *Pseudemys floridana*	Yes	Yes
River cooter, *Pseudemys concinna*	Yes	Yes
Chicken turtle, *Deirochelys reticularia*	Yes	Yes
Florida softshell, *Apalone ferox*	Yes	Yes
Spiny softshell, *Apalone spinifera*	Yes	Yes
Smooth softshell, *Apalone mutica*	Yes	Yes

***G. pseudogeographica* and *G. ouachitensis* were not recognized as distinct species until R. C. Vogt's revision of the group in 1993 (Vogt, 1993).

wildly popular commercial species for a time in the late nineteenth and early twentieth centuries (see account below).

Today in the United States the most widely exploited species for private consumption are the common snapping turtle (*Chelydra serpentina*) and the spiny softshell (*Apalone spinifera*). These species are often common and have wide ranges, are large (at least the female *Apalone*), have good-tasting meat, and can be collected with simple equipment and techniques. In most states with resident populations a fishing license usually allows their legal harvest, although bag and possession limits and seasons vary widely and are based upon little or no knowledge of population dynamics (Levell, 1995). In some states (e.g., Missouri, Wisconsin) a license allows the harvesting of undifferentiated "softshells" which means that the smooth softshell (*A. mutica*) may also be part of the legal catch. Vogt (1981d) expressed concern about the lack of differentiation in Wisconsin as the smooth softshell is much more vulnerable due to its smaller populations and more restricted distribution compared to the spiny softshell. Undoubtedly there is also a certain amount of illegal harvesting of these, and other large, regionally relished species as well, as a result of direct hunting or incidental catch in fin fisheries. The levels of subsistence harvesting of riverine species must be limited in most other temperate locations by the paucity of species, especially of larger forms. Almost certainly there is subsistence use of turtles in such locations as southern Europe, and the temperate regions of South America, Africa, and Australia, but its magnitude is probably small. South central China was another temperate location in the world, besides North America, where subsistence use of river turtles was heavy until recent reduction in stocks (Lau & Haitao, 2000). The presence of large river systems such as the Yang Tze with a resident turtle fauna, and the Chinese proclivity for turtle (see section on Commercial Trade below) for multiple uses virtually assures the veracity of this assumption but quantitative data are scant. Lau and Haitao (2000) pointed out that subsistence use of turtles in China has declined substantially as market value has increased.

While temperate zone subsistence use may normally be considered a dietary supplement, in the tropics and sub-tropics river turtles and their eggs can be important dietary staples of poorer flood plain dwellers and extant aboriginal cultures. For example in India, Chaudhuri (1912) reported that the Santal, Bind, Banpar, Tear, and Gandri tribes along the Ganges River near Rajmahal regularly caught and ate turtles. E. Moll (1990a) reported that the Irulas of Tamil Nadu caught and ate Indian flapshell while usually rejecting the Indian black turtle (*Melanochelys trijuga*) which was also present in the area, although both species may be consumed by coastal, marsh-dwelling people in Sri Lanka (Porej, 2001). Other tribes such as the Kamars and Santhals may reject turtle altogether, believing themselves to be descendants of turtles (Das, 1991). In Malaysia, E. Moll (1976a) and Sharma (1999) noted that indigenous hunter-gatherers known as "Orang Asli" were adept subsistence turtle hunters. Sharma (1999) also stated that native people in the Selangor region of Malaysia eat stream-dwelling Malaysian softshells (also forest-dwelling spiny turtles [*Heosemys spinosa*]). Choudhury and Bhupathy (1993) listed numerous Indian turtle species harvested where subsistence use is practiced. The extensive aboriginal use of Neotropical giant South American river turtles is documented in the appropriate section below. Smith (1975) listed twenty-nine tribes and the way they used this species within various parts of its range. Nearly every species of turtle, aquatic and otherwise, which exists in Venezuela may at least occasionally be used by some group as subsistence fare even if its economic importance is negligible (Pritchard & Trebbau, 1984). Fidenci (2002) provides one example of heavy subsistence use of the yellow-spotted Amazon river turtle by the Yekuana people of Venezuela. They fish for this turtle in the Caura River throughout the year. Carr (1952) related that giant musk turtles (*Staurotypus*) and smaller musk turtle species (*Kinosternon*) are popular food items in Mexico and Central America. Lee (1996) stated that modern rural Maya commonly consume freshwater turtles, including Central American river turtles, giant Mexican musk turtles and sliders. Moll and Legler (1971) noted that Panamanian subsistence hunters harpooned sliders as they slept near floating grass mats and collected nesting females to eat. Mora and Ugalde (1991) also described the use of Costa Rican sliders and their eggs by Guatuzo Indians.

Although generally less well known, a perusal of the species accounts of African turtles by Loveridge (1941), Loveridge and Williams (1957) and Villiers (1958) indicate that turtles are regularly consumed by local people throughout the continent. The Australian aborigines have utilized freshwater turtles as food since earliest times (Goode, 1967; Woinarski, 1993; Cann, 1998). Cann (1998) said that huge numbers are collected for regular feasts in New South Wales. In season, large numbers of aestivating turtles which have been collected may be stored in a pit about 1 meter deep covered with a layer of bark and soil. The turtles continue to aestivate until they are used as food. With the exception of fishes, perhaps, it is hard to think of a more widely utilized subsistence food source throughout the world.

The effects of this type of steady harvest of river turtles are very difficult to assess. As no records of catch are usually kept, we know little of the magnitude of the harvest, which also will often vary regionally and seasonally. For instance, harvest of the Central American River turtle in Belize may be intense in the late dry season when it is most vulnerable due to lowering water levels and increased water clarity (also because of its popularity as an Easter feast dish), but more moderate in other rainier seasons of the year when rivers are swollen and turbid. Also in some parts of Belize the turtle, although available may not be captured and eaten by local people for cultural reasons (D. Moll, 1986; Polisar, 1992; Vogt et al., in press). In addition, we know very little of the tolerance levels of the harvested species for exploitation, although we can assume that for most species it is probably very limited (e.g., Congdon et al., 1993, 1994; Kuchling, 1999). We agree with Klemens and Thorbjarnarson (1995) that this pattern of use of turtles in general—and of river turtles in particular—is in addition to commercial exploitation, worthy of concern by conservationists. As pointed out by Polisar (1992) and Vogt et al. (in press), while fewer turtles are often harvested per unit of time and per hunter by subsistence hunters compared to commercial collectors, there are often many more of the former involved in hunting. As a result, the long-term effects on turtle populations can be significant.

The thought of subsistence use may evoke an initial mental image of naturalness and balance between the consumer and his prey, but even in the most primitive aboriginal societies humans have arguably been able to decimate entire prey species (e.g., Martin, 1966, 1967, 1984; White & Flannery, 1992). Rhodin (1992) suggested that the disjunct distribution of the red-bellied turtle in the northeastern United States is at least partially due to exploitation by prehistoric humans. With increasing human population densities, use of modern more efficient collecting devices and modern transportation capabilities the possibility of such a balance may be illusory in all but the most remote locations and low-density primitive cultures (Altman, 1987). The incredible magnitude of the exploitation of *Podocnemis sextuberculata* nesting females described by Mittermeier (1975) for local use in just one small Amazonian village (Sao Josè, Brazil) is illustrative. Research by Kuchling (1988, 1992) and Kuchling and Mittermeier (1993) on the status and exploitation of Madagascan big-headed turtle on Madagascar provides another case in point. Kuchling and Mittermeier

(1993) and Kuchling (1999) pointed out that subsistence level exploitation of this species is likely to result in its extermination. Even after populations have been reduced by direct exploitation, the turtle is still incidentally caught in fishing operations and, as it is illegal to market it, it is typically slaughtered for private consumption. Kuchling's best-studied population was extirpated between 1987 and 1991 as a result of this pattern of non-commercial exploitation (Kuchling & Mittermeier, 1993; Kuchling 1999). Rose et al. (1982) Georges (1987), and Georges and Rose (1993) expressed similar concern over the effects of aboriginal exploitation of the pig-nosed turtle (*Carettochelys insculpta*) in Papua, New Guinea and tropical Australia where it is still heavily used. Exposure to western culture and technology, specifically the current practice of catching these turtles on baited lines, and in gill nets probably allows greater efficiency in catching these turtles compared to traditional methods. In addition, the availability of refrigeration and modern transportation has encouraged turtle fishermen in Australia's Daly River area to catch more than are required for immediate consumption. The surplus catch was being shipped to Darwin (220 km away) for consumption by friends and relatives. The effects of these practices on the pig-nosed turtle populations are unknown. In Papua, New Guinea, populations are reported to have been seriously depleted over the last 20 years in the Kikori River District and elsewhere that their eggs and adults have been harvested. In Kikori Market over 5000 eggs were sold between October 1980 and February 1981 (Rose et al., 1982), and in the following year over 20,000 eggs were harvested and eaten in Kikori and three nearby villages (Georges & Rose, 1993; Rhodin & Genorupa, 2000). This turtle's predictable, stereotyped nesting patterns render it particularly vulnerable to over-exploitation, and its vulnerability has been exacerbated here in recent times by the introduction of modern technology, especially outboard motors. Human population densities have also increased along the rivers as a result of cessation of clan warfare in the region (Rose et al., 1982; Georges, 1987). Rose et al. (1982) and Rhodin and Genorupa (2000) stated that the traditional hunting and harvesting of eggs of this turtle is the principal threat to its existence in New Guinea (see the section below concerning its use in the pet trade also). Potentially deleterious changes in population structure may also result from some kinds of subsistence turtling practices. Ramo (1982) found that chronic harvesting of female *Podocnemis vogli* as they wandered terrestrially in search of nesting sites (through use of specially trained dogs and horses) caused a much higher level of mortality in females than males in her study population; hence, a sharply skewed sex ratio favoring males resulted. The chronic selective removal of larger individuals, including egg-bearing females, from a turtle population—a very common pattern of subsistence hunting (e.g., the Central American river turtle, Polisar, 1995; Vogt et al. [in press]; Madagascan big-headed turtles, Kuchling, 1999) can also have serious consequences for the populations' future survival (Congdon et al., 1993, 1994; Tucker & Moll, 1997). This subject is discussed in more detail below.

Finally, in modern times the line between true subsistence use and commerce is often a blurry and shifting one based upon supply and demand (Rose et al.,

1982; Georges, 1987). The relationship is exemplified by usage patterns of river softshells and other species in Malaysia (Sharma, 1999) and Central American river turtles in some Belizean settlements where turtles are harvested primarily for personal use, but surplus catch may be sold locally, or sent to market in larger cities (Polisar, 1995; Vogt et al. [in press]). A similar pattern of use was described for pig-nosed turtles, red-bellied short-necked turtles (*Emydura subglobosa*), New Guinea snapping turtles (*Elseya novaeguineae*), and a yet undescribed species of *Elseya* in Papua, New Guinea (Rose et al., 1982: Rhodin & Genorupa, 2000). Subsistence use of Papua, New Guinea's long-necked *Chelodina* species is very limited due to their strong musky odor and snake-like appearance, but their distinctive appearance has made them attractive in the exotic pet trade (see below) so they are collected for this purpose (Rhodin & Genorupa, 2000).

Participants in the 1999 Phnom Penh workshop on the Asian turtle trade (van Dijk et al., 2000a) found a variable amount of subsistence use of turtles in the Asian countries they focused upon, with extensive use still occurring in less well-developed, poorer countries (e.g., Laos, Bangladesh, Myanmar, Papua, New Guinea) and substantially less in rapidly modernizing and/or wealthier nations (although it may remain a regional problem) (e.g., Thailand), or those largely inhabited by people with cultural and/or religious aversions to turtle consumption (e.g., Indonesia). Subsistence use has also generally declined in these and other Asian countries as the market value of turtles has increased and local stocks have dwindled (van Dijk et al., 2000a).

Modern Commercial Exploitation of River Turtles for Food

River turtles and their eggs are still extremely important items of commerce in many parts of the world today. They are normal components of trade at all levels from local through international, and the ever-increasing human population they serve, and modern transportation capabilities continue to intensify, and widen the scope of their exploitation. Modern river turtle stocks are rapidly dwindling in many source locations as a result.

Exploitation Patterns in Asia

Commercial pressure on river turtles is currently greatest in the tropics, and while not insignificant in other areas, is most intense in Asia where vast regional, national, and international export markets exist (van Dijk, 2000; van Dijk et al., 2000a). Local trade is heavy in locations and cultures where riverine species are prized as food (e.g., as among Hindus of West Bengal in northeastern India). River turtles are also under pressure in other areas inhabited by non-turtle-eating cultures (e.g., Muslim) where turtle meat may be shunned but their eggs may be relished, and where turtles may be collected for marketing. The magnitude of this industry is huge, but difficult to define precisely due to the difficulty of obtaining reliable data concerning its participants. Records are usually poorly kept or concealed, as a significant component of the trade is conducted surreptitiously, and species involved are often misidentified.

In spite of these limitations on our ability to analyze the trade, arguably, the worst crisis chelonians have ever faced in their greater than 230 million years on earth is currently unfolding. It is centered in Asia, but has the potential to ensnare almost all of the planet's turtles (van Dijk, 2000; van Dijk et al., 2000a). For a group of animals that has survived and thrived through the major extinction events that marked the beginning, middle, and end of the Mesozoic Era, and the glaciation-related environmental turmoil of the early Quaternary, the current force which threatens them must be awesome. It is—it is us.

Never before have turtles had to face such specific and singular persecution by such a populous and ingenious predator species as ourselves. In addition, we humans overwhelmingly favor the exploitation of the life history stage, the large adults, which turtle populations can least afford to lose (Congdon et al., 1993, 1994; Tucker & Moll, 1997; Gibbs & Amato, 2000). As discussed previously, turtles are remarkably resilient in the face of the extremely high levels of mortality normally encountered by their eggs, hatchlings and small juveniles. Heavily armored and elusive (at least in riverine species), adult turtles typically couple very low mortality rates with considerable longevity that allows them to compensate for the earlier carnage with impressive lifetime fecundities. When even moderate numbers of adults are removed from the population this vital linkage is compromised, and the population declines. When so few adults remain that reproduction will never again be able to balance the continually high levels of attrition of the eggs and young, then the population is functionally extinct—even though the remaining adults may, for a time, present the illusion of viability. Commercial exploitation of turtles in Asia over the last 15–20 years, mainly for food but also for traditional medicines and other uses, has been more instrumental in bringing more turtle species toward this sorry condition than any other force in history. Humans, of course, are capable of overcoming the defenses of adult turtles which evolved long before our species evolved (and even longer before we attained such ubiquity and incredible numbers across the planet's surface), and many of the adaptations which ensured their survival before our arrival (e.g., mass nesting) have now exacerbated their vulnerability to our penchant for over-harvesting them. It is ironic that their greatest modern persecutor, the Chinese, have traditionally revered chelonians more highly than any other modern culture. Based upon the most recent available data at least 13,000 metric tonnes of live turtles (and huge quantities of turtle shell), the vast majority of which are wild caught, are shipped to Eastern Asia (i.e., China and its Hong Kong Special Administrative Regions and Taiwan) from Southeast and South Asia countries annually (van Dijk et al., 2000b).

In addition to the Chinese pressure (discussed further below) and that of continuing subsistence consumption, all tropical Asian supplier countries are involved in varying amounts of internal or regional trade where turtles and their eggs, whether collected purposefully or incidentally, are sold in local markets to middlemen, or directly to restaurants (Compton, 2000). An overview of the dynamics of their colleting exploitation and its effects on turtle populations is provided below.

China has traditionally used its own turtles extensively for food and as ingredients in traditional Chinese medicines (see the relevant section concerning medicinal use later in this chapter—while we have elected to separate our accounts of exploitation of turtles for these uses this decision is somewhat arbitrary since harvested species may often be used for both [e.g., meat for food, and blood and shells for medicine]) (Lau & Haitao, 2000). Zhou and Zhou (1992) note that softshells turtles (still a popular food item throughout China) were so valued that they were used by a Chinese king to reward his army more than 2700 years ago. In spite of laws which have been established in China in recent years to protect native turtles, collecting and trade of wild species is still widely practiced to help supply China's 1.1 billion people with one of their favorite foods. Subsistence use has declined in China as high market value of collected animals favors their sale, directly or through middlemen to local restaurants or city markets (Lau & Haitao, 2000). The inevitable result of such heavy marketing pressure has been the drastic decline of China's native turtle populations (Zhao, 1998 in Lau & Haitao, 2000).

A number of factors beginning in the early 1980s allowed China to continue satisfying its insatiable appetite for turtles in spite of the demise of its native stocks. With economic reforms, China's borders opened to legal trade with other Asian countries, and its currency became more widespread and efficient (Lau & Haitao, 2000). In addition, the human population of southern China—the part of the country where turtle consumption has traditionally been most heavily practiced—became more prosperous, even as the habit of turtle use has continued to spread further north (Lau & Haitao, 2000). With these changes legal turtle export from neighboring Asian countries (in addition to illegal smuggling of turtles which has never ceased) became big business, and the flood of turtles from these countries into China (including the huge volume of turtles specifically entering the Hong Kong Special Administrative Region [valued at U.S. $38.7 million in 1997; Lau et al., 2000] began).

Initially the main providers of turtles for the Chinese market were Vietnam and Bangladesh (Compton, 2000; van Dijk et al., 2000a; Hendrie, 2000; Rashid et al., 2000; Stuart et al., 2000). However, in the typical pattern of unsustainable exploitation of wildlife, as these countries depleted their turtles to levels which could not sustain profitable trade (i.e., "commercially extinct" [Compton, 2000]), they contracted suppliers from *their* neighbors (i.e., India, Myanmar, Laos, and Cambodia), and trans-shipped them to Chinese markets, thereby widening the circle of destruction (Bhupathy et al., 2000; Platt et al., 2000; Stuart & Timmins, 2000; Stuart et al., 2000; Tana et al., 2000). The trade has shifted to the southeast and currently, Indonesia (previously centered in Borneo, but now especially Sumatra), is the "hottest" Asian turtle exporter to the Chinese market (see below for more detailed information from Shepherd [2000]).

The literature cited in relation to the exploration pattern described above is mainly derived from a workshop held in 1999 concerning the Asian trade in freshwater turtles and tortoises (van Dijk et al., 2000a). The reader is referred to these papers for detailed information concerning the characteristics of trade, and the current status of traded species in the countries referred to above. We have selected three of these countries in which the turtle trade has been studied in more detail than the others to serve as case studies which exemplify the general process—Bangladesh and India in particular because knowledge of the river turtle trade has been better documented in these countries also. We have also provided an historical overview of the exploitation of a single species in Southeast Asia (Myammar and Malaysia), the river terrapin, as it represents the best documented case of the destruction of a formerly abundant Asian river turtle species via exploitation. As the Asian turtle trade is largely non-specific (i.e., essentially all turtles caught have trade value [Compton, 2000]) it is sometimes difficult or impossible to differentiate the contribution of riverine species from turtles from other biotopes to overall trade volume. When possible, we have emphasized the role of river turtles in this regard.

Case Study 1: Bangladesh. Bangladesh is a primarily Muslim country with vast freshwater resources (about 8000 km^2 of the country's 147,570 km^2) and a diverse turtle and tortoise fauna (27 species, or about 10% of the world's total) (Rashid et al., 2000). While turtle meat is largely shunned as a dietary item within the country due to religious restrictions, there is no such taboo against profiting from them. As a result, a huge export industry developed here in which river softshells played a significant role. They are favored because they are large and yield more meat than similar-sized hard-shelled species (Fugler, 1984; Das, 1990). The Indian softshell, a resident of the river systems of western Bangladesh, and the peacock softshell of western and eastern Bangladesh were the principal exported species in the legal trade with China (especially Hong Kong), Singapore, Malaysia, South Korea, and Japan (Das, 1990). The government's Export Promotion Bureau listed turtle meat, eggs, and "fins" or "badi," which are the cartilaginous flaps of the plastrons of Indian flapshells (another exported species), and live turtles and tortoises as legal items of trade. The meat, eggs, and live turtles were mainly used as food, of course, and the badi, which was taken from turtles that died in holding centers was used to make soup and/or medicines in Japan and China. Turtle oil was also exported to Japan, China, and Singapore, and bones were used in several countries for poultry feed and fertilizers (Das, 1990). Remarkably, the legal export trade was based upon illegally acquired species, as both the Indian softshell and the peacock softshell are legally protected from capture, killing, or trade in Bangladesh (Das, 1990) (fig. 5.1). This inconvenience was surmounted by assigning aliases to these species which allowed them to be collected, killed, traded, and exported. Leith's softshell and the Asiatic softshell, which do not naturally reside within the country's borders, and "*Trionyx certilsfionus*" (which does not exist) were the only species which could be legally exported from Bangladesh (fig. 5.2). Collection permits were therefore issued to fishermen across the country for the collection of these species, but which in reality are Indian and peacock softshells. The catch was supplied to middlemen who supplied holding

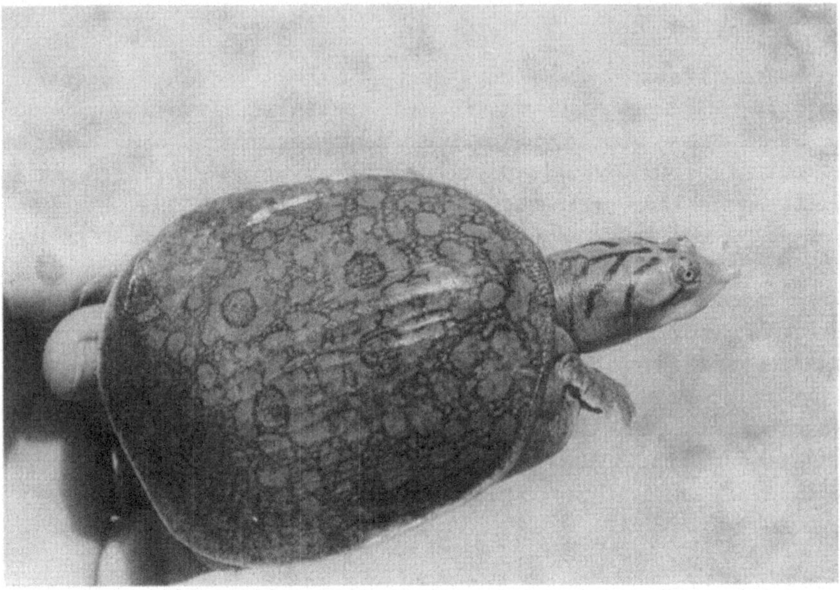

Figure 5.1
A young Indian softshelled turtle (*Aspideretes gangeticus*). Adults of this species are important in the Bangladesh and Indian turtle trade.

Figure 5.2
Leith's softshelled turtle (*Aspideretes leithii*), a species which is ostensibly legal for trade in Bangladesh but in reality does not occur within the country's borders. (See text for further explanation.)

centers near Dhaka (twelve business houses involved in the turtle export trade were operating in 1988–1989 when Das performed the surveys upon which much of this account is based). A "certificate of health" required for export was provided by government veterinary officers, and a "no objection" certificate allowing final clearance for export was obtained from the Office of the Chief Conservator of Forests (Das, 1990). Only turtles below 1 kg in weight were not exportable, and the annual export quota was controlled through the number of permits issued by the government. Through this bizarre, quasi-legal system the export trade thrived in Bangladesh. The number of turtles traded and the financial value of the trade remains cause of concern for conservationists (Oliver, 1979; Khan, 1982; Fugler, 1984; Das, 1990; Rashid & Swingland, 1990; Rashid & Khan, 2000). After increasing in volume and value from the 1970s through the 1980s, the levels of trade fell off sharply from 1989 through 1992 as a result of the tightening of regulations related to the wildlife trade in Bangladesh (Rashid & Khan, 2000). Unfortunately, the turtle trade soon increased to higher levels than ever before in the latter half of the 1990s (Rashid & Khan, 2000). Since 1995 most turtles are exported as frozen meat. It is estimated that about 1500 metric tonnes of wild collected turtles are still exported to the East Asian markets annually from Bangladesh. As sobering as these statistics are, Oliver (1979) stated that the legal turtle trade represents only the "tip of the iceberg." A much larger surreptitious trade, existed in which turtles were smuggled across the border to supply Indian markets (Das, 1990). Das (1990) reported that this trade is largely based upon the Indian flapshell turtle, and large numbers were sold in Calcutta's Sealdah market (now legally closed).

Today, turtles are both collected within the country for export, and Bangladesh also serves as a collection center and trans-shipping point for turtles collected in India, western Myanmar, and probably Nepal (Bhupathy et al., 2000). It is believed that the greatest volume is exported to China with a lesser amount sent to India.

There is also a thriving internal market for riverine turtles supplying the tastes of ethnic minorities in Bangladesh (Das, 1990). The popularity of turtle meat has increased as the price of other meat has risen and availability has declined (Rashid & Swingland, 1990). Species exploited for this trade include Indian narrow-headed softshells, Indian eyed turtles (*Morenia petersi*), river terrapins, crowned river turtles, and larger roofed turtle species. Rashid and Swingland (1990) also found that Indian roofed turtle, a smaller species, and Indian flapshells are consumed locally at a rate of about one individual (either) per week per household. Sylhet is an important marketing center for this trade, with several thousands of turtles sold every winter (Das, 1990). A few river terrapins may also still be traded in markets along the periphery of Bangladesh's Sunderbans region where remnants of a once large population still remain (E. Moll, 1990b; Whitaker, 1982; Bhupathy, 1995).

The consequences of such heavy commercial use of riverine and other turtles in Bangladesh are not known. Although Das (1990) found good wild populations of many of the larger species in the Sylhet and Dhaka areas during his surveys, exporters had begun to complain of diminishing supplies and consequently higher prices charged by suppliers.

Case Study 2: India. In India, a country with twenty-seven resident turtle species, a huge legal commercial turtle industry, based largely upon riverine species, existed until its recent closure by the government. Although demand for turtle meat is regionally localized and restricted to certain cultural/religious groups, other groups—as in Bangladesh—profited from supplying this demand and the current East Asian export trade. Turtle meat is most popular among the Hindu Bengalis of West Bengal in northeastern India. Much of this meat was traditionally marketed at Howrah, just across the Hooghly River from Calcutta (fig. 5.3). Other major turtle markets in West Bengal included those near Malda, Murshidabad, Ragkunathgarj, and Siliguri (E. Moll, 1990a). Wildlife officials have now curtailed the legal trade in turtles in these locations, but a considerable underground market may still exist (Choudhury & Bhupathy, 1993). At any rate, while these markets were operating legally they were the source of a tremendous drain on India's turtle populations, with the majority being harvested from the Ganges and Mahanadi River drainages. As stated above, turtles from Bangladesh were also smuggled in to help supply the demand (Das, 1990).

Figure 5.3
An Indian turtle vendor selling softshells in the market.

Because few written records were kept, and those involved were reticent to discuss the trade due to the utilization of protected species, it is difficult to obtain reliable information on its volume and financial value. E. Moll (1990a) obtained some idea of the magnitude of the trade in one area by interviewing an Uttar Pradesh supplier of the West Bengal markets who obtained turtles from the Chambal, Jumna, Betwa and Ganges rivers. He estimated his monthly trade averaged "5000 to 7000 Indian flapshells, 350–500 Indian narrow-headed softhsells, and 200–300 hard-shelled turtles (mainly three-striped roofed turtles and crowned river turtles)." While no information was provided concerning Indian softshells, which could not be legally traded in Uttar Pradesh, Moll thought that the figures provided for Indian narrow-headed softshells probably included Indian softshells also.

Through a complex marketing system which involved turtle trappers, local jobbers, second-level wholesalers (i.e., in marketing centers) and retailers the turtles made their way to buyers. Most arrived in the major marketing centers by rail, packed alive in wicker baskets with their forelimbs and hind limbs sutured together to limit mobility (fig. 5.4). The volume shipped is difficult to determine. E. Moll (1990a) interviewed a railway official in 1983 who estimated that during the peak winter season (December–March) some 400 to 600 baskets of turtles (each weighing about 100 kg) arrived at his location (Belekoba) monthly. In

Figure 5.4
Softshelled turtles in wicker baskets—a common transportation method in the Indian turtle trade.

May, some 150–200 baskets were still being received. E. Moll and colleagues visited the Siliguri area in May, 1983 to survey the markets and to obtain information concerning the mechanics of the trade system. They found the turtles arriving by rail from Bihar, Uttar Pradesh, Assam, and Orissa states. Most shipments would arrive at the town of New Jalpaiguri where they were divided among local jobbers for distribution in smaller towns in the area. Here, vendors attended daily auctions to purchase stock that they would sell from their stalls in the daily or weekly markets of local villages. Usually, the turtles in the markets were partially butchered, but kept alive to assure freshness until a buyer is obtained by leaving the head, neck, and viscera intact (E. Moll, 1985a).

The largest wholesale market for turtles in India, until its closure at least, was at Howrah. A survey of the Howrah market in 1982 for the World Conservation Union's (IUCN) Freshwater Chelonian Specialist Group (now the Tortoise and Freshwater Turtle Specialist Group) (Vijaya & Manna, unpub.) estimated that 50,000–75,000 Indian flapshells, 7000–8000 large softshells (chiefly Indian softshells, peacock softshells, and Indian narrow-headed softshells), and between 1000 and 15,000 hard-shelled emydids reached Howrah annually. E. Moll (1990a) thought the emydid numbers were probably underestimated as some 300–400 large hard-shelled species were observed at auction in only one day in a later (1983) survey.

As in Bangladesh, softshells are more favored than hard-shelled emydids in commercial trade. Their meat is as good or better testing than hard-shells, they are easier to butcher, and a greater percentage of their body weight is edible. Moll and colleagues surveyed the turtle trade at thirty-five sites (markets and villages) in West Bengal in 1982 (Moll, 1990a) and found at least ten species traded. Softshells were present at thirty-two sites and were the most abundant type at seventeen. Emydids were present at seventeen sites, but were most abundant at only nine. The most widely exploited species were Indian flap-shells and Indian softshells. These two species were also the most abundantly represented species at more sites (nine each) than any other species. Crowned river turtles and three-striped roofed turtles were the most widely utilized hard-shelled species and the latter was the most abundant turtle represented at four sites — the greatest number of sites at which any hard-shelled dominated.

The economic value of the trade is difficult to assess accurately because of the many variables, inadequate records, secrecy, etc. E. Moll (1990a) found the lowest wholesale prices at Howrah market in May to be 10 Rs/kg (~U.S. $0.29) for softshells and 5 Rs/kg (~U.S. $0.15) for hard-shells. Based upon these prices, the annual estimates of turtles arriving at Howrah stated above, and estimates of the mean weight of large softshells at 8 kg, 0.75 kg for flapshells, and 8 kg for emydids, a conservative estimate of 12–13 lakhs (1.2–1.3 million) in annual rupee income from the turtle trade at Howrah could have been generated (~U.S. $35,000.00–$38,000.00).

Choudhury and Bhupathy (1993) conducted more recent investigations of the tortoise and freshwater turtle trade in India from 1991 through 1993 when they visited markets in 55 towns and 13 states. As in E. Moll's earlier surveys, they identified Calcutta's markets as the most important center of the commercial trade, but in both monsoon and winter surveys, they found relatively few markets were involved in turtle trade (6 of 25 and 5 of 23 respectively), and they found fewer species being utilized than in E. Moll's earlier surveys. They thought that the commercial turtle trade, including a significant illegal component, had peaked during the early 1980s. They attributed this upsurge to a combination of factors which included the expansion of traditional fisheries into all kinds of water bodies where turtles could also be easily harvested. Not only were turtles considered detrimental to fish stocks, thereby justifying their removal, but they could increasingly provide an extra source of income for the fishermen. High market demand existed at the time for turtle meat because the price of most other meats and fish had risen, making turtle a cheaper substitute for lower-income groups. In addition, an influx of refugees from Bangladesh, the main turtle buyers at the Calcutta market, had occurred in the 1980s. Turtle collectors could keep turtles (unlike fish) alive and, therefore fresh, for the long periods of time necessary for shipment to this distant marketplace, and the turtle fishery thrived. Choudhury and Bhupathy (1993) thought that the reduction in levels of trade from that observed by E. Moll (1983) might be due both to the increased enforcement of laws which limit the turtle trade, and the increasingly "underground" nature of the trade (including the opening of new smuggling routes between West Bengal and Bangladesh and the Gujarat Coast to the Middle East). It seems likely that reduced turtle stocks due to previous exploitation may have reduced trade levels as well. Choudhury and Bhupathy (1993) found Indian softshells, Indian flap-shells, and crowned river turtles in the markets, but other large softshells seen in trade previously (peacock softshells, Leith's softshells, and Indian narrow-headed softshells) were not observed in their surveys. Choudhury and Bhupathy (1993) found greater exploitation of larger softshells during the monsoon season than the winter. The Indian softshell was the most heavily utilized species during the monsoon, whilst the Indian flapshell was more prevalent in winter.

The rarity or absence of the giant blunt-nosed softshells (i.e., the Indian narrow-headed softshell and the Asian giant softshell) in the major Indian markets during these most recent surveys may indicate that they had declined sharply in the collecting sites which supply the markets. Das (1991) wrote that the meat of the Indian narrow-headed softshell is more coarse and less favored than that of other softshells however, and that it frequently perished in transit from collecting areas to market centers due to its delicacy. It is conceivable that as populations have declined it is less worthwhile to ship the few animals obtained and that they are more likely to be consumed locally. Asian giant softshells have long been considered tasty by some, however. It was caught in fish traps near Penang and relished by the Chinese in early nineteenth century Malaya (Cantor, 1847). It is also considered good tasting in Bangladesh (Khan, 1982, 1987), although opinions differ in Malaysia (Sharma, 1999). Das (1991) wrote that it is not uncommon in northern Orissa state, India and is killed in large numbers for food along the Subarnarheka and Kharasota rivers, and in the vicinity of Gahirmatha. Perhaps its rarity in other parts of the region, and like *Chitra*, its poor transportability preclude its regular appearance in the major market centers.

Choudhury and Bhupathy (1993) found a wide range of prices paid for turtle meat during their survey. In southern India, where turtle meat is held in low esteem it was sold for as low as Rs 15/kg (~U.S. $0.44). In northeastern India, where it is relished, it brought prices ranging from Rs 150 to 200/kg (~U.S. $4.40–5.80). In the Calcutta markets, where volume and competition presumably worked to lower prices they found turtle meat selling for between Rs 30 and Rs 50/kg (~U.S. $0.88 to U.S.$1.47). They were informed by vendors that turtles marketed in Calcutta were collected in the states of Orissa, Bihar, Punjab, Anhdra Pradesh, Jammu, Kashmir, and Uttar Pradesh, and all the major cities and towns along the Ganges and its tributaries in Uttar Pradesh and Bihar were collection centers for turtles to be sent on by rail to Calcutta. E. Moll (1990a) listed these collection areas as well as Madhya Pradesh, Bangladesh, Assam, Rajasthan, and Gujerat.

The major point to make concerning these collection locations is that they are distant from Calcutta and the other West Bengal markets. West Bengal has long been the main center of turtle consumption in India and its citizens' relish for turtle flesh has propelled a now familiar syndrome involving exploited wildlife. As local populations and favored species dwindle from over-exploitation, successively more distant populations and alternate species are exploited to supply demand (sometimes referred to as "the whaling syndrome" after the industry in which the phenomenon was first noted). Biswas and Biswas (in E. Moll, 1990a) stated that in recent years the turtle catch in West Bengal had dwindled to levels so low that it was no longer profitable to attempt to collect them. The FCSG surveys in West Bengal in 1981 led to similar conclusions (Vijaya & Manna, unpub.). They found that few turtles were then being caught in West Bengal and these were usually sold locally. Catches were still high enough at Murshidabad on the Ganges to allow shipment of two to three baskets of turtles to the Calcutta markets during peak months, however. Otherwise, the distant collection locations and extended trade network reflect the limits of the ever-widening circle of depletion emanating outward from West Bengal.

As usual, there are few quantitative baseline data against which the decline of exploited species to modern levels can be compared. There are some statements in the scientific literature which provide an idea of the magnitude of the decline in certain species however. River terrapins for instance, were reported by Blyth in Gunther (1864) to be extremely common in the vicinity of the mouth of the Hooghly River. These were collected and shipped in large numbers to the Calcutta markets at that time. Theobald (1868a) said that it was often used as a substitute for sea turtle meat in turtle soups. Neither Choudhury and Bhupathy (1993) or E. Moll (1990a) observed any river terrapins being traded in their surveys, and the species is almost extirpated from the Hooghly. Only a tiny remnant of the formerly abundant species still occurs in the wild Sunderbans region of West Bengal (i.e., the Ganges Delta), perhaps consisting of as few as ten nesting females (Bhupathy, 1995).

Another comparison of exploited species on the Ganges below Rajmahal, Bihar is possible due to a 1954 status survey by the Zoological Survey of India (Acharji, 1955). Collections and interviews with turtle fishermen in the area resulted in the classification of red-crowned roofed turtles, three-striped roofed turtles, crowned river turtles, and Indian softshells as common to very common (fig. 5.5). In 1982 surveys by E. Moll and colleagues, three-striped roofed turtles were found to be still common in the area, but the Indian softshell, though present, was no longer common. Crowned river turtles and red-crowned roofed turtles were seldom seen or caught, and were regarded as rare by local fishermen (E. Moll, 1990a). E. Moll (1990a) also accumulated anecdotal evidence from various other sources that crowned river turtles and red-crowned roofed turtles had also undergone serious declines elsewhere. For instance, market vendors complained of receiving fewer crowned river turtles in shipments every season, and that the red-crowned roofed turtle was only rarely received. Also Falconer (in Gunther, 1864) referred to red-crowned roofed turtles as being common in the Hooghly River near Calcutta where it, and all other turtle species, are extremely rare today. The 1982 surveys by Moll and associates support these findings as they found only small numbers of crowned river turtles, and only rarely encountered red-crowned roofed turtles (E. Moll, 1990a) (fig. 5.6). Gupta (2002) has also noted that the pace of turtle

Figure 5.5
A pyramid of river turtles awaiting auction in an Indian market.

Figure 5.6
Adult female red-crowned river turtle (*Kachuga kachuga*). This was formerly important in the Indian market trade, but is now severely depleted in numbers.

collection for both subsistence and commercial use has recently increased in northeastern India (Assam), and includes the harvest of rare, endemic forms such as *Kuchuga sylhetensis*. While other factors than commercial exploitation could have contributed to the decline of these species (e.g., pollution, deforestation, habitat alteration, etc.), it is highly likely that commercial trade has been the single most damaging factor in their decline (e.g., E. Moll, 1990a; Gupta, 2002).

Today, in spite of existing legislation protecting many turtle species, a thriving trade leads many Indian species to markets outside the country. Three or four trade routes are recognized with all but one involving river turtles (a route through Bombay for smuggling star tortoises [*Geochelone elegans*] for the international pet trade is the exception) (Bhupathy et al., 2000). Turtles, principally softshells, collected in southern India are transported to Madras for air shipment to Singapore, or to Colombo, Sri Lanka for adding possible other turtles from Sri Lanka, before all are sent to Singapore (where most of the turtles [>90%]) are usually trans-shipped on to other countries; Shepherd [2000]). Turtles collected in central and northern India are transported to Calcutta, and most are flown to Singapore and Hong Kong to supply the food markets of southern China (some turtles which reach Calcutta may be trucked into Bangladesh, however). A fourth route leading from India and Nepal into southwestern China is also suspected to exist (Bhupathy et al., 2000).

Case Study 3: Vietnam. An enormous commercial trade involving hundreds of thousands of tortoises and freshwater turtles annually was centered in Vietnam during the 1980s and 1990s (Le Dien Duc & Broad, 1995). At least eighteen of Vietnam's twenty-three non-marine species—including several river dwellers—were identified as being traded. In addition, species collected from Laos and Cambodia were also funneled into Vietnam. Baird (1964 in Le Dien Duc & Broad, 1995) reported that softshells from the Mekong River and its tributaries in southern Laos were exported illegally into Vietnam in large numbers. They were bought by Vietnamese traders for as much as U.S. $3/kg live weight. These traders reportedly preferred animals under 5 kg body weight as larger animals were more difficult to transport. Mundkur et al. (1995 in Le Dien Duc & Broad, 1995) reported that at least several tons of turtles were exported into Vietnam each month. The vast majority of animals originating from these three countries (perhaps as much as 90%) were destined for food and medicine markets in China (Le Dien Duc & Broad, 1995). The rest supply more local consumptive use and the pet trade (Le Dien Duc & Broad, 1995).

Le Dien Duc and Broad (1995) endeavored to understand the details of the trade routes, volume and species composition of the trade, and its financial value. They found that Ho Chi Minh City and Hanoi were the largest trade centers. Traders at Ho Chi Minh City received turtles from collection and staging areas in the Mekong Delta region of southern Vietnam and Cambodia, then (with some lesser number accumulated at Phan Thiet) trucked them to upper central Vietnam. Here they were joined by turtles supplied from the High Plateau of Laos for transport to Hanoi. From Hanoi, most turtles were moved to Lang Son or Dong Dang on the Chinese Border where they were trucked to Bang Tuong (China), and eventually to Guiyang (China). A smaller percentage of Hanoi's turtle holdings were sent to Mon Cai where they were joined by turtles brought in from Thai Binh for collective sea transport to Dong Hung (China).

Le Dien Duc and Broad (1995) were hindered in their efforts to determine the volume of the trade due to its complexity and lack of official documentation. Their monitoring of the major four turtle dealers at border export locations allowed them to estimate a minimum combined daily export of around 450 kg of live turtles to China (which suggests an annual export quota of some 165 metric tonnes). The actual volume traded may be substantially more based upon other indicators. For instance, Le Dien Duc and Broad (1995) state that Ho Chi Minh City dealers claimed a daily turnover of some 3000 kg or over 1000 metric tonnes per year. Although precise estimates of numbers of turtles represented by these volumes is impossible to calculate. Le Dien Duc and Broad (1995), using estimated mean turtle weights and total weights traded, thought that at least 200,000 individual tortoises and freshwater turtles were traded each year, but they conceded that the actual number could be several times higher than their estimate. Similarly, although the financial value of the chelonian trade was conservatively estimated at U.S. $1 million, it could be valued considerably higher. The unofficial nature of the transactions and export process preclude more exact figures. Regardless of the actual volumes, numbers, and values attached to the commercial chelonian trade, all indications from Le Dien Duc and Broad's (1995) surveys are that it was of truly staggering proportions.

Le Dien Duc and Broad (1995) identified several species commonly associated with lotic environments in the trade, although many of these are generalized enough to be found in other aquatic habitats as well (e.g., stripe-necked leaf turtles, giant Asian pond turtles). The stream specialist big-headed turtle was reported as common in the trade, and the Asiatic softshell, the wattle-necked softshell (*Palea steindachneri*), the Asian giant softshell, and the Chinese softshell were also traded. Asiatic softshells and Asian giant softshells are river dwellers of particular concern, the former being quite common in trade. Although it was usually encountered in low volume, it is a species considered in need of conservation action by the IUCN's Tortoise and Freshwater Turtle Specialist Group. van Dijk et al. (2000a) consider the Asiatic softshell to be vulnerable due to its demand in trade, and the Asiatic giant softshell (*Pelochelys*) as critically endangered in Vietnam, so even though relatively low volumes of trade were observed in these species they must still be considered at risk. Higher volumes of trade were observed in the other two softshells (*Palea* and *Pelodiscus*), and van Dijk et al. (2000a) recommend that they be considered as "endangered" and "vulnerable" respectively.

Le Dien Duc and Broad (1995) thought that most of the commercially traded turtles were destined for consumption as food in China, but turtles and turtle products are also widely used there for medicinal purposes (as discussed later in this chapter). A small proportion of the trade and local turtle collection also served internal consumptive needs which, locally, can be heavy. At Cau Mong, about 30 kg of turtle meat was being consumed every day in the form of grilled turtle, casseroled turtle, etc. Freshwater turtle eggs were also collected and sold as food in Vietnam and Cambodia. A localized, but thriving pet trade also exists within Vietnam (Le Dien Duc & Broad, 1995; Hendrie, 2000). Today, turtle export to China and internal trade and consumption, especially of softshells, continues within Vietnam (Hendrie, 2000). Although difficult to quantify, the level of use has severely depleted the countries' turtle fauna, and is almost certainly unsustainable (Hendrie, 2000).

Case Study 1: **A History of Exploitation of the River Terrapin in Southeast Asia.** The river terrapin was highly valued for its meat and eggs, and was treated as an inexhaustible resource in many areas of its formerly vast range across tropical Asia. A widespread trade network involving this species may have existed since ancient times. Its remains have been identified at Sarnath in Uttar Pradesh, India from the twelfth century, and fragments—possibly from trade animals—have been identified at Mohenjodaro and Harappa, archaeological sites containing ruins of cities in Pakistan's Indus Valley (c. 3000 to c. 1500 B.C.) which are far from the species' modern range (Nath [1959 in Groombridge & Wright, 1982]). Tandy in 1833 (in Parsons, 1962) referred to a sandbank in the Ava River, Burma (Myanmar) from which enough eggs and live turtles of an unidentified communally nesting species (probably the river terrapin) were taken "sufficient to supply a great part of the kingdom." River terrapin and possibly a second species (*Kachuga trivittata*?) were formerly abundant and widespread in the Irrawaddy Delta of Myanmar, producing an estimated 70,000 eggs per year (Maxwell, 1911; Morrison, 1915). Herds of up to 500 river terrapin would emerge from the sea every afternoon during the nesting season to bask on the sand bars and beaches of the Irawaddy Delta (Maxwell, 1911). They were thought to be already in decline by early in the twentieth century due to the heavy exploitation of eggs, increased cultivation of inland nesting areas, and heavy steamer traffic (Maxwell, 1911), and now are absent from areas where they were formerly abundant (Salter, 1983). Salter (1983) reported that the river terrapin was then known to nest only rarely on Kadonlay Kyun Island off the Irawaddy Delta, but no other nesting sites were known to be used at that time. The current status of this last known site is not known, but van Dijk (1997) and Platt et al. (2000) reported declines in nearly all Burmese turtle species in their surveys conducted within that country.

In Malaysia, nesting aggregations of thousands of river terrapins were still known in the Perak River at the end of the nineteenth century (Roff, 1967). While the turtles were not eaten due to restrictions of the Islamic religion, their eggs were under no such taboo, and were considered a delicacy. The collection of eggs here was traditionally the prerogative of the Sultan of Perak, a custom formalized by the "River Rights enactment of 1915" which officially gave the Sultan the power to control the collection of their eggs and prohibited the killing of adults (E. Moll, 1978a). The collection of the eggs was described by Sir Frank Swettenham, a British official of Perak state (Roff, 1967). Of three laying periods on the Perak's major nesting beaches, the first two sets of eggs were harvested under the auspices of the Sultan (the harvest of the second set of eggs was the focus of a festive holiday for the Sultan's family and invited guests), and the third set was left to hatch and maintain the population (E. Moll, 1978a,b). Due to such restrictions and conservation efforts, the Perak's population of river terrapins remained strong through the first three decades of the twentieth century. Cohorts of several hundred females would nest on a beach in a given night; and egg harvests on the Perak as recently as 1937 ranged from 5000 to 7000 eggs per night during the laying season or around 375,000–500,000 (by some estimates 650,000 eggs) (E. Moll, 1980a, 1987b, 1990b). The number of laying turtles this represented is obscured by uncertainty concerning the number of clutches laid in a year by female river terrapin. Maxwell (1911) stated that Burmese river terrapins laid three clutches of eggs totaling 50–60 eggs, and a marked female laid at least two clutches in Malaysia's Kedah River (E. Moll, 1987b). Based upon a mean clutch size for Perak River females of 22.5 eggs (E. Moll, 1990b), the number of females may have ranged between 6482 (if all laid three clutches) and 19,444 (if all laid only one clutch) (fig. 5.7). The likelihood of multiple clutching would suggest that something near the lower figure would be closer to the actual number. As E. Moll (1980a) found adult females to comprise 33% of the juvenile and adult river terrapin, he collected during surveys of the Perak River, the river's *Batagur* population might have been somewhere between 19,444 and 58,332 adults and juveniles at that time (fig. 5.7).

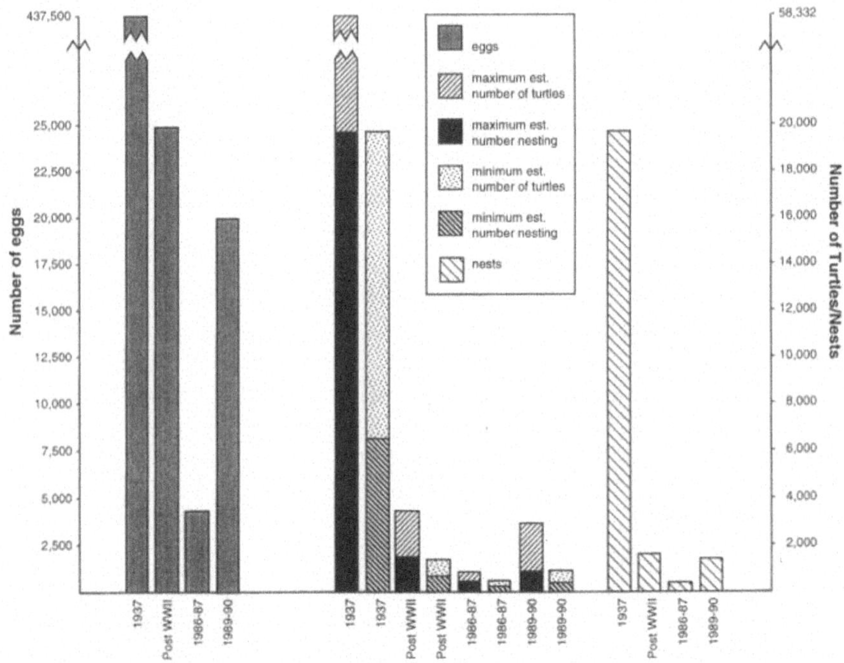

Figure 5.7
Changes in Malaysia's Perak River, river terrapin nesting populations, nests, and egg production during the twentieth century. (Refer to the histogram key and text for further explanation.)

During the Japanese occupation of Malaysia during World War II, the system of controls which protected the terrapin populations broke down and widespread food shortages resulted in heavy exploitation of both eggs and adults. As a result, the population plummeted. Following the war, egg harvests dropped to 20,000–30,000 in especially good years (fig. 5.7), and nesting cohorts rarely exceeded fifty females. Licensed egg collectors now harvested the eggs (fig. 5.8). The licensees hold title to particular nesting sites along rivers which they keep clear of vegetation and construct conical sand hillocks thought to attract nesting terrapins to the site. Odiferous substances may be burnt, whiskey may be poured over the sand, and fruits may be anchored in the river near the site to increase its attractiveness to the terrapin. At night, the men may pound the sand to duplicate the sounds made by nesters as they rhythmically force their plastrons upon the sand in completion of the nesting process. The licensees do not disturb the nesters, but at dawn they expertly probe the nesting area with pointed sticks to find the nest chambers and dig up the eggs (fig. 5.9). They are supposed to give a third of their harvest to the Sultan and rebury another third as a conservation method, while selling the remainder. In practice, however, few egg collectors typically reburied any more than a token

Figure 5.8
A licensed egg collector at his river terrapin nesting site on the Perak River, Malaysia. The sand hillock in the picture is believed to attract nesting females.

Figure 5.9
Probing for eggs laid by nesting river terrapins the previous night on a Perak River, Malaysia nesting beach.

amount and the population continued to decline (Khan bin Momin Khan, 1977). In spite of the establishment of three river terrapin hatcheries, and later breeding facilities on the Perak, Terengganu, and Kedah rivers by the Malaysian government, by the early 1980s, populations in these rivers remained low (E. Moll, 1990b; Sharma, 1999; Sharma & Tisen, 2000). Surveys of major nesting beaches in 1986–87 on the Perak River found only 190 nests containing 4137 eggs (E. Moll, 1990b). At maximum, no more than 190 females nested, and probably fewer, as multiple clutches and clutches divided into more than one nest were probable in some proportion of the population (E. Moll, 1980a, 1984a, 1987b, 1990b) (see fig. 5.7). The largest nesting cohort observed consisted of only nineteen females. E. Moll (1990b) considered this to be evidence that the number of nesting females had declined by at least half since his initial surveys in 1975–76. Surveys in 1989–90 were slightly more encouraging, resulting in the discovery of 923 nests containing 20,768 eggs—the work of perhaps 308 to 923 females (see fig. 5.7).

Another indicator of the river terrapin's steady decline on the Perak in modern times comes from the pattern of nesting on one of the most heavily used nesting sites on the river—Mohammed Shah Beach near Bota Kanan (E. Moll, 1980a) (fig. 5.10). Game Department records indicate a 38% decline from the average

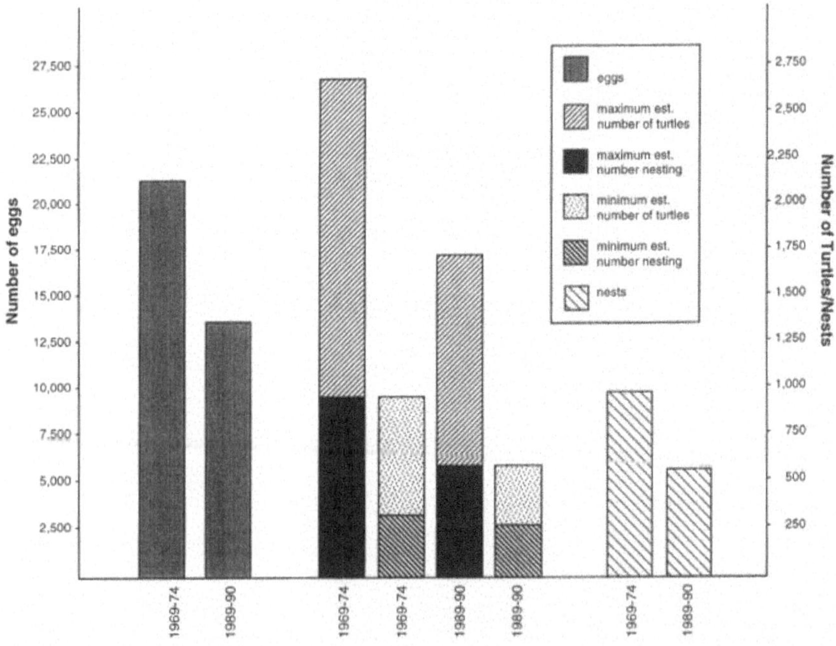

Figure 5.10
Changes in Mohammed Shah Beach (near Bota Kanan on the Perak River, Malaysia) river terrapin nesting populations, nests, and egg production between 1969 and 1990. (Refer to the histogram key and text for further explanation.)

annual total of nests here between 1969 and 1974 (904 nests) and 1989 and 1990 (567 nests) when the last surveys were conducted (fig. 5.10). The much-depleted state of this turtle in Bangladesh and India has already been addressed, and the status of this species across its range in Asia today is further addressed below.

Favored River Turtle Species in the Modern Asian Food Trade

Softshelled turtles are probably the most widely traded species across Asia as they are often considered the most palatable, and yield the most meat by weight. The trade typically occurs at high volume, regardless of whether protective legislation has been enacted or not (Jenkins, 1995). Jenkins (1995) pointed out that there was a substantial volume, of softshells exported from Indonesia for consumption in the Far East. van De Bunt (1990) considered that Sumatra was the major center of Indonesia's turtle trade, with the provinces of Aceh, Riau, West Sumatra, South Sumatra, and Tanjung Balai, North Sumatra the major areas from which freshwater species were exploited. He stated that populations in Aceh province had already been depleted significantly ten years previously. The Asiatic softshell turtle was the most important freshwater species in the trade here (which also includes tortoises and sea turtles), followed by the generalist emydid, Malayan box turtle (fig. 5.11). Softshells are killed in huge numbers and their eggs are collected at several locations. These products are transported to trade centers such as Medan where they are sold locally and the meat is exported abroad. Softshell meat is mainly consumed by the Chinese population within Sumatra, but Moslems eat their boiled eggs which are available during the November to January nesting season. van De Bunt (1990) estimated that all but 2% of the 66,500 kg of softshell flesh exported from Indonesia in 1988 came from Sumatra, with the remainder derived from West Kalimantan. Shepherd's (2000) surveys found that staggering numbers of softshelled turtles and other hard-shelled turtles (see below) were still being exported from Sumatra's North Sumatra and Riau provinces (mainly through the cities of Medan and Tembilihan) to China, Hong Kong, and Singapore. He estimated that more than 25 metric tonnes of live turtles were air-shipped each week from just two of several equally active exporters present on the island, and that export by sea was as frequent, though tonnages of turtles shipped were unavailable (softshells were usually exported by air, hard-shelled species usually by sea). The most common softshell in the trade was the Asiatic softshell, but Asian giant softshells, Malayan softshells and even a few Thai narrow-headed softshells may be utilized also. These two provinces exported about half of the softshells exported from the entire country in 1998 (Shepherd, 2000). These numbers suggest a decline in stocks supporting the Medan export market (a large area of northern Sumatra and its satellite islands), as does a decline in the number of full time trappers, and middlemen in the turtle trade (Shepherd, 2000).

The huge illegal trade in softshells collected in the Mekong River in Laos and destined for Vietnamese and ultimately Chinese markets has already been

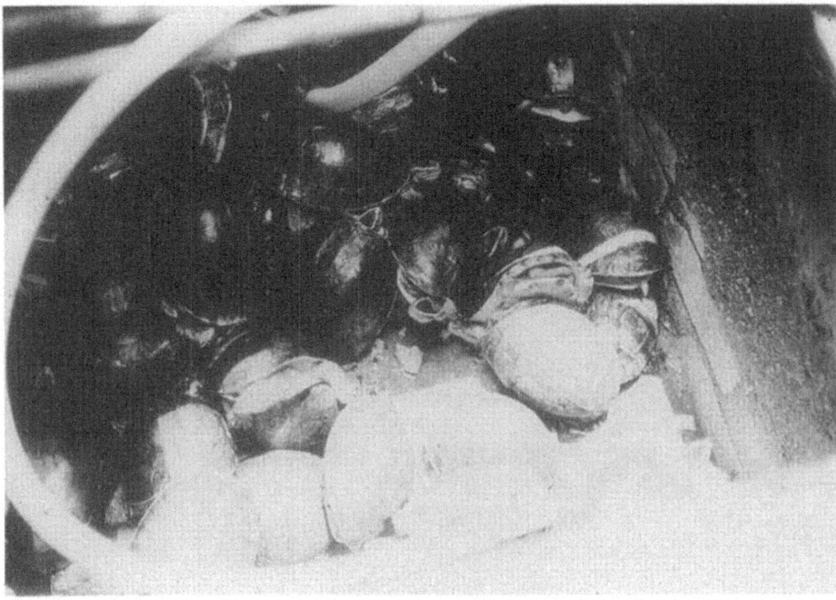

Figure 5.11
A tub of Malayan box turtles (*Cuora amboinensis*) for sale in a Chinese "pet" shop in Kuala Lumpur, Malaysia.

addressed. Jenkins (1995), Thirakhupt and van Dijk (1995 "1994"), and van Dijk and Thirakhupt (1997) have discussed the intensive exploitation and subsequent depletion of Thailand's softshells. Although laws protect all species except the Asiatic softshell covert trade still occurs (Jenkins, 1995; Thirakhupt & van Dijk, 1995 "1994") (fig. 5.12). The narrow-headed soft-shell, which is mainly restricted to the Mae Klong River basin, has been particularly hard-hit by hunting pressure and habitat degradation (Jenkins, 1995; van Dijk & Thirakhupt, 1995), and large specimens are now seldom captured (fig. 5.13). Jenkins (1995) reported that a fish dealer obtained a 76-kg specimen from a local fisherman for U.S. $120 in August 1993. The dealer subsequently sold the turtle to a softshell farmer for U.S. $640. In Malaysia there is evidence that the Asiatic softshell has become less available, and this species and the Malayan softshell (*Dogania subplana*) (a stream dweller) remain in demand for food (Jenkins, 1995). E. Moll (1976a) said many Asiatic softshells (about U.S. $1.78 per kg) and some Asian giant softshells were mainly sold in markets on Malaysia's west Coast (such as Telok Anson [now Teluk Intan]) where most of the country's Chinese population had settled. Jenkins (1995) stated that softshells are caught in Myanmar for both subsistence use and trans-border trade with Thailand. Kuchling (1995) reported on the presence of large Asiatic softshell and/or Burmese peacock softshell turtles (*Nilssonia Formosa*) (which are sometimes difficult to distinguish from one another), Burmese flap-shell turtles (*Lissemys*

Figure 5.12

A harvested Asiatic softshell (*Amyda cartilaginea*) in Thailand. This is the only freshwater turtle which may be legally caught in Thailand (a widely ignored restriction).

Figure 5.13

A huge Thai narrow-headed softshell of unverified weight caught on a bank line in Thailand's Srinagrind Reservoir. (Photo by Wachira Kitimasak)

scutata) and several other hard-shelled species in the markets of Ruili in Yunnan, China, near the Myanmar border. He thought that, based upon the good condition of the marketed animals and other signs, it was likely they were collected nearby in the Longshuan Jiang River which forms the border between China and Myanmar for about 30 km. Kuchling (1995) also noted that he had observed large numbers of the foreign soft-shells (i.e., Indian softshells, peacock softshells, and Indian flapshells) with numerous other exotic hard-shelled species in Canton and Shenzhen, China markets in 1993—obviously the result of exploitation from distant Asian localities. More recent surveys of southern Chinese turtle markets continue to record the presence of foreign softshelled and hard-shelled species (e.g., Salzberg, 1998a; Artner & Hofer, 2001; Haitao, 2002).

Altherr and Freyer (2000) summarized the magnitude of the trade in Asian softshells, and its effects on the targeted species, all of which are steeply declining as a direct result of heavy trade. The rarest softshell of all is probably the East Asian giant softshell or "Shanghai softshell" (*Rafetus swinhoei*), a species on the verge of extinction (Hendrie, 2000; Pritchard, 2001).

Although all of the softshells referred to in the commercial trade summarized above represent wild-caught specimens, the Chinese softshell has long been propagated as a source of food in China and Japan. At the beginning of the twentieth century, Mitsukuri (1905) reported that over 7000 kg were sold annually in Japanese cities at about U.S. $3.25 to U.S. $3.75 each. Subsequently, they (with wattle-necked softshells) were exported to Hawaii where they were marketed in the 1940s and 1950s). Today there is a growing enterprise of captive breeding and rearing of the Chinese softshell in Asia. These current operations are known from Japan, China, Taiwan, Thailand, Malaysia and, to a limited extent, in Vietnam (E. Moll, 1990a; Le Dien Duc & Broad, 1995; van Dijk, 2000; Chen et al., 2000). Their degree of promise in shifting pressure from wild populations is an intriguing, but controversial, issue which is discussed in more detail in chapter 7 (E. Moll, 1991; van Dijk, 2000). Thirakhupt and van Dijk (1995 "1994") have pointed out the possible disruptive effects of feral Chinese softshells on native Asian species as one danger resulting from these ranching operations.

The large, riverine hard-shelled species of the Bataguridae family are among the most threatened species in the region as a result of commercial exploitation. The large-roofed terrapins (e.g., red-crowned roofed terrapin and Burmese roofed terrapin) of the Indian subcontinent and Myanmar, the river terrapin, widespread across southern Asia, the painted terrapin of Thailand, Malaysia and western Indonesia, and the Malaysian giant turtle (*Orlitia borneensis*) of Malaysia and western Indonesia have all been heavily traded within the region, and still are to the extent their depleted populations allow (Jenkins, 1995). Their eggs have traditionally been important commercial products as described in particular for the river terrapins above, but their meat is (or was) extensively used in numerous locations as well (Jenkins, 1995). Jenkins (1995) reported that there was little or no export of larger river hard-shells from Kalimantan for food until the late 1980s, but subsequently exports increased substantially. Large numbers of Malaysian giant turtles and painted terrapins were also being exported from Medan in northern Sumatra in

1994 (Jenkins, 1995). Although prohibited by law, the former species is still one of the most commonly exported hard-shelled turtles from Medan and Tembilahan, Sumatra, while the capture of painted terrapins is restricted to 300 animals from South Sumatra and 200 animals from West Kalimantan annually (450 of these may be exported) (Shepherd, 2000). As discussed in detail above, the river terrapin has declined to very low levels, even in its last stronghold of recent abundance, West Malaysia's Perak River, chiefly as a result of egg harvesting. The nesting population here has declined by more than 90% this century and has collapsed completely in Kedah state (Jenkins, 1995). The species is even rarer today in other areas of former abundance across its vast range such as the Indian Sunderbans region, and in Thailand where it faces imminent extinction from chronic exploitation for meat and eggs (E. Moll, 1990b; Bhupathy, 1995; Jenkins, 1995). A breeding population may yet occur in Bangladesh's Sunderbans (Groombridge & Wright, 1982; Whitaker, 1982) and viable populations may also exist in Sumatra (E. Moll, 1990b). In 2001, a small breeding population of river terrapins was discovered in Cambodia's Koh Kong Province (New York Turtle and Tortoise Society Website, 2002). Unfortunately, the river terrapin is under pressure from local use and commercial trade. Although legal trade in river terrapins has been banned since 1975 when it was placed on CITES, Appendix I (Jenkins, 1995), some occasionally appear in shipments to China which have been sent from Medan, Sumatra (Shepherd, 2000). The commercial importance of painted terrapins has also waned in Malaysia due to depleted populations. As with the river terrapin, its populations were largely reduced by over-harvesting of its eggs by licensed collectors (Sharma, 1999; Sharma & Tisen, 2000). These eggs were preferred over those of sea turtles and, due to their large size and good taste, fetched prices four to five times higher than chicken eggs (Groombridge & Wright, 1982) (fig. 5.14).

Altherr and Freyer (2000) summarized the extent of commercial trade (regional and export to China) in the roofed turtles (genus *Kachuga*). They found evidence of serious population declines in all six traded species.

Former Commercial Exploitation of River Turtles for Food in the United States

River turtles have long been of commercial value in the United States. Perhaps the earliest explorers and colonists did not initially realize the potential value of this inland food resource as they came, by and large, from turtle-poor home countries. As Nicholls (1977) pointed out, however, they undoubtedly quickly learned from native Americans, who had been catching and consuming turtles for thousands of years, that native species were good to eat. Nicholls (1977) provided an interesting quote from a 1588 book by Thomas Hariot, "Brief and True Report of the New Found land of Virginia" which indicated this recognition was not long in coming: "Tortoises both of the land and sea varieties, are more than a yard in breadth, with thick shells on their backs and bellies. Their heads, feet and tails look very ugly like those of a venomous serpent. Nevertheless, they are very good to eat, as are their eggs."

Figure 5.14
Painted terrapin (*Callagur borneoensis*) eggs for sale in a Malaysian market.

As stated earlier in this chapter, the first non-marine species to be encountered was probably the estuarine diamondback terrapin (*Malaclemys terrapin*). It was tremendously abundant, easily caught, and had long been a staple in the diet of coastal tribes (Nicholls, 1977). Destined to be served to slaves and, eventually, to gourmets, it presumably became an item of commerce in the earliest cities of colonial America. Pope (1939) stated that it was already well known in Philadelphia markets 150 years ago—indicating that, as a strict coastal dweller, it was already an object of long distance inter-city commerce by the late eighteenth century. Here in Philadelphia, which has traditionally served as a major United States turtle market, the diamondbacks were joined by other, inland freshwater species. These latter species increased in commercial importance, and found many other markets in the developing cities of the eastern seaboard (e.g., Boston, New York, Washington, D.C.), the Midwest (e.g., St. Louis, Detroit, Chicago), and the deep south (e.g., New Orleans), as roads, railways, and buyer's tastes evolved through the nineteenth and early twentieth centuries. While these other forms never reached the zenith of epicurean regard attained, rather mysteriously, by the diamondback during this time period, they enjoyed (if that is the appropriate word) a steady demand which in some species, continues to the present day. Native freshwater species which Carr (1952) indicated were utilized in commercial trade in the United States are listed in (table 5.1).

The Snapping Turtles

The common snapping turtle is the most important freshwater species in past and present commercial trade in the United States. Most of the individuals utilized in the trade have been collected from rivers. Clark and Southall (1920) provided a detailed account of its utilization in the early twentieth century and, unless otherwise noted, the information on this fishery provided below is from this source.

The central states of Wisconsin, Minnesota, Iowa, Indiana, South Dakota, Michigan, and Kentucky were major suppliers of snappers at this time. Turtles were collected from the rivers of the region usually when they concentrated in hibernacula in the autumn and winter (see chapter 4 for collecting techniques utilized). Not only could turtles be collected in large numbers at these times (perhaps 250–500 kg (500–1000 pounds) per day, or 5 tons in a season), but they were also sluggish and easily handled, and were destined primarily for use in soups—mainly a cold-weather dish. There were a few warm-weather turtle fisheries, however. Clark and Southall mention one operating on Michigan's Grand River in the summer of 1913 which supplied snappers and softshells to Detroit and Chicago markets. The rivers of the Mississippi Basin were major cold season-collecting areas with such rivers as the Mississippi itself, the Illinois, and the Cedar River (Iowa) yielding huge numbers and poundages of snappers to the market annually. Kofoid (1903) estimated that the Illinois and its backwaters alone provided some 15,000 dozen turtles (probably both snappers and emydids) annually.

The catch was bought by middlemen in river cities along the Mississippi (e.g., Winona, Minnesota and Guttenburg and Muscatine, Iowa) and the Illinois (e.g., Peoria, Havana, and Grafton, Illinois). Here, the turtles might be shipped on to the bigger city markets in small batches, or held in impoundments which might contain thousands of turtles before shipments were made. Clark and Southall referred to a storage pen at Clear Lake, Illinois which had formerly held 25,000–30,000 turtles. One impoundment at Grafton held some 2 tons of snappers. (Clark and Southall point out that terrapin were usually designated and priced by number, and snappers and softshells by weight in the commercial trade.) Another impoundment held 4000–5000 terrapin (i.e., *Graptemys pseudogeographica/ouachitensis* and *Trachemys scripta*) for eventual shipment to the markets of Boston and Philadelphia. (The sibling, sympatric species *G. pseudogeographica* and *G. ouachitensis* were not recognized as separate species until recently [see Vogt, 1993]. They were designated as *G. lesueurii* at the time.) These turtles would usually be stacked in barrels with holes bored through the bottom and sides for drainage and ventilation, and with burlap covers nailed on top for rail transport to the big city markets.

Huge volumes of snappers were processed in these market centers. The Chicago wholesale market was estimated to handle about 10,000 snappers per year worth about $5000 in 1920 dollars ($39,575 in modern dollars), with its buyer handling about a ton every week. A census in 1908 indicated that the Mississippi River Basin produced 713,000 pounds (320,000 kg) of turtles and terrapin with a value to fishermen of $25,000 ($439,750 in today's currency).

Snappers would sell for anywhere from 5 to 18 cents a pound, depending upon such variables as whether it was live weight or dressed, wholesale or retail, season, market location, etc. Roughly half the weight of the live snapper could be dressed out (if done carefully, and especially if eggs and liver were included), and this proportion and the effort involved compared favorably to that of processing fish and poultry. Snapper was mainly purchased as a source of cheap meat in soups prepared for restaurant and saloon customers in these large cities. In smaller river cities, however, snapper was often purchased by individuals for private consumption.

Market demand could fluctuate considerably. Clark and Southall (1920) referred to a fish company in La Crosse, Wisconsin which had accumulated 29,609 pounds (~13,300 kg) of Mississippi River snappers between November, 1917 and May, 1918. They were having trouble selling off this stock as their main New York City market had declined from being able to absorb "carloads" of animals to practically nil within a decade. (Philadelphia was generally considered the best market for snappers by Grafton dealers.) In contrast, Lagler (in Carr, 1952) said that in Michigan in the summers of 1937 and 1939 market demand far exceeded supply. Trappers could sell their entire catch wholesale for 5–8 cents/lb live weight, and most of these would eventually be shipped off to Chicago and cities in northern Indiana and Ohio where they supplied the restaurant trade and private individuals. Their retail price was 10–20 cents/lb live weight or 20–30 cents/lb dressed. Lagler knew of a trapper and his helper who worked about five dozen traps in Michigan who had made about $1000 ($11,200 in modern dollars) profit for each of the last three years in supplying this market.

In other areas, a general century-long decline in the commercial turtle fishery mainly based upon snappers has occurred. An example is provided by the commercial turtle catch statistics between 1899 and 1971 (after which commercial turtling became illegal in Illinois) in the Illinois River. Bellrose et al. (1977) summarized the situation as follows:

> Softshell and snapping turtles were, and still are, considered an excellent food by some people, and the commercial turtle fishery flourished in the Illinois River at the turn of the century, landing a reported 547,000 pounds worth about $12,000 [about $228,840 in modern dollars] in 1899. However, the catch had declined to only about 60,000 pounds in 1922, valued at only about $450 [about $4,184 in modern dollars]. The decline is probably attributable to a decline in the economic value of turtle meat, coupled with a reduction in catch due to pollution and extensive drainage of bottomland areas. By 1931, the catch had been further reduced to only 14,000 pounds, although the value of turtle meat had increased from one to five cents per pound, making the 1931 catch worth 50% more than the 1922 catch. Since 1957, the turtle catch from the Illinois has fluctuated between highs of 9,700 and 8,700 pounds to lows of 1,900 and 1,300 pounds. Large changes often occur from one year to the next. The most recent reported catches (1970, 1971) were less than 500 pounds each—a marked decline from 9,700 pounds in 1969. The price of

> turtle in recent years (25 cents per pound in 1971) suggests few commercial fishermen would have thrown away snapping or softshell turtles that were caught in their fishing gear, as was often the case in the earlier years of the century. (Bellrose et al. [1977])

In the second half of the twentieth century the demand for common snapper meat underwent a general decline as the United States became more affluent and dietary patterns changed. Bans on commercial turtle fisheries were also imposed in some states (e.g., Illinois), thereby further stifling commercial trade. Snapper meat was always available in many areas where it traditionally has been coveted, however, such as in the south, in the fish markets of river towns, and in Philadelphia where Nicholls (1977) stated that 500 pounds of snappers were still being processed for soup each week. It is interesting to note that, in contrast, in some regions where snappers exist they have never been particularly popular fare, however. The Florida snapping turtle (*Chelydra serpentina osceola*) has never attained the commercial popularity of its northern relative elsewhere throughout the species range, for instance. Carr (1952) believed this was probably due to the abundance of better looking, more easily prepared, and as good or better tasting species (such as softshell and "Suwannee chicken" = Suwannee river cooter) available to Floridians.

The common snapper's much larger relative—the alligator snapper—has also been an important commercial species, especially in the southern states. It has never been as widely popular commercially as the common snapper because of its inferior taste (Carr, 1952), its proportionately lower yield of meat to total body weight (close to one-third in larger specimens versus about one-half in the common snapper) (Clark & Southall, 1920; Pritchard, 1989), and its more restricted distribution. Due to its great size, and the relative ease with which animals may be collected by experienced turtle trappers, it has not been ignored however. Clark and Southall (1920) referred to an active fishery for this species during certain seasons in Louisiana where it was then considered to be especially abundant. Its sale was mainly confined to southern markets such as New Orleans, but Carr (1952) observed that it had also been regularly sent as far as St. Louis. It was eaten locally along the Suwannee (Carr, 1952), and undoubtedly was regularly featured in other local markets within its range.

The Terrapin Trade

The most famous turtle of commercial trade in the United States is the aforementioned diamondback terrapin. Only marginally a "river" turtle as it inhabits estuarine and other coastal habitats of the Atlanta and Gulf Coasts from Massachusetts to Texas, it was the main impetus propelling other inland riverine forms to commercial importance along with it. As stated earlier, the diamondback has been an item of commerce for a long time in the eastern states. This is not particularly surprising as it is a good-tasting turtle, although small by comparison with the chelydrids discussed previously. Its rather meteoric rise to the heights of gustatory prominence in the late nineteenth century (and equally precipitous descent in the early twentieth) is more difficult to comprehend. Nicholls (1977) saw its rise as a possible result of opportune promotions (shrewd or desperate) to "a market always anxious for culinary novelties" with a product which was abundant, cheap to obtain, and could be guaranteed to be served fresh as it was shipped alive to its market destination. Once it became a fixture in the menus of the elite, it undoubtedly acquired a certain "snob" quality which increased its appeal to the always much larger group who, as Nicholls put it, "were intent on copying the whims and lifestyle of the elite." Carr (1952) speculated that its decline might have been associated with the advent of prohibition, as alcohol was both an important component of most terrapin recipes, and a near-essential adjunct of its consumption. The Depression might also have played a role in its overall effect of toning down "high living" in general. Both Carr and Nicholls agree that much of the phenomenon had unmistakable faddish qualities which, by definition, are largely unexplainable. The terrapin equivalent of the hula-hoop, its time came and passed for a variety of poorly understood reasons, leaving much diminished populations in its wake.

Carr (1952) eloquently described the particulars of the diamondback's exploitation, the real or imaginary palatability differences of its various races, and the development of captive propagation programs to supply the industry's requirements as natural populations dwindled. He considered Baltimore to be the center of terrapin cookery (as the names of various defunct and extant local sports teams still suggest), but Philadelphia was always the best market. The Delaware Bay terrapins, and subsequently the "Chesapeakes" were most favored in the markets, with the other more southern Atlantic and Gulf forms considered distinctly inferior fare—the "Biloxi terrapin" (*M.t. pileata*) considered the best of the southern lot. In 1891 the Chesapeake Bay terrapin fishery yielded 89,150 pounds valued at $20,559 ($361,610 in 2003 dollars). Nicholls (1977) stated that early in the twentieth century a single specimen of the favored races could sell for as much as $10 (about $175 in 2003), and a dozen adults would fetch from $90 to 120 (about $1583–2110 in 2003). By 1920 the total catch in Chesapeake Bay had plunged to 823 pounds worth $1000 ($7915 in 2003) as a result of the thirty-year binge (McCauley, 1945 in Carr, 1952).

Inland "terrapins" (several hard-shelled species in the family Emydidae were utilized) came to prominence in national commercial trade largely as an adjunct of the elevation of the diamondbacks to the zenith of their popularity (and accompanying prices), and subsequent decline of their populations as a result. A primary reason for their coincident popularity in the eastern turtle markets is that they were not markedly inferior in flavor to the diamondbacks (if at all), and they could be obtained much more cheaply. Clark and Southall (1920) quoted one Chicago market price for these species at 25–30 cents apiece ($1.98–2.38 in 2003), and another at 75 cents to $3 per dozen ($5.94–23.75 in 2003). The most desirable specimens for the Philadelphia markets were the larger egg-bearing females, called "queen terrapin," which were sold to buyers for this market at $1.00 each in 1918 ($10.48 in 2003). Smaller males were discarded (some of the species utilized were sexually dimorphic and males are substantially smaller than females). Clark and Southall (1920) reported that in Philadelphia and Washington, D.C. (and presumably elsewhere), these cheaper terrapin were often used as substitutes or partial substitutes for diamondbacks in "diamondback" soup and other dishes. Given the price differential, the profit motive for such substitution is clear, and it is doubtful that even the most discriminating turtle gourmets would realize they had been duped. As diamondback populations declined it is likely that these inland species played an increasing role in this regard until the terrapin fad eventually faded for good.

The demand for these inland "terrapin" was largely met with shipments from the rivers of the Mississippi Basin. Three species were most important in this regard, although only two were probably recognized at the time. The most important of these terrapin in the trade in the early twentieth century was then known as Lesueur's terrapin (*Graptemys lesueurii*). Only recently has it become apparent that this "species" was actually composed of two very similar, sibling species, the false map turtle, and the Ouachita map turtle, which are widely sympatric across much of their Mississippi Basin range (see Vogt, 1993 and chapter 3 this volume). The other species utilized extensively was the red-eared slider, or elegant terrapin (*Trachemys scripta elegans*). Cahn (1937) ranked these turtles as third and fourth in importance as commercial food turtles in Illinois (after common snappers and softshells), but Clark and Southall (1920) point out that although a few people were aware of their excellent taste they were seldom utilized as such locally. They suggested that the main reason for their lack of popularity within their range was probably due to their unfamiliarity to the general population and their small size (even large females average only about 2 lb [1 kg]) compared to the larger and locally more favored snappers and softshells. In addition, they have smaller meat to bone/shell ratios, and the meat is more difficult to access than in the larger species. Insulated by distance, and Midwestern common sense from the diamondback fad, the citizens of the Mississippi Basin were happy to take the dollars received from shipping their queen terrapin east and buy with it some substantial snapper or softshell to fry. Clark and Southall (1920) found that the main buyers of inland terrapin in the Chicago and St. Louis markets, as well as diamondbacks which were shipped in from the east, were members of the Chinese community. Following longstanding traditions in China they utilized the terrapin for medicinal purposes (see the appropriate section later in this chapter for more detail) as well as for food. Although they considered the diamondback the best species for these purposes, its expense often forced them to utilize the cheaper red-eared slider (then about 20 cents each) as a substitute.

Terrapin collectors caught most of their animals in rivers with traps of varying sophistication set among basking sites during the warm season. Basking turtles would fall into or entangle themselves in the traps and nets as they dropped into the water to leave the basking sites. In this way thirty to forty turtles could be taken per day. Terrapin would be housed in impoundments holding hundreds or thousands of animals (as described above concerning common snappers), presumably until the cold, "soup season" returned and they were shipped off to the eastern turtle markets.

Other hard-shelled emydids also at least occasionally, and in some regions regularly, served the terrapin trade as well (see table 5.1). Conant (in Carr, 1952) said that large numbers of Blandings' turtles (*Emydoidea blandingi*) from Ohio were caught and sent to market at one time, and even wood turtles (*Clemmys insculpta*) from Michigan were once sent to eastern markets (Harding, 1990) (both species are often more aquatic and associated with lotic habitats in the Midwest than in other parts of their range). McCauley (in Carr, 1952) mentioned that a fisherman was able to sell painted turtles to a Pennsylvanian for $1.00/dozen, but whether the purchaser thought he was obtaining diamondbacks at a bargain price or desired a cheap substitute was unclear.

An eastern species of more importance in the trade at one time was the red-bellied turtle. Pope (1939) said it supported a profitable fishery between Washington, D.C. and the sea, and that in 1883, 240 individuals were taken in one seine haul. During the latter part of the nineteenth century, Chesapeake Bay region animals increasingly replaced the declining diamondbacks in the commercial trade (Carr, 1952). McCauley (in Carr, 1952) found it still selling for 75 cents and up depending on size in Washington D.C. markets in the 1940s.

On the West Coast, where a supply of eastern diamondbacks for San Francisco epicures must have been very expensive and discontinuous at best, the Pacific pond turtle (*Clemmys marmorata*) (actually a stream and river dweller in many areas) filled the commercial terrapin niche. Lockington (in Carr, 1952) stated that as early as 1879 this turtle, formerly abundant, had become scarce in the San Francisco area because of commercial demand. Turtle dealers of the day were able to distinguish the relative degree of palatability of the yet to be described northern and southern races, and the fishery provided a nearly constant supply for San Francisco markets during the 1880s. The price in the 1920s here ranged from $3 to $6/dozen (about $27–54 in 2003 dollars) (Carr, 1952).

A few other North American aquatic emydid species have occasionally reached restaurant markets, but they have been mainly of local significance. One river form which is particularly delicious is northern Florida's Suwannee River cooter (*Pseudemys concinna suwanniensis*) or famed "Suwannee chicken." Carr (1952) expressed surprise that it had never been marketed in any organized manner, as it was superior to nearly any other species as a component of soups and stews. Due to its habit of foraging for turtle grass far offshore from Gulf river estuaries, it was occasionally picked up by Cuban trawlers and featured in Havana markets at times, however (Carr, 1952). Other races of the Suwannee River cooter have also been locally marketed in Gulf Coast markets that mainly catered to small restaurants and the housewife. The cooter races, *Pseudemys floridana floridana* and *P. f. peninsularis* were also used for soup. The former was sold in some numbers in Charleston and Savannah, and an occasional barrel or two of the latter also reached even northern markets (Carr, 1952).

The Softshells

The native softshells of North America, the Florida softshell (*Apalone ferox*) the spiny softshell, and the smooth softshell are large, excellent tasting turtles but they were not significant in trade at the national level. Clark and Southall (1920) saw none in

Washington or Chicago markets where, in the latter, they were told "they could not be given away, much less sold". They attributed this contradictory situation to the fact that the table qualities of these turtles were too well known locally, but too little known at the great market centers. Clark and Southall considered softshells to be the favorite food turtles of the small towns along the Mississippi and Illinois Rivers, where they were prized for soup and fried turtle. Evermann and Clark (1920) considered the spiny softshell the finest food turtle of all the Indiana species. As they were usually incidentally collected in small numbers, and were more delicate and difficult to ship than hard-shelled species anyway, they were usually just consumed by their collectors or sold in local riverside markets. Nevertheless, Ditmars (in Pope, 1939) and Carr (1952) saw Florida softshells for sale in southern city markets, and Pope (1939) saw the Texas spiny softshell (*A. spinifera emoryi*), which extends across the border into Mexico, for sale in Ciudad Juarez markets. Clark and Southall (1920) reported that market men found that softshells could be easily shipped on ice after beheading and rough cleaning, without loss of flavor. Fortunately for American species, their popularity never escalated nationally to such an extent that this shipping technique was widely employed.

Vogt (1981d) noted that spiny and smooth softshells (along with common snappers) were commercially collected in Wisconsin and sold to restaurants for soup. They sold for 45–75 cents per pound live weight in Wisconsin markets in 1975.

Current United States National and Export Food Trade in Turtles

Unfortunately for freshwater turtles there has been a resurgence of interest in turtles as food in recent years in the United States. This interest has been associated with an increase in general popularity of American regional cuisines (such as "cajun cuisine") which feature turtle. Coupled with the steady demand for turtle in some locales, and by certain ethnic groups within the country which have continued to support a modest commercial trade, the number of animals traded locally and nationally has sharply risen from what it was two or three decades ago. Although significant, this national trade is dwarfed by the immense export trade in United States freshwater species destined for foreign nations. As an arguably significant proportion of these exported turtles are potentially used as food, and as many river dwelling species and/or riverine populations of more generalized species are harvested for the export trade, the dynamics, value, and impact of this export trade, will also be discussed here.

Gilbert (1993) suggested that commerce in snappers within the United States has greatly increased in recent years, with hundreds of thousands of pounds of dressed turtle meat sold annually in this country. Gilbert attributed the recent increase in this trade to a continuing preference for turtle meat coupled with the declining accessibility of other sources such as sea turtles and terrapins. At any rate, there are still some big suppliers of snapper products operating in the country. Gilbert profiled one of the largest—Fred Millard of Birmingham, Iowa—who annually sells 180,000 pounds (81,000 kg) of live snappers and meat to restaurants and specialty markets (at $4.75/lb for the latter). He buys his turtles form a network of suppliers across the snappers' U.S. range. Gilbert also stated that a Louisiana turtle buyer processed 15,000–20,000 pounds a year for sale within Louisiana, where turtle meat remains very popular (some of this poundage might be composed of alligator snapper meat also; see below). Warwick and Steedman (1988) reported another trader in Louisiana who estimated that 100,000 pounds (45,000 kg) of common snapper used for human consumption passed through his establishment in one year. Other states with a high volume of commercial trade in snappers include Minnesota (80,000–91,000 lb [~36,000–40,800 kg] or 6000–6800 average-sized adult snappers annually) (Brooks et al., 1988), and Virginia (1350 snappers processed annually) (J. Mitchell, pers. com). In southern Ontario the estimated annual catch is 66,000–111,000 lb (30,000–50,000 kg), or 5000–8300 adult snappers (Brooks et al., 1988).

Pritchard (1989) and Sloan and Lovich (1995) provide a modern perspective on the alligator snappers' commercial exploitation for food. Pritchard (1989) found this species to have been the subject of intensive commercial exploitation in Florida, Louisiana, Georgia, Mississippi, Arkansas, and Oklahoma. The late 1960s and early 1970s were times of particularly heavy harvesting from many areas—perhaps to supply the needs of the Campbell Soup Company, which produced a frozen snapper soup at that time (Pritchard, 1989). Turtles obtained from northern Florida rivers such as the Appalachicola and Ochlocknee were sold for $1.00/lb in New Orleans in the early 1970s, while Georgia animals were also being shipped to northern states by the thousands to supply the restaurant trade at that time. Boneless meat sold for about $3.00/lb (Pritchard, 1989). By 1982, Georgia animals were selling for $3.00–4.50/lb depending on location and dealer. If no local market could be found, turtle fishermen would keep their catch alive until they had accumulated enough to make a trip to the New Orleans market profitable (Pritchard, 1989).

Pritchard (1989) was informed that the most active Louisiana markets for alligator snappers in the early 1980s were at Manchac (Tangipahoa Parish) and on Hwy 90 in St. Charles Parish near the Jefferson Parish line. Turtles were trucked in from locations throughout the state, and from bordering states also. One informer said turtles at Manchac and Bayou de Allemands markets were usually sold alive for $1.50/lb live weight, or if butchered and cleaned, the meat sold at $4.50/lb. One turtle market in Manchac (Reno's Fish House) was reported by Pritchard (1989) to have purchased 7000 lb of alligator snapping turtle between May and August, 1981, mainly trucked in from distant locations, for 85 cents/lb; they were then sold for $1.30–1.40/lb live weight. Meat (with bones) could be sold locally for $4/lb or elsewhere in the state for $5–6/lb at that time.

Sloan and Lovich (1995) reported that heavy commercial usage of alligator snappers continued in Louisiana. They monitored the catch processed by a single, anonymous turtle buyer in south central Louisiana from 1984 through 1986. The buyer purchased 17,117 kg (7771 lbs) of live alligator snapping turtles, and the mean weight processed each year was 5706 kg (2590 lb). Most came from

Louisiana, but significant importing from Arkansas and Mississippi also occurred. Turtles were supplied by between two and seventeen trappers per month, and a direct positive relationship between rate of harvest and number of trappers was determined. Sloan and Lovich estimated that the live weight received probably represented some 1223 turtles or around 408 killed annually by just this dealer. As about thirty-five major dealers are currently (1995) known to be operating in Louisiana, and an unknown number of smaller dealers and private consumers are also legally taking alligator snappers in Louisiana, the annual mortality is undoubtedly considerable.

Sloan and Lovich (1995) applied Pritchard's (1989) price schedules obtained from Reno's Fish House to estimate their dealers' mean annual harvest as worth $10,693 at wholesale price, and between $16,353 and $17,611 at local retail value. Sloan and Lovich's figures indicate that the trade in alligator snappers remains profitable in Louisiana.

Other species may also reach specialty and regional markets and restaurants where demand exists. Asian Americans have a cultural heritage of turtle eating, and Chinatown fish markets may often have various species for sale. Eley (1984) found Florida softshells, smooth softshells, spiny softshells, Chinese softshells, common snappers, wood turtles, western pond turtles, chicken turtles, sliders, river cooters, common cooters, diamondback terrapin, and green turtles (*Chelonia mydas*) available for sale in San Francisco, and Oakland California markets. Garber (1990) found hundreds of freshwater turtles for sale in New York's Chinatown food markets as well. Many species were represented, but diamondback terrapin predominated. Garber (1990) estimated that at least 10,000 diamondbacks were being sold annually in New York City during the late 1980s. Garber (1990) calculated that the Fulton Fish Market alone was selling more than 30,000 pounds (67,000 kg) of diamondback females annually. The effects of the resurgent trade on natural populations of diamondback terrapins are difficult to separate from several other significant sources of mortality which currently plague the species, including incidental drowning in crab traps, road kill, and habitat loss. A workshop on the species' ecology, status, and management in 1994 identified commercial trade as one of the factors which have collectively caused the decline of diamondbacks in seven states within their range (Seigel & Gibbons, 1995).

Warwick and Steedman (1988) provided information concerning a food trade in red-eared sliders (mainly), which supplies both national and international markets. They assert that adult turtles for the food trade are usually wild-caught, although ranch stock (later replenished from the wild) may also occasionally be used. They quote a source as estimating that some 25,000–30,000 wild-caught adults collected by about 200 professional collectors were shipped from Louisiana hunter/traders every one to two weeks during the warmer seasons when turtles are active (April–November). Adjusting for a somewhat slower catch rate probably occurring later in the season, they estimated an annual catch of 637,000 per year. The turtles came from Louisiana and other states, but were sent to California for local sale and also exported to Far Eastern countries. McCloud (1985) reported the weekly arrival of up to several hundred turtles at the San

Francisco airport each week (often inhumanely packed) destined for various food markets. Warwick and Steedman (1988) estimated that 60–100% of these shipments were composed of red-eared sliders. McCloud (1985) also reported that softshells, map turtles, and perhaps other emydids were commonly received as well. It is assumed that the adult turtles shipped on to other countries are destined for use as food, or possibly for medicinal purposes as discussed below, but that vast numbers of exported hatchlings supply a pet trade abroad. However, Warwick and Steedman (1988) suggested that some percentage of these hatchlings may actually be fried, or used in stocks and stews also. The percentages of large and small turtles actually exported are unknown, as the sizes of exported turtles are not usually recorded by government inspectors. It should be noted that there is some disagreement with Warwick and Steedman's (1988) conclusions concerning the magnitude of an overseas food trade in American sliders. Hemley and Gaski (1989) found little evidence of such a trade based upon interviews with shippers, and from contacts with TRAFFIC (Japan) and WWF-Hong Kong. There is certainly a huge export trade in baby red-eared sliders, and other United States species for pets which will be considered below.

The possible effects of heavy collecting of red-eared sliders for direct shipment to markets and for supplying ranch stock include the depletion of many populations accessible to collectors and the need for harvesting ever more distant population to supply demand (Warwick, 1985a); deleterious ecological effects on the habitats from which they have been removed (Warwick et al., 1990), and the diminution of average body sizes in populations subjected to harvesting (Close & Seigel, 1997).

One other species known to be subject to heavy regional harvesting for food is the Florida softshell (fig. 5.15). Enge (1991) estimated that some 3600 softshells were purchased for meat in south Florida between July 1990 and July 1991. Unknown quantities of this species from southern Florida are also being shipped to Chinatown markets in northern United States cities and in Toronto (B. Sawlaw, pers. com.).

Overall Effects of Exploitation on United States River Turtles

Despite the heavy exploitation of the native species described above, none is currently in danger of extinction. Many have certainly been reduced in population size, sometimes drastically in various parts of their range, and some of their populations and described races (e.g., Suwannee cooter) are threatened. The magnitude of their declines and even the extent to which the depletion is due to their commercial use is usually difficult to quantify, however. The reasons for these uncertainties concerning exploitation effects are first, that we have only anecdotal information concerning former pre-exploitation population sizes for comparison with modern populations; second we still have no particularly reliable methods for accurately estimating abundance of lotic (especially large river) turtle populations (Vogt & Villareal-Bentez, 1993); and third, populations of these species have also

Figure 5.15
The Florida softshell (*Apalone ferox*), a marketed U. S. species. (Photo by Indraneil Das)

been negatively affected by other factors such as habitat elimination, degradation and alteration—making assessment of the population effects solely due to commercial exploitation (and the species' ability to recover from it) difficult to assess. Collectively, the circumstantial evidence available concerning the linkage between commercial exploitation and depleted populations is compelling, however. The familiar patterns related to the excessive harvesting of other commercially valuable species are recognizable in the United States commercial turtle fishery also. The ever-widening geographic range of collecting activity in order to meet market center demands as more local stocks are depleted, the rising market prices as supplies dwindle, and the substitution of less-favored species for depleted primary target species have proven to be reliable indicators of over-exploitation of other wildlife, including turtles, on other continents. As these signs are—or were in the case of defunct fisheries—also earmarks of various United States turtle fisheries, there is no reason to interpret them differently in this case. In addition, surveys of population densities or other relevant characteristics compared between unexploited and exploited populations such as those conducted on the red-eared slider by Warwick et al. (1990) and Close and Seigel (1997) indicate likely over-exploitation consequences. Coupled with the Congdon et al.'s (1993, 1994) and Tucker and Moll's (1997) conclusions concerning the negative ramifications of removal of critical older juvenile and adult cohorts from turtle populations in regard to their future viability, the concern for harvested species is surely not misplaced.

Patterns of Commercial Exploitation in Mexico, Central America, the Caribbean, and South America

Case Study 5: Exploitation of Giant South American River Turtle in South America. The abundance of these giant river turtles in South America's Amazon and Orinoco River drainages as European explorers began to reach the continent centuries ago is difficult to comprehend today. In 1745, Padre Dumilla considered these turtles as common as the grains of sand along the river shores (Parsons, 1962). Although resident aboriginal groups had undoubtedly harvested the turtles and their eggs for millenia, a number of factors worked together to ensure that the turtles were not exploited beyond their capacity to replace themselves. The humans of the region were largely hunter-gatherer groups with a poorly developed market economy, and their populations were small and widely dispersed. In addition, in some cultures an appreciation of the role of the turtle in their diet led to its acknowledgment and portrayal in their ceramic art and mythology (Smith, 1975). Some Orinoco tribes believed that a female spirit guarded over laying turtles and would harm anyone who disturbed them (Smith, 1975). Another tribe believed the species was protected by a huge spirit turtle (Smith, 1975). Aspects of the turtles' ecology also afforded it some protection. It is a powerful swimming and diving species which is present in large concentrations (for egg laying) for only a limited time each year. At other times, the populations are widely dispersed through the rivers and flooded forests where they feed, and are much less accessible. The species as a whole also has a vast range spanning much of northern South America. These combined biological and cultural checks on over-harvesting allowed both species to flourish while coexisting through the ages.

Excessive restraint in harvesting turtles was apparently not necessary to ensure its well-being in this harmonious balance, however. When the Orellana expedition traveled down the Amazon in 1542 they encountered a village with more than 1000 captive turtles, housed in wells, and aquatic corrals, which had been captured during dry season nesting concentrations (Medina, 1934 in Smith, 1975). In 1560, Pedro de Ursua encountered another settlement holding more than 4000 river turtles (Vasquez de Espinosa, 1942 in Smith 1975), and an observation by Acuna in the mid-1600s that there was not a village which did not hold a hundred or more giant river turtles provoked his comment that hunger must be virtually unknown among the people of the Amazon river bank settlements (fig. 5.16) (Smith, 1975). Typically, the captive turtles would be kept for slaughter through the wetter parts of the year when river turtles and fish were dispersed and less easily captured in large numbers (Smith, 1975). The term "river cattle" was applied to these turtles, and provides an idea of their role as a dietary staple. The eggs were also an important staple food for river tribes throughout the range of the turtle.

Parsons (1962) related one early account of collected eggs which were piled in heaps up to 20 feet high that resembled "stacks of cannonballs such as are seen at Navy yards."

In spite of this relatively heavy use, Alexander von Humboldt (1819, 1852) estimated in 1799 that more than one million female giant South American River

Figure 5.16
A turtle corral in a Conibo village, Peru (St. Cricq, 1874).

turtles were still nesting annually in the lower Orinoco River (Mittermeier, 1975). The influx of missionaries and early Spanish and Portuguese traders and settlers soon began to change the balance, which had existed between man and turtle through all previous time, however. The early Jesuits were particularly fond of turtle meat, and it had the advantage of being considered a type of fish so it could be consumed during Lent (Smith, 1975). Smith (1979) summarized the pressures on the giant South American River turtles' populations in the Itacoatiara area of Amazonian Brazil beginning early in the eighteenth century as representative of a typical pattern of exploitation which evolved throughout the region. The turtles were particularly large and abundant here, and numerous nesting beaches were scattered through the area. The nesting concentrations quickly attracted Portuguese exploiters who made annual expeditions to collect the turtles as they finished nesting. At least one nesting beach was designated a royal beach, and was reserved to supply the militia stationed along the Rio Negro, especially the garrison at Barcelos (Lacerda & Almeida, 1944 in Smith, 1979). Despite heavy exploitation of this kind, the turtles were abundant enough that they remained an important component of diet and commercial trade at Itacoatiara for a century and a half, until at least 1859 (Ave-Lallemant, 1961 in Smith, 1979).

As important as turtle meat and eggs for direct consumption were in the commercial trade, they were eclipsed by the great demand for turtle-derived oil. The manteiga or azeite de tartaruga was used for cooking and as lamp oil (a heavier oil was mixed with resin and used for caulking boats), and was usually derived from eggs. The oil produced a nearly smokeless, long-burning flame when properly wicked, and it was used all over the region for this purpose until kerosene became widely available. Despite its fishy flavor the oil was also very popular as a substitute for butter or lard (Parsons, 1962). The finest grade of oil came from boiling fresh turtle fat, however, and once the fat was obtained the rest of the turtle was often discarded unused (Ferreira, 1786 in Smith, 1975)

Initially there was concern, based upon the great value of the oil resource, that the turtles would be quickly depleted by this kind of overzealous exploitation. This concern led to the development of strictly enforced conservation regulations on heavily used Orinoco nesting beaches which were imposed by church and state officials (Smith, 1975). von Humboldt (1852), referring to his observations from 1799 through 1804, stated that "The Jesuits did not suffer the whole beach to be searched, they left a part untouched, from the fear of seeing the breed of arrau tortoises (i.e., giant South American River turtles), if not destroyed, at least considerably diminished." Nevertheless, Humboldt thought that the egg harvest was already beginning to decline by the end of the eighteenth century and recommended a three-to-four-year moratorium on egg harvesting to allow populations to recover. Similarly, Ferreira criticized the wasteful exploitation of hatchlings for food and the destruction of nests for production of turtle oil in Brazilian Amazonia as early as 1786 (Mittermeier, 1978).

During the nineteenth century, Amazonian populations were the focus of similar intense exploitation which was recorded by many observers including S. Cricq (1874), Agassiz (1879), Spruce (1908) and, most eloquently, by Bates (1863). As Bates's account, which is based upon observations made between 1848 and 1859 on the upper Amazon near Tefé, provides the most detailed account of the process and magnitude of the egg (and hatchling) harvest, and the oil-making procedure of this time period, it is quoted here in full:

> The last of these minor excursions which I shall narrate, was made (again in company of Senhor Cardozo, with the addition of his housekeeper Senhora Felippa) in the season when all the population of the villages turns out to dig up turtle eggs, and revel on the praias. Placards were posted on the church doors at Ega, announcing that the excavation at Shimuni would commence on the 17th of October, and on Catua, sixty miles below Shimuni, on the 25th. We set out on the 16th, and passed on the road, in our well-manned igarite, a large number of people, men, women, and children in canoes of all sizes, wending their way as if to a great holiday gathering. By the morning of the 17th, some 400 persons were assembled on the borders of the sandbank; each family having erected a rude temporary shed of poles and palm leaves to protect themselves from the sun and rain. Large copper kettles to prepare the oil, and hundreds of red earthenware jars, were scattered about on the sand.
>
> The excavation of the taboleiro, collecting the eggs and purifying the oil, occupied four days. All was done on a system established by the old

Portuguese governors, probably more than a century ago. The commandante first took down the names of all the masters of households, with the number of persons each intended to employ in digging; he then exacted a payment of 140 reis (about four pence) a head, towards defraying the cost of sentinels. The whole were then allowed to go to the taboleiro. They ranged themselves round the circle, each person armed with a paddle, to be used as a spade, and then all began simultaneously to dig on a signal being given—the roll of drums—by order of the commandante. It was an animating sight to behold the wide circle of rival diggers throwing up clouds of sand in their energetic labors, and working gradually towards the center of the ring. A little rest was taken during the great heat of mid-day, and in the evening the eggs were carried to the huts in baskets. By the end of the second day, the taboleiro was exhausted; large mounds of eggs, some of them four to five feet in height, were then seen by the side of each hut, the produce of the labors of the family.

In the hurry of digging some of the deeper nests are passed over; to find these out the people go about provided with a long steel or wooden probe, the presence of eggs being discoverable by the ease with which the spit enters the sand. When no more eggs are to be found, the mashing process begins. The egg, it may be mentioned, has a flexible or leathery shell; it is quite round, and somewhat larger than a hen's egg. The whole heap is thrown into an empty canoe and mashed with wooden prongs; but sometimes naked Indians and children jump into the mass and tread it down besmearing themselves with yolk and making about as filthy a scene as can well be imagined [fig. 5.17]. This being finished, water is poured into the canoe, and the fatty mess then left for a few hours to be heated by the sun, on which the oil separates and rises to the surface. The floating oil is afterwards skimmed off with long spoons, made by tying mussel-shells to the end of rods, and purified over the fire in copper kettles.

The destruction of turtle eggs every year by these proceedings is enormous. At least 6000 jars, holding each three gallons of the oil, are exported annually from the Upper Amazons and the Madeira to Pará where it is used for lighting, frying fish, and other purposes. It may be fairly estimated that 2000 more jars-full are consumed by the inhabitants of the villages on the river. Now, it takes at least twelve basketsful of eggs, or about 6000, by the wasteful process followed, to make one jar of oil. The total number of eggs annually destroyed amounts, therefore, to 48,000,000. As each turtle lays about 120, it follows that the yearly offspring of 400,000 turtles is thus annihilated (Mittermeier (1978), again using reproductive data from modern populations, suggested that this amount of oil and eggs was probably the result of the reproductive output of closer to 600,000 females). A vast number, nevertheless, remain undetected; and these would probably be sufficient to keep the turtle population of these rivers up to the mark, if the people did not follow the wasteful practice of lying in wait for the newly-hatched young, and collecting them by thousands for eating; their tender flesh and the remains of yolk in their entrails being considered a real delicacy. The chief natural enemies of the

Figure 5.17
Conibos squashing giant South American river turtle eggs for oil (St. Cricq, 1874).

> turtle are vultures and alligators, which devour the newly hatched young as they descend in shoals to the water. These must have destroyed an immensely greater number before European settlers began to appropriate the eggs than they do now. It is almost doubtful if this natural persecution did not act as effectively in checking the increase of the turtle as the artificial destruction now does. If we are to believe the tradition of the Indians, however, it had not this result; for they say that formerly the waters teemed as thickly with turtles as the air now does with mosquitoes. The universal opinion of the settlers on the Upper Amazon is, that the turtle has very greatly decreased in numbers, and is still annually decreasing (Bates, 1863).

In spite of the continuing harvests of this magnitude and early concerns over declining populations, giant South American River turtles's populations were still enormous in many locations during the mid-19th Century. Even as late as the 1850s it was still common enough to impede river traffic during nesting aggregations on the Madeira River, and the clacking of their shells as they collided with one another could be heard for great distances (Parsons, 1962; Coutinho, 1868 in Smith, 1975). Another observer (Mathews) on the same river in 1868 is quoted thus: "For miles as far as the eye can see, which hereabouts runs straight for some six or seven miles, were continuous rows of turtles at the water's edge: the rows being eight or ten deep, many thousands must have collected together" (Smith, 1975). In 1874, 4000 giant South American river turtles were collected from only one beach (Tamanduá) on the Madeira River (Keller, 1874 in Smith, 1975).

Resilient as the populations obviously were, the effects of such excessive, unrelenting exploitation inevitably led to the collapse of the populations upon which it was based. Smith (1975) provided an overview of the oil traded annually and an estimate of the eggs destroyed in the process from thirty-four Amazonian locations over an approximately 203-year time span (1700s to 1903). The 1860s harvest from the Madeira River populations referred to above apparently represented the zenith of destruction, with 48 million eggs destroyed to produce some 8000 pots of oil annually during this decade. Although the egg harvests and oil production continued at impressive levels since, the backbone of the industry had been broken and nothing approaching the 1860s slaughter would ever be seen again. Smith's (1975) figures indicate that some 214,400,000 giant South American River turtles eggs must have been destroyed during these two centuries.

There were still enough turtles in the mid-twentieth century to support a brisk but ever-declining industry. Mosqueiro Manso (1945, 1960 in Pritchard & Trebbau, 1984) described its organization on the Orinoco's Pararuma nesting beaches. The structure and divisions of labor of turtling here were summarized by Prichard and Trebbau (1984) and recounted in detail in chapter 4.

Although the collection of eggs on Orinoco beaches was prohibited in 1946, the collection of nesting females was legal until 1962 in Venezuela (Pritchard & Trebbau, 1984). Ramirez (1956) estimated that 11,407 turtles (or nearly 30% of all nesting females) were taken annually between 1950 and 1956. After the ban, illegal harvesting continued at high levels until placement of armed guards on critical beaches reduced the poaching to a trickle (Pritchard & Trebbau, 1984).

The effect of the ban on harvesting of nesting females and their eggs is difficult to determine, but indications of continued declines in nesting on the major Orinoco beaches are not encouraging (Pritchard & Trebbau, 1984). Ojasti (1967) reported a decline in number of nests on Playa del Medio from 34,300 in 1963 to 13,800 in 1965, while another observer (Rastillo, in Pritchard & Trebbau, 1984) reported only 4695 nests at this beach and only thirty at Pararuma in 1981. While annual fluctuations in numbers of nesting females are often observed in colonially nesting species, such as sea turtles, there is no doubt that over the long term the Orinoco populations have declined precipitously (fig. 5.18). Today, there is no organized commercial fishery for giant South American River turtles in Venezuela, although many are still killed for subsistence and local market use (Pritchard & Trebbau, 1984; Licata & Elquezabal, 1997).

Elsewhere within the species' range the populations are similarly depleted, and with some exceptions they are rare or absent from vast areas of their former Amazonian range. Although Brazil, Colombia, and Peru have long had legislation on the books protecting giant South American River turtles, there is still much subsistence harvesting and illegal commerce occurring (Mittermeier, 1975, 1978; Smith 1975, 1979; Shoumatoff, 1978; Groombridge & Wright, 1982; Klemens & Thorbjarnarson, 1995). Fortunately, several factors have contributed to the continued survival of the species as a whole, and though much reduced in

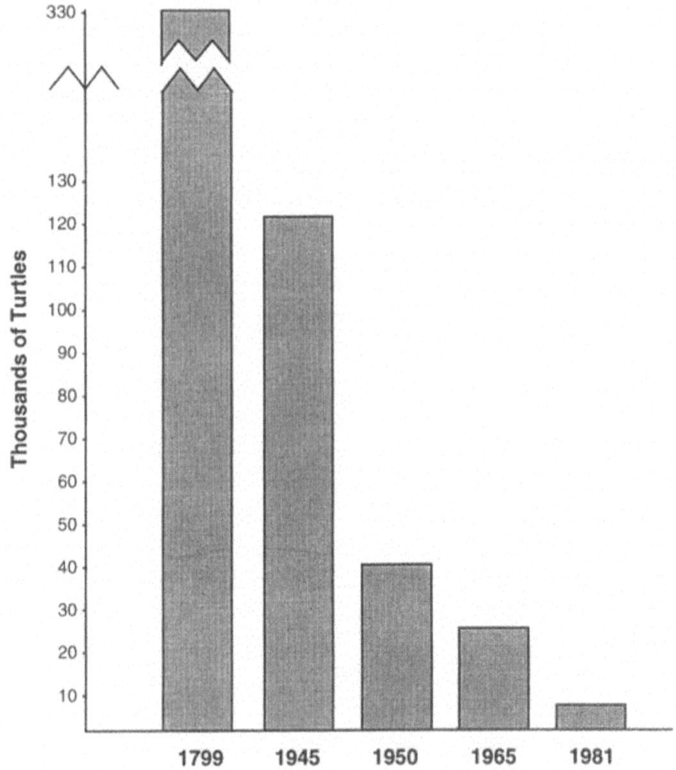

Figure 5.18
Reduction of giant South American river turtle nesting populations on Orinoco River nesting beaches since 1799. Data from Groombridge and Wright (1982). (See text for further explanation.)

numbers, the likelihood of immediate extinction is small (Groombridge & Wright, 1982). Smith thought the increasingly widespread availability of kerosene and vegetable oils as substitutes for declining and increasingly expensive turtle oil during the latter nineteenth century probably saved giant South American River turtles from complete extinction at that time. Across its huge range several large nesting colonies—such as Brazil's Rio Trombetos, Tapajós, Xingu, Purus, Jurua, Guaporé, Branco and elsewhere—still exist which collectively produce millions of hatchlings (Groombridge & Wright, 1982; Vogt, in press). Viable populations also exist in Colombia, Peru, the Guianas and perhaps other peripheral parts of its range (e.g., Bolivia). In Ecuador the species is present but rare as a result of exploitation and habitat damage (Velasco, 1995). The increased protection of a number of these important areas by governmental agencies such as IBAMA, the conservation arm of the Brazilian government and by designated reserves (e.g., Cahuinari National Park in Colombia, Cayabeno National Park in Ecuador, Pacaya-

Samiria National Reserve in Peru) provide encouragement concerning their prospects (Groombridge & Wright, 1982). It seems unlikely, in the face of modern pressures on the Neotropics, that the giant South American river turtle will ever be able to regain its former dominance in these great rivers, however. The once incredible abundance of this species could not protect it from the equally incredible levels of exploitation sustained for centuries, and its modern diminution is solely attributable to the hand of man.

River turtles have traditionally been of commercial importance in much of Latin America. They remain so today, although as previously discussed in reference to South America's giant South American river turtles populations, the magnitude of trade in many areas has dwindled as the turtle populations supporting it have declined.

In southern Mexico and northern Central America (Belize and Guatemala), the Central American River turtle has been the primary riverine species in the trade. Like the river terrapin and giant South American River turtles, it is a large, herbivorous species with succulent meat and eggs which can be very abundant where not heavily persecuted. It has been heavily exploited over much of its range and is now rare in most areas as a result, however (Alvarez del Toro et al., 1979; Groombridge & Wright, 1982; Polisar, 1995; D. Moll, 1986; Vogt et al., in press).

In Mexico overexploitation for restaurant use is the greatest problem today (Vogt et al., in press). In 1991, Central American river turtles sold in Villahermosa for $4.00/kg live weight or about $40 for an average-sized female, and prices continue to rise (Vogt et al., in press). With such high prices, exploitation intensity has continued to rise accordingly and expand geographically (across state and probably national borders into Belize and Guatemala) to supply demand as populations have declined.

In Belize, commercial exploitation is also the species' main threat, although their market price is considerably cheaper than in Mexico (approx $10–12/kg) (Moll, 1986; Polisar, 1995; Vogt et al., in press). Their populations' status here ranges from poor to good depending upon their location and accessibility, degree of protection, and cultural variables (D. Moll, 1986; Polisar, 1995; Vogt et al., in press). In both Mexico and Belize the meat is most sought after for feasts associated with Lent and Easter, respectively. As both occasions occur during the spring dry season when the Central American River turtle are most concentrated and easily accessible, very large numbers can be harvested for the market at this time. However, the numbers seen in city markets may not necessarily reflect the true harvest, as the animals sent to market may actually reflect only surplus animals remaining after subsistence and village commerce needs have been satisfied (Polisar, 1995).

Although the species is commercially traded in Guatemala (Lee, 1969; Vogt et al., in press), the species and trade characteristics have not been studied as thoroughly as in Mexico and Belize. It is known that many Guatemala City restaurants feature a soup made from *Dermatemys*, and that soldiers stationed in the Petén region utilize it for meat (Vogt et al., in press). There are indications that it is being seriously overexploited here also, and surveys to determine its status and trade levels are urgently needed.

Other freshwater turtles may also be sold in Mexican and Central American markets as food items. Mittermeier (1971) found various mud turtles, narrow-bridged mud turtles, sliders, common snapping turtles and Mexican giant mud turtles for sale mainly in Veracuz and Tabasco markets, in Mexico. Mittermeier (1970) also observed various mud turtles, *Rhinoclemmys* (i.e., a neotropical batagurid genus), sliders and giant mud turtles sold as food in Nicaragua, El Salvador, and Guatemala markets. Carr (1952) noted that mud turtles were relished by Nicaragua's Mosquito Indians. D. Moll (1986) observed sliders and Mexican giant mud turtles commonly for sale as food in Belize City in 1983 and 1984 (fig. 5.19a,b). While in general, the status of these genera and species is thought to be good compared to that of the Central American river turtles, there has been no comprehensive study of the effects of harvesting on any of them to date. There is some concern that exploitation for food in conjunction with other factors may be threatening some slider populations in parts of its vast range in Mexico and Central America (Smith & Smith, 1979; Moll et al., in press).

In the Caribbean, several of the Islands contain sliders composed of seven or eight species closely related to the mainland slider (*Trachemys scripta*) (see chapter 3, this volume). On the Greater Antilles (Cuba, Jamaica, Hispaniola, and Puerto Rico), the resident species are often river dwellers (Barbour & Carr, 1940; Seidel, 1988). All of these have undoubtedly served as subsistence food at one time or another (see Barbour & Carr, 1940), but several have also been considered gourmet food and have been briskly marketed. As early as 1756, Browne (in Barbour & Carr, 1940), said of Jamaica's resident species the Jamaican slider (*Trachemys terrapin*), "It is often served up at gentlemen's tables in that island and looked upon as delicate wholesome food by many people." Danforth (1925 in Barbour & Carr, 1940) said that egg-bearing females of the Puerto Rican slider (*T. stejnegeri*) were collected on Puerto Rico as they emerged to lay, and were sold as food. Barbour and Carr (1940) noted that Cuban sliders (*Trachemys descussata*) were surprisingly rare in some suitable habitats where they had long been hunted for food. They were then most abundant in the deep pools of the island's many sluggish rivers. More recently Seidel (paraphrased by Moll, 1984b) found that sliders were being heavily commercially exploited for food on Hispaniola, and becoming rare as a result. Indeed, some persons made their entire living from selling these turtles in the market. Seidel (1988) noted that both Jamaican slider and Puerto Rican sliders appeared to be uncommon in most areas, but whether exploitation was a factor or not was unclear.

In South America the past exploitation of the large riverine species giant South American River turtle has already been recounted in detail. Although not in danger of extinction, and in fact common or even abundant today in a few areas (Pritchard & Trebbau, 1984; Alho, 1985; Pritchard, 1986; Vogt, in press), it has been severely depleted or eliminated in much of its former range. The main commercial effect of this change in status, and in some areas due to better protection, is its

(a)

(b)

Figure 5.19
(a) Live tropical sliders (*Trachemys scripta venusta*) and (b) a Mexican giant musk turtle (*Staurotypus triporcatus*) awaiting sale in the Belize City Market, Belize.

transition from a dietary staple of the common people to an expensive delicacy reserved for special occasion meals and wealthy gourmets (Smith, 1975; Alho, 1985). Although giant South American River turtles and the other *Podocnemis* species are nominally protected throughout much of their Amazonian range, enforcement is often inadequate and smuggling to market is widely practiced and lucrative (Smith, 1975). Smith (1975) said that in Manaus fifty years ago large turtles could be bought for one cent each, but at the time of writing smuggled animals could be sold for $30 each. Shoumatoff (1978) and Smith (1979) quoted its value at U.S. $60 (Itacoatiara) to U.S. $100 (Manaus) at that time. Undoubtedly prices are even higher today. Johns (1987) said that giant South American river turtles fetched prices between U.S. $40 and U.S. $130 (mean = $60) in Tefé, Brazil in 1985, dependent on size, while yellow-spotted Amazon river turtles might sell for U.S. $7 to U.S. $14 per turtle dependent upon supply. Turtle eggs sold for between U.S. 13 and U.S. 34 cents. Johns said prices for both turtles and eggs were two-to-three-fold higher in Manaus than in Tefé, and those in southern Brazil were higher still.

Other predictable effects of the over-exploitation of giant South American River turtles have been the expansion of collecting areas away from marketing centers as local stocks have declined (Shoumatoff, 1978; Smith, 1979), and the substitution of smaller, less desirable size classes or species, or formerly less easily collected species such as yellow-spotted Amazon river turtles and six-tubercled Amazon River turtle as giant South American River turtles declined in availability (Mittermeier, 1975, 1978; Smith, 1979; Pritchard & Trebbau, 1984; Johns, 1987; Klemens & Thorbjarnarson, 1995). Smith (1979) estimated some 8000 yellow-spotted Amazon river turtles were being caught annually within a 60-km radius of Itacoatiara, with about 6000 being landed in town, and half of these smuggled on to Manaus where they fetched higher prices. In Itacoatiara a 7-kg female sold for about U.S. $10, while a 2-kg male might bring U.S. $1.00 at that time. The same-sized animals were worth four or five times that in Manaus (Smith, 1979). Smith (1979) thought that six-tubercled Amazon River turtles were being caught in about the same numbers, but were not traded much because they were small (about 1 kg) and worth less money (then about U.S. $1.00). The majority caught were consumed in rural areas and small villages (Mittermeier, 1975; Smith, 1979). Klemens and Thorbjarnarson (1995) stated that near Tefé, Brazil both giant South American River turtles and yellow-spotted Amazon river turtles have been reduced to small populations of little commercial importance by chronic exploitation. Now, the smaller six-tubercled Amazon River turtle is being regularly captured nearby in large numbers for sale in Tefé. One shop owner reportedly sold some 12,000 roasted six-tubercled Amazon River turtles a year.

Other *Podocnemis* species may also be captured for food, and occasionally for commercial trade within their South American ranges. Savanna side-necked turtle of the Llanos region of Colombia and Venezuela may serve as an important protein source for humans (Mittermeier, 1978). Even the small red-headed side-necked turtle may be exploited heavily in some areas (Mittermeier & Wilson, 1974; Alho, 1985).

The chelid side-necks of South America tend to be smaller than many of the podocnemids and not generally as attractive for commercial purposes. They have not been entirely ignored, however, and some may be heavily exploited and traded in certain areas (e.g., Geoffroy's side-necked turtle in Colombia, Medem [1960]; common toad-headed turtles (*Phrynops nasutus*) near Iquitos, Peru, Dixon and Soini [1977]). Smith (1979) stated that the mata mata and the twist-neck turtle (*Platemys platycephala*) may also be used as food in Amazonia, but they are seldom marketed.

Of the continents' emydid genera, the slider is the most heavily exploited. In Venezuela and Colombia, the Colombian slider (*T. s. callirostris*) is in decline due in part to its intense utilization (especially egg-bearing females) for feasting during Holy Week celebrations (Pritchard & Trebbau, 1984). The Venezuelan slider (*Trachemys s. chichiriviche*) is also moderately to heavily exploited in Venezuela. Maracaibo wood turtles (*Rhinoclemmys diademata*) and spotted-legged turtles (*R. punctularia*) are eaten, but are probably of negligible commercial importance (Pritchard & Trebbau, 1984).

The small scorpion mud turtle (*Kinosternon scorpioides*) is a popular food item in restaurants in Belém, on Marajo Island, and elsewhere in the Amazon Delta region where it is common (Smith, 1979; Alho, 1985). In Pará restaurants it is often sold as "Casquino de mucua," where it is cooked in the shell (Freiburg, 1981).

Africa and Australia

We have little information concerning the commercial use of river-dwelling aquatic turtles for food in Africa, although there is little doubt that it occurs here as on the other continents discussed. The absence of really large mass-nesting river turtles in Africa may inhibit human dependence on such food sources to the extent we have seen in other geographic areas (Iverson & Richmond, 1992; Thorbjarnarson et al., 2000). With the notable exception of Kuchling's (1988) surveys for *Erymnochelys* in Madagascar, there has been little recorded in recent years concerning the status of the African aquatic turtle fauna. Loveridge (1941), Loveridge and Williams (1957), Villiers (1958), and Pritchard (1979) provide some information concerning the utilization of several native softshells, African side-necks, and pond turtles as food (and by whom), but the distinction between its use as subsistence fare and/or commercial product was seldom provided. Branch (1988), Boycott and Bourquin (1988, 2000), and Pritchard (2001) provide the most recent accounts of the status of some of these species. Loveridge and Williams (1957) referred to an old observation by Strauch of hundreds of *Mauremys leprosa* on sale in Algiers fish markets, however. The level of its modern use is not known. Surveys of the current food trade in turtles are obviously required everywhere across the continent in order to assess the level of exploitation and its effects on its resident species.

In most other locations besides those discussed above, the commercial food trade in resident river turtles is limited by a paucity of turtles (e.g., northern Eurasia) or by effective laws and/or cultural considerations (e.g., Australia). Australians of European descent once utilized the Murray River short-necked tortoise (*Emydura macquari*) in the restaurant trade, but the duration of the trade was limited (Goode, 1967). Waite (1929 in Goode, 1967) stated: "I am told that though both tortoises [*Chelodina longicollis* and *Emydura macquarii*] are eaten by natives, *Emydura* only is relished by Europeans, and at one time when the chelonians were more plentiful in the vicinity of our towns than they are today, the Murray tortoise was regularly caught and consigned to Adelaide where it was sold to restaurants and served as turtle soup. *Chelodina*, was on the other hand not utilized in any way because 'it stinks horribly'."

Exploitation for Purposes Other than Food

Medicinal Use

It is not widely recognized in the West that turtles, including many river dwellers, are valued in other parts of the world (especially Asia) as the source of an impressive variety of medicines and tonics. Nevertheless, a huge number of turtles are traded, imported, and exported each year primarily for this purpose. Much of the food trade cited in the previous chapter may actually serve this apothecary function instead of, or in addition to, its listed primary function. Records of actual use patterns of purchased turtles are incomplete, and purchased turtles could be used to fulfill both ends simultaneously, thereby blurring the distinction between the two. Choudhury and Bhupathy (1993) pointed out that since Indian turtle remedies are often derived from the shell, turtles collected and sold for meat may often serve both functions. Alternately, there is some evidence of collection of Asian turtles for shell only, with the meat discarded or used as food in crocodile farms (Jenkins, 1995).

The medicinal use of turtles has a long history, and turtle-derived drugs have been used by the Chinese for treating ailments since at least 2737 B.C. (E. Moll, 1982). Zhao (1995) stated that the plastron of hard-shelled turtles and carapace of softshells has been used in traditional Chinese medicine (i.e., TCM) since at least the Han Dynasty (202 B.C. to 9 A.D.). Rome's Pliny the Elder in *Natural History* provided sixty prescriptions based upon tortoise flesh and organs (Goode, 1967). *Unani*, the Greco-Arabic system of medicine which forms an important foundation for medicinal practice in India and many other modern countries, prescribes the use of various turtle body parts for curing human ailments (Vohora & Khan, 1979, in Das, 1991). Das (1991) listed numerous turtle body parts, eggs and fluids which were believed to have healing properties, including ash from burnt shells (for cancer, tuberculosis, infantile diarrhea, and respiratory ailments), bone ash (for rectal prolapse, ringworm, and uterine discharge), flesh (for fighting inflammation, as an aphrodisiac, and for curing menorrhagia), turtle fat (applied externally for treating convulsions and improving sexual performance), blood (for rheumatism, gout, and epilepsy), and turtle bile mixed with honey (for eye diseases and epilepsy, and by itself for wound healing and diphtheria). Turtle eggs were eaten to cure chronic coughs in children. Folk medicines of indigenous peoples of India and other countries often involved turtles also. Das (1991) listed the uses of gallbladder (for intense stomach aches), ovaries (to lower high blood pressure, especially during childbirth), powdered bones (to exorcise spirits), and ash from burnt plastrons (sprinkled on wounds to hasten healing).

Choudhury and Bhupathy (1993) thought that there was a very low level of exploitation of turtles in India for medicinal use exclusively (the turtles used having been primarily captured for food), but that secondary medicinal use of turtle shell by various indigenous groups was widespread across the country. They stated that the riverside people and fishermen along the Narmada, Tapi, Mahi, Godavari, Krishna, Mahanadi, and Brahmaputra rivers used turtle eggs medicinally also. Choudhury and Bhupathy (1993) listed five aquatic species, three of which are the exclusively riverine species *Aspideretes gangeticus*, *A. hurum*, and *A. leithii* (the others are the generalist Indian flapshell and *Melanochelys trijuga*) which are used medicinally by various tribal groups (fig. 5.20). The shells (burnt and ground with

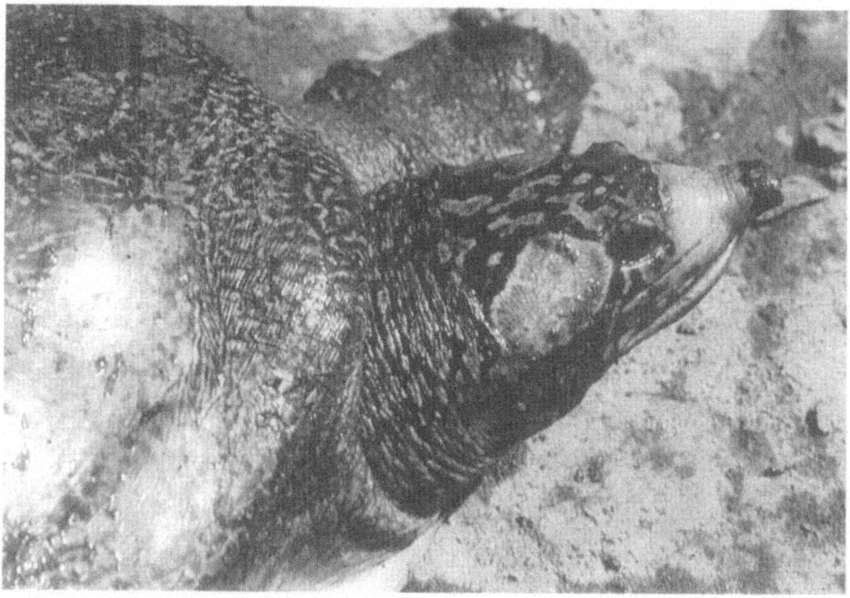

Figure 5.20

The peacock softshell (*Aspideretes hurum*), a heavily marketed Asian species used for both food and traditional medicine.

oil), blood and gallbladders from these species are particularly important in the treatment of such ailments as skin diseases, stomach disorders and diarrhea, and cuts and burns. E. Moll (1982) and Das (1991) also noted that Indian flapshell is widely sold in India for use in treatment of tuberculosis.

The greatest consumers of turtles for medicines and tonics are the Chinese (E. Moll, 1982; van Dijk et al., 2000a). As resident species have been depleted, and increasingly protected by laws (Zhao, 1995), China has become an importer of turtles in huge quantities from many other countries including the United States, as discussed above. It should also be noted that other countries with large Chinese populations (e.g., Taiwan, Singapore) and other Oriental countries with a tradition of turtle use for food and medicines (e.g., Japan, South Korea) are large importers (E. Moll, 1982; Warwick, 1986; Das, 1990; van Dijk et al., 2000a; D. Moll et al., in press). As mentioned previously, there is often incomplete knowledge of the exact proportions of the imported animals that are utilized for medicinal purposes (or as food or pets), but the use of many primarily for the former purpose is indisputable (E. Moll, 1982; Le Dien Duc & Broad, 1995; van Dijk et al., 2000a). Zhao (1995) listed sixteen species of turtles and tortoises which were imported into China as "recently living" specimens or shells (the latter almost certainly destined for medicinal use) from neighboring countries. All but three of the species listed were aquatic or semi-aquatic forms, and five of these are largely or totally lotic species (crowned river turtle, Indian roofed turtle, Asiatic softshell, peacock softshell, and Malayan softshell).

The issue of usage patterns is clearer in Taiwan where few wild caught turtles are used as food. Here, almost all turtle products imported are destined for medicinal use (Chen et al., 2000). Taiwan may be the largest consumer of turtle shells for medicinal use with 958 metric tonnes of hard-shelled turtle shells and 228 metric tonnes of softshelled turtle shells imported from mainland China and Southeast Asia (mainly from Indonesia) between 1992 and 1998, an estimated value of U.S. $3.2 million (Chen et al., 2000). An unknown, but obviously huge number of wild turtles continue to be slaughtered to supply this demand. The proportion of river turtles meeting the demand for hard-shells is unknown, but essentially all of the softshell shells are from riverine species.

E. Moll became concerned about the problem of the apothecary trade in turtles after reading an article by Shibata (1975 in Moll, 1982) which indicated that five non-Chinese turtles (Asian leaf turtles [*Cyclemys dentata*], giant Asian pond turtles [*Heosemys grandis*], Malayan snail-eating turtles [*Malayemys subtrijuga*], elongated tortoise [*Indotestudo elongata*], and impressed tortoise [*G. impressa*]—probably from Thailand) were being imported into China to provide meat, gall, blood, and urine for various medicines, but primarily for plastral scutes used to make "Gui Ban." This is a balm or jelly produced from heating the plastral scutes of chelonians and is used to treat a variety of afflictions including cancer, tuberculosis, leukorrhea, menorrhagia, chronic malaria or dysentery, lethargy, piles, rickets, and failure of a child's skull bones to grow together. A similar drug called "Biejia" is produced from the carapace of *Pelodiscus sinensis*. E. Moll

(1982) recorded that, over a six-month period, 154,146 kg (approximately 30,000 turtles) were imported into Hong Kong from Thailand (mainly) and Bangladesh presumably to mainly supply medicinal demand. Moll noted that Gui Ban is also sold in Japan (under the name "Kiban") and is probably an ingredient of pep pills made from South-East Asian turtles that are marketed in that country. Lovich et al. (2000) noted that the Chinese refer to the Malaysian giant turtle as the "thousand year turtle," and it is specifically sought and eaten by people in order to extend their longevity. As discussed above, huge numbers of this species are exported from Sumatra to meet Chinese demand.

Le Dien Duc and Broad (1995) also listed additional diseases which are treatable with turtle-based medicines. These include acting as a tonic to yin (such as kidney-yin deficiency), faint and weak voice, afternoon heat spells, nocturnal sweats, lumbago, weak sinews, swollen or infected pancreas, and tumors. In addition they noted that Chinese athletes and swimmers consume turtle blood as part of their preparation for competition. Le Dien Duc and Broad (1995) considered that the fate of a significant but unknown proportion of the turtles imported into China and Vietnam was to be used for medicinal purposes. They also noted internal marketing of chelonians (especially tortoises) in Vietnam in far lesser amounts. In Hanoi, plastrons used for medicines sold for U.S. \$3–4, and carapaces (used as an additive to plastral remedies) for U.S. \$1.

The use of turtles and their eggs for various elixirs is prevalent in many other parts of the world also. The bizarre mata mata, for instance was used near the confluence of the Amazon and Tapajos Rivers as a remedy for rheumatism (Orton, 1870 in Smith, 1979). In many tropical countries, the eggs of many species are eaten for their reputed aphrodisiacal and/or medicinal qualities, as well as their food value. The decline in Malaysia's river terrapin and painted terrapin populations is largely due to the unrelenting harvesting of their eggs for these purposes (e.g., Moll, 1980a,b; Sharma, 1995, 1999) (fig. 5.21). Sharma (1999) provided a summary of peninsula Malaysian species and their use as food or medicine.

Wherever Chinese communities exist, the sale of turtles for food and medicine will probably occur, as noted by Garber (1990). The medicinal use of turtle in the United States Chinese communities probably extends back to their beginnings. Clark and Southall (1920) wrote that the only local buyers of the expensive diamondback terrapins shipped to Chicago and St. Louis from the East were the Chinese, who used them mainly for medicinal remedies designed to clear the blood and to treat tuberculosis and other internal problems. When diamondbacks were unavailable or too expensive, the local "Texas terrapin" (*Trachemys scripta elegans*) was substituted for diamondback in the tonic recipes (Clark & Southall, 1920).

Another North American species, the common snapping turtle, was once thought to be medicinally useful by residents of Massachusetts who collected its oil for treatment of bruises and sprains (Evermann & Clark, 1920). In a rather dubious endorsement, it was considered to be a therapeutic agent as worthy as goose, rattlesnake, and skunk oil.

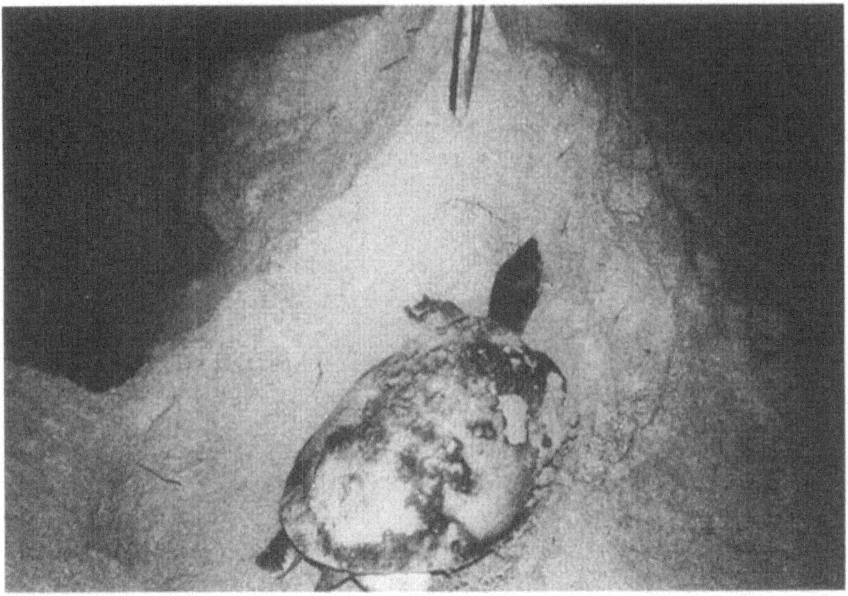

Figure 5.21
A river terrapin nesting on a Perak River nesting beach. Its eggs will be harvested for food and for their supposed aphrodisiacal value.

The Pet Trade

Turtles, especially hatchlings and small colorful species, have long been popular pets. Young children in Belize keep locally collected black turtles and baby Central American river turtles, while in Tanzania small helmeted turtles (*Pelomedusa subrufa*) are housed in tanks made from cut-out oil drums (D. Moll, pers. obs.) (fig. 5.22). The appeal is universal, and in more developed countries where extra spending money is more often available (and perhaps local stocks of turtles less so) the sale of pet turtles, often imported from distant countries, is big business. River turtles have figured heavily in this trade in the past and continue to do so today in many parts of the world (e.g., van Dijk et al., 2000a).

The United States has traditionally been the largest supplier of pet turtles for the international pet trade, and is itself an important hobbyist market for attractive native species (currently those above hatchling size) and exotic imports. The roots of the industry stem from the establishment of so-called turtle farms in the south (especially Louisiana) during the 1950s to supply the growing domestic demand for hatchlings as pets. The "farms" (which actually depended upon annual restocking of reproductive stock from wild populations) raised and shipped tens of millions of red-eared slider hatchlings (*Trachemys scripta elegans*) to the United States, and later, to overseas markets until 1975, when the U.S. Food and Drug

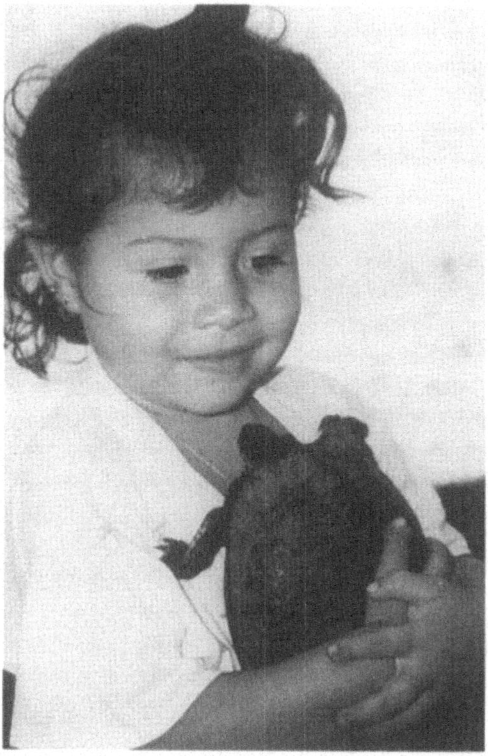

Figure 5.22
A young girl with her pet black turtle (*Rhinoclemmys funerea*) in Belize. (Photo by Kevin P. Jansen)

Administration banned the domestic sale of turtles under 4 inches (10 cm) due to their role in spreading salmonellosis (Kennedy, 1969; Tauxe et al., 1985; Feehan, 1986). Although slider hatchlings always dominated this trade, it is our recollection (having admittedly been enthusiastic buyers ourselves during this unenlightened period) that numerous other species' hatchlings—both native and exotic—were also increasingly available for sale. Painted turtles, various map turtles, false map turtles, river cooters, spiny softshells, snapping turtles, and occasionally other native species were often sold. During the 1960s and early 1970s a flood of *Podocnemis* hatchlings (mainly yellow-spotted Amazon river turtles, giant South American River turtles and red-headed side-necked turtles) entered the United States, European, and also Japanese pet markets (Weaver, 1973; Mittermeier, 1978; Pritchard & Trebbau, 1984). Weaver (1973) provided a detailed account of the dynamics and inhumanity of the trade (accompanied by photographs of thousands of dead and dying yellow-spotted Amazon river turtle hatchlings which were confiscated and subsequently incinerated). He described three shipments of yellow-spotted Amazon river turtles of 28,000, 20,150, and 33,275 hatchlings, respectively (the latter two shipments destined for Japan), which were detained by U.S. wildlife officials in just one month (April) in 1971. Other exotic hatchlings, such as South American sliders and Chinese three-keeled pond turtles (*Chinemys reevesi*), were also regularly available. It should be emphasized that all of the species referred to above were routinely available to the most casual pet keeper—sold cheaply in large numbers in the "dime stores" and local pet shops to any child with about 79 cents in his or her pocket. Almost all were destined to die swiftly (and be replaced) from malnutrition (the commercial turtle food of the day consisted essentially of ants' eggs or dried flies), cold, and neglect. The more sophisticated turtle hobbyists of the day were able to select from a much more diverse international menu of species (in 1970–71, as many as eighty; Humane Society of the United States [HSUS], 1994) sold as pets through mail order from an impressive number of well-stocked reptile-supply companies.

Australia was another location with a significant pet trade in native species during the 1960s (Goode, 1967). Goode (1967) noted that adult *Chelodina longicollis* was the commonest species for sale in the eastern states, and occasionally *Emydura macquarii* was available also. In tropical north Queensland, *Elseya latisternum*, *Emydura krefftii*, and *Chelodina novaeguineae* were also offered for sale. Goode observed that babies were not often offered for sale, but occasionally several hundred would appear at once which were "snapped up" for the then exorbitant price of A. $2.50–4 each. Depending on a number of variables, adults usually sold for less. Interestingly, some of Australia's yet undescribed or newly described freshwater tortoises, such as the Mary River tortoise (*Elusor macrurus*), were first recognized as unique from animals which turned up in the pet trade (e.g., Cann & Legler, 1995). Laws now limit the collection and sale of turtles in most of Australia (Cann, 1978, 1998; see Lunney & Ayers, 1993).

In spite of the United States ban on the sale of small turtles (also instigated in several other countries), the turtle farms have rebounded to supply an enormous overseas market for pet turtles worth many millions of dollars. The magnitude, dynamics, and effects of the trade have been summarized (Warwick, 1985a,b, 1986; Feehan, 1986; Hardy, 1988; IWC, 1988; HSUS, 1994; Hoover, 1998; Salzberg, 1998a; Thorbjarnarson et al., 2000; Telecky, 2001; D. Moll et al., in press). Once again, the trade is based principally upon red-eared slider hatchlings which are shipped by the millions (i.e., 52,122,389 between 1989 and 1997 [Salzberg, 1998a; Telecky, 2001] or more than 80% of the total number of live reptiles leaving the United States [Hoover, 1998]) to pet markets mainly in Europe and Asia, but also to many other lesser markets. For example, the 1993 export destinations for 6,308,938 red-eared sliders listed by the U.S. Fish and Wildlife Service were South Korea (21.9%), Italy (12.3%), Japan (10.6%), Spain (8.9%), France (8.3%), Singapore (8.3%), Mexico (7.6%), Hong Kong (7.5%), and 31 other smaller markets combined (14.6%). Salzberg (1998a) also summarized the distinctions of 7,852,050 red-eared turtles exported from the United States between 1994 and 1997. Markets continually shift as countries become wealthier and pet-keeping becomes fashionable, and close down as countries develop their own turtle forms to supply internal pet markets (e.g., Taiwan) or restrict further importation (e.g., South Korea) (van Dijk et al., 2000b; Chen et al., 2000; Ranson, 2002). While we are concerned with the humane issues involved with the shipping of such huge numbers of turtles to an almost certain early death (often in transit), there are several other even more serious consequences associated with the trade. First, the "turtle farms" are still not closed operations and should be more correctly thought of as ranching operations where depleted stock is regularly replaced from wild populations on a regular basis. Warwick (1986) estimated that 100,000 or more wild adults (mainly adult females) are collected annually to replace breeding stock that die annually in the fifty or so existing farms. This chronic harvest has apparently reduced populations severely in collection areas (Warwick, 1986; Warwick et al., 1990; Close & Seigel, 1997), reduced the proportion of large adult females (Close & Seigel, 1997; Warwick et al., 1990), and forced collectors to exploit ever more distant habitats to supply demand (D. Moll et al., in press). The effects of the removal of such large numbers of turtles from their respective habitats in terms of future species survival and ecosystem function are still unclear, but worrisome (Warwick et al., 1990; Congdon et al., 1993, 1994; Close & Seigel, 1997; Tucker & Moll, 1997; Moll & Moll, 2000; chapter 2, this volume). Another area of concern is the effect of feral red-eared sliders, and potentially other introduced turtle species (e.g., Chinese softshells) on native turtles and ecosystems outside of their natural range in America and in countries which have imported them. Many countries around the world now host thriving populations of red-ears, and surveys and studies are in progress to determine the consequences of these introductions (e.g., Branch, 1988; Boycott & Bourquin, 2000; Green, 1991; Dupre, 1995; Ferri & Di Cerbo, 1995; Chen & Lue, 1998; van Dijk et al., 2000a; chapter 6, this volume). In Asia, feral Chinese softshells derived from the pet trade and farming operations are also of concern (van Dijk et al., 2000a).

Numerous other species of American aquatic turtles, including exclusively or mainly river-dwelling forms have been exported in large numbers from American locations (Hoover, 1998; Telecky, 2001). Based upon export records available from 1989 through 1997, map turtles (almost all composed of common and false map turtles) are particularly common in the United States export trade, as are American softshelled turtles, painted turtles, common snapping turtles, alligator snapping turtles, cooters and river cooters, red-bellied turtles (genus *Pseudemys*), and musk turtles (genus *Kinosternon* and *Sternotherus*). As the United States is also a major re-exportation center for live animals, it plays an important "middleman" role in the world-wide redistribution of exotic river turtles and other wildlife in the international pet trade (Hoover, 1998). As with the red-eared slider, it is assumed that almost all are destined for the foreign pet trade, but it is possible that some may also be used for food or potentially medicines (Warwick & Steedman, 1988). As the size of turtles shipped from United States ports is not regularly recorded, and the possibility exists that in the Far East even small turtles may occasionally be used for food, room for doubt exists concerning the fate of at least some of the consignment. The proportion of these other exported species which are ranched or which are obtained directly from wild-caught stock (as in box turtles, *Terrapene*, which have also been exported in large numbers for the pet trade) is not clear (HSUS, 1994). Very few or none of the aquatic turtles in the trade come from captive bred or truly farmed populations (HSUS, 1994).

There is also currently a very high level of interest in keeping turtles and other reptiles as pets in the United States (Hoover, 1998; Thorbjarnarson et al., 2000; Telecky, 2001). Consequently a multi-million dollar support industry in reptile housing, care and feeding and a thriving retail and mail order supply system exists to support it (HSUS, 1994). River turtles which enter the trade are virtually all wild-caught individuals (often illegally) and are obtained both domestically and from abroad. Particularly attractive species in the domestic trade are wood turtles (primarily a stream dweller in some seasons and parts of its range) and the several species of colorful riverine map turtles. Many of these are subject to "laundering"—they are illegally collected in one area then smuggled to another location where they may not be protected by law (perhaps because they do not naturally occur there) and then sold. Natural populations of these and other North American species are known or suspected to have been seriously reduced by such activity (Muir, 1984; Ernst, 1995; Thorbjarnarson et al., 2000). A 1994 study of the reptile pet trade listed thirty-eight exotic species of turtles and tortoises being sold by mail-order dealers in the United States—a drop from almost eighty different species imported into the United States between 1970 and 1971 when restrictive laws were less stringent (HSUS, 1994). The trade remains active, however. Hoover (1998) listed thirty-seven species of wild-caught tortoises and turtles (both domestic and exotic species), including numerous riverine forms available for sale in the United States between 1980 and 1996. Telecky (2001) listed the total number of turtles and tortoises imported into the United States between 1989 and 1996 as 5,709,972 animals. Currently a brisk trade in chelonians and other reptiles and amphibians is conducted over the internet (e.g., see www.kingsnake.com and www.chitraindica.com) (Salzberg, 2001).

Aquatic turtles outside the United States may be sold as pets internally or exported abroad as well. Choudhury and Bhupathy (1993) reported a modest commercial trade in India involving the aquatic species the Indian roofed turtle, Indian black turtles (*Melanochelys trijuga*), and possibly Assam roofed turtles (*Kachuga sylhetensis*). They thought the pet trade in India was on the increase and should be closely monitored.

In eastern and southeastern Asia, turtles have not traditionally been kept as pets to any significant degree, but with increasing affluence the trend is growing in several countries (van Dijk et al., 2000a,b). In Vietnam for instance, Le Dien Duc and Broad (1995) observed juvenile Malaysian snail-eating turtles (*Malayemys subtrijuga*) commonly sold to children as pets, and one American spotted turtle (*Clemmys gutatta*) for sale (at U.S. $100) in Ho Chi Minh City's Cau Mong turtle market (now closed). More recently, Hendrie (2000) found that some people in rural areas and urban centers such as Hanoi keep turtles, and Dong Xuan market in Hanoi carried several turtle species which were sold as pets. Sharma (1999) found twelve of twenty-six pet shops surveyed in several peninsular Malaysian cities to sell a few species of small native turtles and red-eared sliders as pets (these were not the so-called Chinese pet shops which sell turtles primarily for food and medicine). In Thailand, pet turtles are available in Bangkok and other cities' aquarium shops, and especially in Bangkok's vast Chatuchak market. D. Moll saw a variety of Thai and exotic species (including hatchling pig-nosed turtles and alligator snapping turtles) for sale in the latter part of 1998. A few wealthy collectors in Thailand are also willing to pay thousands of dollars for large specimens of Thai narrow-headed softshells—these are a status symbol as a result of their rarity (W. Kitimasak, pers. com.). At the other end of the spectrum, the pet trade is a huge business in Hong Kong and some other parts of China, and in Taiwan. A year-long East Asia TRAFFIC survey of Hong Kong markets in 1998–99 recorded eighty-four species of chelonians which were used as food and/or pets (Lau et al., 2000)—up from the fifty-eight species recorded in the trade in Hong Kong, Macau, and southern China in 1993–94 (Lau et al., 1996). Excluding red-eared sliders and Chinese softshells, which came from farms, eight of the top ten traded species were of Asian origin, while two were American imports (common snapping turtles and false map turtles). Of the eighty-four species as a whole, thirty were Asian species, and the other fifty-four species were mainly American species imported for the pet trade (Lau et al., 2000). While turtles are popular pets in Hong Kong, the practice is just beginning to catch on elsewhere in mainland China. There are many large farms which raise Chinese softshells for the food market, and some of these also raise other native and foreign species presumably for the (export?) pet trade. One example is the huge Tunchang turtle farm in Hainan Province which may house as many as fifty species and 50,000 individual turtles (Haitao & Parham, 2001). Although the Chinese softshell is most abundant here (about 30,000 turtles), other lotic species being bred include the endangered wattle-necked softshells (*Palea steindachneri*), Chinese big-headed turtles (*Platysternon megacephalum*), and American common and alligator snapping turtles.

In Taiwan, pet turtles are commonly kept, and a survey of Taipei pet shops conducted in 1995–96 found twenty-four species of tortoises and turtles for sale (including numerous riverine species from Asia, North America, and South America) (Chen et al., 2000). While large volumes of exotic chelonians are imported into Taiwan (Salzberg, 1998a), there are at least five turtle farms established here that also supply the pet trade with production capabilities of more than 30,000 hatchlings propagated per year (mainly red-sliders and Chinese striped-necked turtles [*Ocadia sinensis*]) (Chen et al., 2000).

Some Asian countries export native river turtles and other species to other regional and overseas markets for use as pets. A small number of rare and/or undescribed species are exported from China via Hong Kong for the international pet trade, and Malaysia contributes chelonians to the trade also (Sharma, 1999).

A huge illegal export trade in hatchling pig-nosed turtles originates from the southern Irian Jaya–Papua, New Guinea (PNG) border region where pig-nosed turtles nest in large numbers (Samedi & Iskandar, 2000; Rhodin & Genorupa, 2000). Local people harvest and incubate the eggs (estimated to be between 1,5000,000 and 2,000,000 per year in the Merauke regency of Irian Jaya), on both sides of the border, then transport the hatchlings to Merauke, where they are smuggled via Jakarta and other export centers to international pet markets (Rhodin & Genorupa, 2000; Samedi & Iskandar, 2000). PNG also has a number of other distinctive and/or endemic native riverine species such as the red-bellied short-necked turtle (*Emydura subglobosa*), New Guinea snapping turtles (*Elseya novaeguineae*), and red-bellied snapping turtle (*Elseya sp. 1*), Siebenrock's snake-necked turtle (*Chelodina siebenrocki*), Parker's snake-necked turtle (*Chelodina parkeri*), the New Guinea snake-necked turtle (*Chelodina novaeguinea*), and Pritchard's snake-necked turtle (*Chelodina pritchardi*), which are smuggled from PNG into the international pet trade also.

The Trade for Biological Specimens

Collecting turtles as biological specimens remains a big business in the United States. This also occurs in other countries (e.g., India), but its volume is poorly known.

In the United States a number of biological supply companies, including Quivira, Wards, Carolina Biological Supply, Frey Scientific, Fisher Scientific, Nebraska Scientific, Connecticut Valley Biological Supply, Nasco, Southern, Sargent-Welsh, and Wm. A. Lemberger Co., recently sold or currently sell live, preserved, and skeletonized and otherwise treated aquatic turtles by mail-order. The trade is large based upon red-eared sliders, but other sliders (e.g., *T. s. scripta*), river cooters and cooters, painted turtles, common snapping turtles, common map turtles, false map turtles, Ouachita map turtles and others are used also (D. Moll, pers. obs.). Universities and other training and research institutions are the main customers. Purchased turtles are utilized for laboratory instruction (e.g., physiology experiments, comparative anatomy dissections and demonstrations) and personal research. It is estimated that 400–500 turtles per year are utilized by each supply house (IWC, 1988). Turtles may be supplied by local collectors or from southern populations and turtle farms as more local supplies are depleted (Warwick & Steedman, 1988). Vogt (1981d, 1989) noted that many of Wisconsin's turtles (and many other vertebrates and invertebrates also) have been decimated by some of the big in-state supply companies without any sort of restrictions or reporting procedure required. He estimated that about 40,000–50,000 turtles were moved out of some of these places annually in the 1960s and 1970s. By the late 1970s they were importing turtles from South Dakota and the southern United States because depleted Wisconsin populations could no longer satisfy their demand. The strength of the companies' lobby in the state prevented regulation of the industry.

Besides the obvious drain on wild populations related to the trade, there are humane considerations concerning the widespread use of these specimens for vivisection, and the often filthy conditions and neglect which characterize their internment. In addition, surplus animals are often released into the nearest aquatic habitat whether or not it is suitable, and without concern whether it is within the natural range of the species. Choudhury and Bhupathy (1993) listed studies of reproductive biology and physiology, genetics, morphology, aging, and blood chemistry as laboratory uses for which riverine turtles have been exploited in India. They identified fourteen universities which utilized a wide range of riverine, semi-aquatic and terrestrial chelonians for research purposes (there were many other in institutions which were not contacted, however). They considered the level of exploitation for laboratory purposes, mostly with permission from wildlife authorities, to be justified and non-detrimental. They similarly considered the level of exploitation for zoos and captive breeding programs to be small, acceptable, and well regulated. The same description is probably applicable also to major United States zoological parks.

Religious and Cultural Influences on River Turtles

In North and South America, Australia, and Asia, especially, turtles and humans have been traditionally linked by religious and cultural, as well as nutritional bonds. While it is beyond the scope of this work to delve into these fascinating relationships for their own sake, it is relevant to relate how certain practices have influenced river turtle conservation. In South America, for instance, the Cavina and the Tumposa tribes believed a huge spirit-turtle protected its species (giant South American River turtles) (Smith, 1975). Some Orinoco tribes believed that a mystical lady stood guard over nesting female giant South American River turtles and would harm anyone who disturbed them during the process (Smith, 1975). Such beliefs encouraged restraint on turtle harvesting by these aboriginal groups in the pre-European colonization days of the Neotropics.

Cann (1998) observed that when some Australian aborigines in the Daly River region catch gravid females to eat they will incubate some of the eggs and release the hatchlings back into the wild. In Asia, the turtle is featured prominently in the mythology and doctrines of the Hindu, Buddhist, and Muslim religions, and also in many of the beliefs of indigenous tribal peoples. The Karnars and Santhals of India will not eat turtles, which they consider their ancestor (Das, 1991), and Gupta (2002) noted that traditional beliefs and taboos of some native Assamese groups in northeastern India formerly provided restraints against over-harvesting turtles.

In the Buddhist religion a belief persists in which a person who finds and delivers a turtle to a turtle temple (where it may or may not receive subsequent care) is entitled to good fortune and long life as a reward (E. Moll, 1976a; van Dijk et al., 2000a). Many species which are featured in Malaysian Chinese "pet shops", including riverine forms such as painted terrapins, river terrapins, Asian giant softshells, Asiatic softshells (and many other smaller aquatic and terrestrial species), in other Asian countries may be purchased for this purpose or for food, medicine or pets (van Dijk et al., 2000a). E. Moll (1976a) saw river terrapins, painted terrapins, Malaysian giant turtles, and several smaller aquatic species (including imported red-eared sliders) at two Malaysian turtle temples in Penang and Ipoh, respectively. An even greater diversity of native and non-native species are currently held at Bangkok's famous Turtle Temple of Wat-Prayoon and in many other Thai Temple ponds also (fig. 5.23a,b).

Turtles are also protected in Muslim and Hindu sanctuaries in other Asian countries (van Dijk et al., 2000a). In some of these (but not all), turtles are fed and protected in ponds associated with temples. One of the most famous is the Muslim shrine of Hazrat Bayazid Bostami in Chittagong, Bangladesh, which is

(a)

(b)

Figure 5.23
(a) The famous Wat Prayoon Turtle Temple, Bangkok, Thailand; (b) one of its tenants, a giant Asian pond turtle (*Heosemys grandis*) being fed papaya-on-a-stick.

Figure 5.24
The black giant softshell (*Aspideretes nigricans*) in the Bostami Shrine Temple pond, Chittagong, Bangladesh.

inhabited by the huge enigmatic black softshell (*Aspideretes nigricans*) (fig. 5.24). These turtles are thought to have restorative properties for pilgrims visiting the shrine, whilst eating the mud from the back of the turtles is thought to make barren women conceive (Das, 1991). The origins of these temple populations are unknown, but they may be very old individuals of peacock softshells (as younger individuals in the pond suggest) (Rashid, 1987) or possibly Indian softshells (Khan, 1987). In any case, they are certainly of riverine origin. Indian Hindu temples with turtle ponds also often contain large trionychids of various species (Annandale & Shastri, 1914; Das, 1991, 1995; E. Moll, 1990a). E. Moll (1990a) observed many Leith's softshell turtles eating puffed rice and hibiscus flowers provided by local villages at a shrine near the mouth of the Godaveri River in Andhra Pradesh. Softshell turtles were also once used in Vedic sacrifices (Annandale & Shastri, 1914).

Of greater significance in terms of river turtle protection is the practice of protecting and feeding turtles in a stretch of river adjacent to certain temples (see E. Moll, 1985b, and Case Study in chapter 7, this volume). The Muslim practice of shunning the meat of turtles while relentlessly harvesting and consuming their eggs ultimately provides a zero-sum return. While the rate of decline of turtle populations may be slower than if the adults were taken for food as well (as in reality they will be if a commercially valuable species), the decline of the populations is eventually assured if there is no relief from the chronic egg harvesting. The declines of Malaysia's river terrapin and painted terrapin populations are largely attributable to this practice, as described previously.

Turtles and their parts may also serve as charms or talismans of various kinds. In eastern India and Bangladesh, hanging carapaces on human dwellings, cowsheds, and livestock is thought to provide wealth for the humans and greater milk production for the cattle and buffalo so adorned (Das, 1991; E. Moll, pers. obs.). Vijaya (1982b) stated that *Lissemys* are tied to stakes near village ponds in West Bengal to guard against evil spirits. The shells of species such as common snapping turtles in North America (e.g., used as ceremonial rattles; Carr, 1952) and New Guinea giant softshells, Asian giant softshells and New Guinea snapping turtles (for ceremonial masks; Rhodin et al., 1993; Rhodin & Genorupa, 2000) in Papua, New Guinea also have significance in rituals of aboriginal groups. The New Guinea species' shell masks are also widely sold for the tourism industry (Rhodin & Genorupa, 2000).

Other Uses of River Turtles

River turtles are the source of a variety of other products and materials, and have uses which are secondary in the sense that the turtles had been primarily collected and killed for use as food. The concave, overturned carapaces of hard-shelled species have been commonly used as stepping stones, basins, fruit and cassava bowls, agricultural tools, planters, and containers for animal feed and other products (Agassiz, 1868; André, 1902; Smith, 1975, 1979; Pritchard & Trebbau, 1984). Smith (1975, 1979) also stated that giant South American river turtle shells were used to make combs, and burned for ash which could be mixed with clay in pottery making. In addition, skin from their necks was used for making tobacco pouches and stretched for making tambourines. As previously described, the giant South American River turtles' oil was once widely used for cooking and lighting, while its fat, when mixed with resin, was also used for caulking boats (Smith, 1979). Das (1991) stated that the Sahnsias of India's Punjab use the fat of Indian softshells for making oil and lotions. E. Moll saw trionychid shells in India used as boat bailers, and as receptacles for carrying materials such as mud and sand. He also saw an Indian flapshell's shell used as a sounding drum for a stringed instrument called a "sarangi." Similarly, *Cyclanorbis* carapacial disks are used as basins by Bari tribesmen of Sudan and elsewhere in Africa *Trionyx triunguis* carapaces with untanned fish skins stretched over them served as guitars (Loveridge & Williams, 1957). Das (1990) observed that the exported bones of Bangladesh turtles are used in several countries to make poultry feed and fertilizers. Anderson (1898, in Loveridge & Williams, 1957) recorded that the carapaces from large Nile softshells were used as war shields by *Nubian* tribesmen in Africa. The carapaces of alligator snappers have been used as a garden border (i.e., fencing) in the United States (Pritchard, 1989).

Another parallel use of turtles in both Asia and Africa is to keep them alive in wells as a sort of living filter system. Indian flapshells are commonly used for this purpose in India, keeping the wells free of algae and insect larvae (E. Moll, 1985b;

Das, 1991). *Cyclanorbis senegalensis* are used in African wells to keep them free of putrefying matter (Anderson, 1937 in Loveridge & Williams, 1957).

In addition to all of the reasons for killing turtles which have been presented and discussed in these chapters, the last two are in some ways the most frustrating. First, caught turtles incidentally by commercial and sport fishermen are routinely killed in the (usually erroneous) belief that they are competitors for fish resources. Mahmoud (1969) provided an example of this type of senseless killing where two fishermen on Oklahoma's Blue River caught and killed fifty-one adult razorback musk turtles (*Sternotherus carinatus*) in 2 hours. In reality, very few aquatic turtles are well-adapted for capturing living, healthy fish as prey in their natural habitat, and there is no evidence that fish populations are limited in any way by turtles. In a variation on this theme, Prashad (in Das, 1991) wrote that in the Indus River, the Indian softshell is baited with kneaded flour by fishermen who then break their necks and throw them back into the river in retaliation for damaging their lines and scaring away fish. Finally, the most senseless reason for killing turtles is to have no reason at all. A widespread practice in some parts of the United States is to use basking turtles for target practice as the shooter walks along shore or canoes through a stretch of river (Lovich et al., 2000) (fig. 5.25). The practice is common enough in Missouri rivers that the Missouri Department of Conservation created a poster to discourage the practice which was posted at many canoe put-in locations around the state (fig. 5.26).

Figure 5.25
A "plinked" wood turtle (*Clemmys insculpta*) (Photo by James H. Harding)

Figure 5.26
A poster from the Missouri Department of Conservation designed to educate the public about the stupidity of shooting river turtles.

Revenge of the River Turtles?

Unlike the similarly valuable crocodilians which can—and do—occasionally pay humans back in kind for the heavy levels of exploitation to which they have been subjected, river turtles, with few exceptions, are limited in their capacity to inflict all but minor damage on humans. One way, however, may be their capacity to act as a vector for *Salmonella* bacteria which can be transmitted to humans, causing the serious disease salmonellosis. To their benefit, this resulted in legislation which eliminated the domestic sale of small turtles within the United States, but remarkably the huge export trade still flourishes—unconscionably in light of the known serious, even fatal effects the disease may have on afflicted humans (although now some exporters claim to sell only salmonellafree hatchlings [e.g., Louisiana Cypress Turtle Farm, LLC: www.laturtles.com]). Nevertheless, even the United States hatchling sliders are still frequently sold by street vendors (Ypsilanti, 1993) and in pet stores—sometimes with a posted sign "for research or educational purposes only" upon their aquarium. The degree of documentation often required to purchase these animals is nil, however.

Vogt (1981d) has called attention to the dangers posed to canoeists by falling common map turtles, along Wisconsin's St. Croix River, where they bask far out on tree trunks leaning over the water. Another exception might be provided by several large species of softshells which inhabit Asian and African rivers. These are powerful, aggressive species which may regularly attain lengths over 1 meter and weights well above 50 kg (see also chapter 2, this volume, concerning the ferocity of Nile softshells in consuming large prey). Annandale (in Pritchard, 1967) reported that an Indian narrow-headed softshell could attain a carapace length of 2 meters and that it was capable of attacking and sinking small boats. Chaudhuri (1912) also stated the capacity of *Chitra* to damage fishing boats with blows from its head. While the likelihood of this occurrence was doubted by Minton and Minton (1973) because of the probability that it would damage itself in the process, both it and the other blunt-nosed softshells (*Pelochelys*) are known to shoot their heads out rapidly while keeping their mouths closed when disturbed (Das, 1991; Pritchard, 1979, 2001). Cantor (1847) considered the Asian giant softshell to have a ferocious temper, but Pope (1935) thought it had a mild disposition. Holland (in Rhodin et al., 1993) reported a captured turtle which Rhodin et al. (1993) believed from Holland's description to be a New Guinea giant softshell (*Pelochelys bibroni*) that bit and twisted off a piece of a wooden canoe paddle. An indisputably vicious species, the Nile softshell should be treated with caution. Baker (1867, in Loveridge & Williams, 1957) described how a large specimen caught on a baited hook was able to bite through the triple brass wire attached to the hook "as with shears." Durrell (1953) claimed that the jaws of a decapitated specimen kept on snapping and had enough force to splinter a piece of wood. Another reason for caution around the large Asian softshells may stem from their long habit of consuming partially burnt or unburnt human bodies which are placed in the Ganges River in large numbers annually as the culmination of Hindu funeral rites (Das, 1991). The association of human parts with food by a large softshell encountering an unwary bather could potentially have serious consequences. Burton (1928 in Das, 1991) observed that large softshells of the Mahe River were aggressive enough to rush out of the water to tear off the limbs of corpses on funeral pyres, even though the heat from the burning pyre burnt the soft skin of their back during the process. Acharya (1949 in Das, 1991) claimed to have seen a man eaten by a large softshell at the tank at Dakor, Gujarat. Whether true or not, the capability of these huge animals to inflict damage and their aggressive tendencies deserve respect and caution for those within their immediate proximity.

Also worthy of respect are the American snapping turtles *Chelydra* and *Macroclemys* which have extremely sharp jaws and attain large (in *Macroclemys*' case, huge) sizes. While *Chelydra* is more agile and able to shoot out its neck to produce a painful bite, the crushing and shearing power of *Macroclemys*' jaws is even more impressive. Pritchard (1989) presented a photograph of a turtle trapper whose finger tips were removed when they were carelessly placed in the open mouth of a boated alligator snapper.

Other large aggressive species, such as Australia's snapping turtles (Cann, 1998) and Amazonian big-headed turtles can also inflict damage, but an unknown species which bit off the penis of a bather in Malaysia (E. Moll, 1984c) perhaps provides the most memorable reminder to many of us that caution in the vicinity of river turtles is not misplaced.

Indirect Factors Contributing to Extinction

6

Profile 6. *Slowly with the moonlight reflecting off her massive, high vaulted shell, a female river terrapin or tuntung in local vernacular, cautiously emerges from Malaysia's Kedah River and onto the hard clay-packed beach. In former times—better times—this turtle would have emerged en masse with a hundred or more other females on what was then a deep sand beach. According to the old tuans, there were nights when more than 700 females nested and in excess of 20,000 eggs were dropped into crater-like nests on this very beach. At the peak of nest excavation, sand would fly in a myriad of directions. Then, as eggs were laid and covered, characteristic thumping sounds would begin, as females raised their bodies on extended limbs then dropped the huge shell onto the nest, compacting the sand over the eggs. Slowly at first and barely audible, the sound of shells dropping on the sand would grow progressively, as more and more females joined in. Eventually, a somewhat rhythmic "tun tung, tun tung" reverberated over the entire nesting beach, advertising the annual return of the terrapin. Sadly, this scene is now only history. It depicts a spectacle that will likely never be repeated on the Kedah River.*

The curtain began to fall on the tuntung extravaganza in the early sixties. Alor Setar, capital of Kedah, was experiencing a growing binge. Dams were being constructed on the river, both above and below the nesting sites. Sand and gravel were needed for this construction and the sand banks along the river offered an obvious and plentiful source. Jalur (river boats) and sagur (dugouts), piled high with sand, congested the river. One beach after another was depleted but the sand miners continued until no sand remained. With an upstream dam preventing new sand from moving downstream, the beaches no longer could be replenished during the monsoon floods. In turn, river terrapins and painted terrapins could not or would not reproduce without the sand. Further downstream a tidal barrage obstructed the terrapins' access to the productive estuary near the sea. Now tuntung numbers dwindle further with each passing year. The sand that formerly incubated and protected their eggs now forms the walls and foundations of the buildings comprising Alor Setar.

For now, this large female, along with a scattering of others, still returns annually to former nesting beaches seeking the deep blankets of sand that may have once protected the very nest where she hatched so very long ago. But it is a futile quest; she wanders aimlessly over the hard clay pan, searching for a commodity that no longer exists. Suddenly the female steps into space and disappears over the edge of a steep sided trench tumbling almost a meter to its floor. Although she struggles to escape, the sides of the trench are too steep and the depth a little too great. She is still there an hour later, when Fuad, the local egg collector who dug the trench, arrives to claim his quarry. The terrapin is placed in a nearby sand pit surrounded by a bamboo enclosure. Despite the disturbance, the reproductive drive of the female is still strong and the sand substrate seems to trigger the necessary nesting behavior. In just under an hour the nest is excavated and a clutch of 25 eggs laid. In nature another hour and a half would be spent covering and concealing the nest site, but Fuad aborts the remainder of the process and returns the female to the water.

The turtle was never in danger; Fuad, being a devout Muslim, would not eat an amphibious animal. Eggs, however, are a different story. Believed to have aphrodisiacal properties, they can be eaten as a delicacy or sold at local markets for several times the cost of a chicken's egg. As for the female, later in the month or perhaps next year she will likely return again and deposit another clutch. Fuad has learned to recognize many of the females and knows them to be creatures of habit. With one last check of the pit fall trap, Fuad heads home with the precious clutch of eggs ensconced in a plastic bag. It has been a productive night; one turtle came to nest.

E. O. Moll

Damming and sand mining are examples of factors that indirectly damage or destroy populations of river turtles. Direct factors such as human exploitation are typically more obvious causes of population decline and often serve as stimuli to incite conservation action by a concerned public or government. While direct factors typically kill animals outright or at least remove them from the gene pool, indirect factors can surreptitiously reduce their chances for survival by altering habitat or reducing food supplies. As such, they may decimate a population before it becomes obvious that something is wrong. Though less conspicuous than the direct causes, indirect factors are at least of equal importance in determining the ultimate survival of a species. Table 6.1 summarizes the types of indirect factors affecting selected species. Two important types of indirect factors, habitat alteration and species introduction, are discussed below.

Habitat Alteration

Habitat alteration implies any change in an animal's environment, but herein we will consider human or anthropogenic alterations. Anthropogenic habitat changes are not necessarily harmful to every type of turtle. Riverine specialists are more vulnerable to such changes than are eurytopic generalists that occupy a variety of lotic and lentic habitats. Generalists are by their nature adaptable and thus are less likely to be harmed by changing conditions.

A study by D. Moll (1980) on the Illinois River illustrates this principle well. The original environment of the Illinois River has been greatly altered as a result of clearing and draining land for agriculture, dumping of municipal sewage (particularly by the Chicago Sanitary District), and the construction of a series of locks and dams by the Corps of engineers to facilitate barge traffic. Moll found that while these alterations had reduced or eliminated populations of Blanding's turtles, yellow mud turtles and smooth softshells, generalist species such as the common slider, false map turtles, spiny softshells and common snapping turtles were thriving in the altered environment (see also Mills et al., 1966; Bellrose et al., 1977). Similarly, Anderson (1965) reported that commercial fishermen of the Mississippi and Missouri Rivers noted increases in softshells (spiny?) and snappers in areas having moderate sewage pollution.

Chelonian generalists are usually extremely resilient vertebrates that resemble weeds to the extent that they often do better in human altered environments than they do under more natural conditions. Examples in literature include the following: In tropical Asia, the Malayan box turtle (*Cuora amboinensis*) and the black pond turtle (*Siebenrockiella crassicollis*) appear to be more common in human-altered environments than in the pristine forested regions of Malaysia (Moll & Khan bin Momin Khan, 1990). Similarly, the Asiatic softshell, (*Amyda cartilaginea*) and the Malaysian snail-eating turtle (*Malayemys subtrijuga*) seem to be the only Thai turtles benefiting from the conversion of wetlands into agricultural areas (Thirakhupt & van Dijk, 1995 "1994"). Snapping turtles and painted turtles have been the chief beneficiaries of the proliferation of human-built farm ponds throughout Iowa. Both species now occur in large numbers in almost every square mile of the state (Christiansen, 1981). These generalized turtles (e.g., Caspian terrapin [*Mauremys caspica*], common sliders, and snapping turtles) tolerate a variety of environmental contaminants including many chemical pollutants such as pesticides, polychlorinated biphenyls (PCBs) and heavy metals and can be useful monitors in assessing the bioavailability of toxicants in the environment (e.g., Stone et al., 1980; Helwig & Hora, 1983; Olafsson et al., 1983; Bryan et al., 1987; Lamb et al., 1995, Meyers-Schone & Walton, 1994; Yawetz et al., 1983; Watermolen & Amrhein, 1995). Herald (1949) recorded a case of probable DDT poisoning in a striped musk turtle (*Kinosternon baurii*); however, Cann (1993) reported the occurrence of *Elseya* turtles with clouded and degenerated eyes in creeks and rivers receiving agricultural run-off in Australia.

Not only do these generalist chelonians tolerate polluted conditions, some seem to thrive. Without specialists to compete for resources, generalists may greatly increase their densities. For example, the Caspian terrapin (*Mauremys caspica*), the only vertebrate found in most sewage-polluted water bodies in Israel, seem to be prospering, attaining densities of 500 per ha (Gasith & Sidis, 1984). European pond turtles (*Emys orbicularis*) were reported to tolerate highly polluted

Table 6.1
A selection of indirect factors that actually or potentially impact riverine turtle populations throughout the world.

Species	Locality	Factors*	Source
Batagur baska	Malaysia	2,3,5,6,7,8,9, 10,11,12,13	E. Moll, 1976b,c, 1984a, 1985c, 1997; Sharma, 1995, 1999
Callagur borneoensis	Malaysia	2,3,5,6,7,8,9, 10,11,12,13	E. Moll, 1985c, 1997; Sharma, 1995, 1999
Kachuga dhongoka	Nepal, India	3,13	Shrestha, 1997; E. Moll, 1997
K. kachuga	Nepal, India	1,3,9,11	Shrestha, 1997; E. Moll, 1997
K. smithi	Nepal	6,9,13	Shrestha, 1997
K. tecta	Nepal	2,13	Shrestha, 1997
K. tentoria	Nepal	9,13	Shrestha, 1997
Clemmys insculpta	Northeastern, USA	1,3,5	Harding, 1990
C. marmorata	Western USA	3,9,13	Bury & Holland, in press
Emys obicularis	Europe	7,9, 13	Ferri, 1995; Mascort, 1997
Graptemys barbouri	Florida	1,3,9,12	Ashton & Ashton, 1985
G. ernsti	Alabama, Florida	1,3,9,12	McCoy & Lovich, in press a
G. flavimaculata	Mississippi	1,9,11,12	Anon., 1991; LaClaire, 1996
G. gibbonsi	Mississippi	1,9	Buhlmann & Gibbons, 1997
G. nigrinoda	Alabama, Mississippi	1, 10	McCoy & Vogt, 1980b
G. oculifera	Mississippi, Louisiana	1,3,9,11,12	McCoy & Vogt, 1980a; Stewart, 1988
G. pseudogeographica	Mississippi River Drainage	6,9,13	Anderson, 1965; Vogt, 1993

G. pulchra	Alabama, Georgia	1,3,8	Shealy, 1976; McCoy & Lovich, in press b
Malaclemys terrapin	Eastern & SE Coastal U.S.	1,6,9,10,13,14	Seigel & Gibbons, 1995; Wood, 1997; Wood & Herlands, 1997
Pseudemys floridana	North Carolina	9	Pope, 1939
P. nelsoni	Florida	9,13	Jackson, in press.
P. rubriventris	Pennsylvania	9,13	Ernst, 1985
Trachemys scripta callirostris	Columbia, Venezuela	4	Pritchard & Trebbau, 1984
Kinosternon herrerai	Northeastern Mexico	2	Carr & Mast, in press.
Sternotherus depressus	Alabama	2,8,12,13	Dodd, in press.
Carettochelys insculpta	Australia	2,7,8,10	Georges & Rose, 1993
Apalone mutica	Iowa, Missouri	1,9,12	Williams & Christiansen, 1981; Johnson, 1987
A. spiniferus	Iowa, Missouri	1,9,12	Williams & Christiansen, 1981; Johnson, 1987
Aspideretes gangeticus	India, Nepal	1,3,5,9,11	Sinha, 1995; Shrestha, 1997
Chitra chitra	Thailand	3,9,11	Thirakhupt & van Dijk, 1995, "1994", 1997
Chitra indica	India, Nepal	1,11,12	E. Moll, 1997; Shrestha, 1997
Nillsonia formosa	Myanamar (Burma)	3,11	Meylan & van Dijk, in press
Pelochelys cantorii	Malaysia	2,3,5,6,7,8,9,10,11,12,13	E. Moll, 1997; Sharma, 1995
Rafetus euphraticus	Turkey	3,11	Taskavak, 1995
Trionyx triunguis	Africa, E. Mediterranean	6,7,9,10,11	Atatür et al., in press
Elusor macrurus	Australia	7,11	Cann, 1998
Erymnochelys madagascarensis	Madagascar	3,6,13	Kuchling & Blanc, in press
Podocnemis expansa	Brazil	2,3,8,10	Mortimer, 1992; Alho, 1985; Vogt, 1995; in press.

*Factors defined as follows: 1 = Channelization; 2 = Clearing/deforestation; 3 = Dams/impoundments; 4 = Fires; 5 = Fragmentation; 6 = Incidental catch; 7 = Introduced species; 8 = Mining; 9 = Pollution; 10 = River traffic; 11 = Sand mining; 12 = Siltation; 13 = Wetlands/habitat destruction; 14 = Road kills.

habitats near Frankfurt, Germany (Ernst & Barbour, 1989). More recently, Souza and Abe (2000) recorded extremely high densities and biomass of the South American generalist species Geoffroy's side-neck in a badly sewage-polluted urban river in Brazil. Vermersch (1992) reported that the Texas cooter (*Pseudemys texana*) was very tolerant of pollution in the San Antonio River and that they were more numerous today than they were 25 years previously. Other studies have found that growth rates of certain generalists are relatively rapid in polluted environments. Painted turtles in polluted portions of Michigan's Kalamazoo River grew faster than populations in the southwestern part of the state (Knight & Gibbons, 1968). False map turtles feeding on a chiefly carnivorous diet of midge fly larvae in polluted flood plain backwater on the Illinois River grew markedly faster than members of a largely herbivorous population from a less-polluted site on the Mississippi River at Lake Co., Tennessee (D. Moll, 1976a). Snapping turtles in a polluted section of Michigan's Kalamazoo River (Gibbons, 1968) and from a polluted marsh in Massachusetts (Graham & Perkins, 1976) grew at a similarly rapid rate, averaging 32 and 26 mm per year respectively from the first through the sixth year of life.

Thermally enriched environments have favored another generalist, the slider, which has faster growth rates, grows to larger size and has greater reproductive potential in such habitats studied in South Carolina (Gibbons, 1970b; Gibbons et al., 1981; Gibbons & Greene, 1990; Dunham & Gibbons, 1990), and in Illinois (Thornhill, 1982).

Despite these examples, even tolerant species may reach their limits as habitats become further degraded. As noted by Nelson (in D. Moll, 1980) in reference to the late nineteenth century's Chicago River, "In July and August when the water is worst even the mud turtles leave the river in disgust and seek less odorous homes."

The above demonstrates that habitat alteration can benefit some (usually) generalized species, but in this chapter we are more concerned with the negative aspect of habitat alteration—loss of biodiversity. For the most part the generalists favored by habitat alteration are common species and not of conservation concern. It is the riverine specialists that are most negatively impacted by habitat alterations and which are dominating the threatened and endangered lists world wide. Herein, we review types of habitat alteration that have been significant in the declines of these species and assemblages.

Terrestrial Alteration

It might appear that factors typically associated with the terrestrial environment would have little significance to an aquatic turtle, but this reasoning is fallacious. Every herpetologist has witnessed the carnage that results when highways are built between rivers and turtle nesting grounds (reviewed by Mitchell & Klemens, 2000). This is mortality of the most destructive variety in that adult females—the group viable populations can least afford to lose—are being killed in wholesale quantities. Wood and Herlands (1995, 1997) have implicated road building and coastal development as key factors in the decline of diamondback terrapin populations along the Atlantic Coast of New Jersey. Using modeling, Gibbs and Shriver (2002) established that road networks typical of the northeastern, southeastern and central United States have the capability of significantly contributing to the decline of semi-aquatic, terrestrial, and large aquatic turtle species in their vicinity. Other types of development may occasionally result in the immediate loss of entire populations. Cann (1993) witnessed the destruction of hundreds of turtles which were buried under tons of sand during land reclamation for an oil terminal constructed in Australia in the 1950s.

Removal of shade-generating plants at a nesting site can alter nest temperatures, thus adversely affecting hatchling sex ratios of species with temperature-dependent sex determination (TSD). Wibbels et al. (1991) warned that any alteration of beach temperatures would be a particular concern for a species with limited distribution such Cagle's map turtle. At this point we can only speculate on the effects of the world-wide phenomenon of global warming on the genetics and sex ratios of TSD species, which include many turtles and crocodilians (Gibbs & Amato, 2000).

Deforestation and mining operations on land impact river turtles in various ways, but chiefly through causing increased flooding and siltation in the water (fig. 6.1). Forests help to hold the rainfall on land and to keep soil from eroding away. Removal of plant cover then causes streams and rivers in the watershed to become

Figure 6.1
Deforestation along the Perak River, Malaysia.

deluged with silt and excess water. Mining operations often dump spoil and tailings on the watersheds of rivers or directly into rivers—actions that correspondingly increase turbidity and silt (Cann, 1993; also see Mitchell & Klemens, 2000 for turtle-related examples).

Among the resulting problems, silt lowers the productivity of lotic habitats for river turtles through turbidity (reducing the light available to aquatic plants), by covering and destroying the riffles needed by a variety of aquatic invertebrates, and by destabilizing substrates and congesting the gills of mollusks (both important prey groups for river turtles). Silt accumulation in the channel decreases the river's capacity to carry off water, and this—along with the added run-off of water from the cleared land—leads to increased flooding. Not only does increased flooding cause turtle nesting beaches to deteriorate through erosion, but unseasonal floods during the nesting season can destroy part or all of the annual reproduction of a river turtle population. Eggs of most species tolerate only short periods of submergence (but see Kennett et al., 1993a,b; Polisar, 1996) and thus are quite vulnerable to inundation.

Plummer (1976) found significant mortality in smooth softshell eggs submerged for two days, and complete mortality in eggs submerged for four days. He also noted that smooth softshells from a Kansas River abandoned a popular nesting site on a sand bar following a brief flood that left the sand covered with silt. No hatchlings emerged from those clutches laid prior to the flooding, indicating that either development was terminated or hatchlings could not break through the hard silt crust. D. Moll (1980) believed that the increasing silt load carried by the Illinois River in the latter half of the twentieth century was probably the main factor which caused the decline of its smooth softshells for similar reasons.

Reproduction of the endangered river terrapin, on the Perak River of Malaysia, has also been severely affected by flooding and increased siltation (fig. 6.2). Factors contributing to these problems include large-scale deforestation and tin mining in upstream areas. Floods that wipe out portions or all of the river terrapin annual reproductive output are not unusual. Khan bin Momin Khan (1977) described an unseasonal flood in 1967 that destroyed some 2000 eggs on a major river terrapin nesting beach. E. Moll (pers. obs.) has observed flooding which destroyed portions of the river terrapin reproductive potential in each of three nesting seasons that he studied river terrapins on the Perak River (including a man-made flood in which water was let out of an upstream dam for boat races). Flood-related mortality of South American podocnemid river turtle eggs was discussed previously in chapter 2 (see also Pezzutti & Vogt, 1999).

In addition to direct mortality on turtle eggs, each flood lays down layers of silt over the sand banks where turtles nest. This additional organic layer has promoted the growth of lalang, a tropical grass (*Imperata cylindrica*), on Malaysian nesting beaches. As river terrapins require open sand for nesting (E. Moll, 1980a), the grass covering the sand banks has effectively destroyed the ancestral nesting sites of the river terrapin on this river. The only open sand banks presently available for nesting are those that are cleared annually by the local turtle egg collectors and by the Malaysian Department of Wildlife to attract nesting females. Ironically, Perak

Figure 6.2
The Perak River, Malaysia—an important habitat for river terrapin, painted terrapin, and other riverine species that is threatened by pollution and numerous alterations to the river and its floodplain.

River terrapins are now almost totally dependent on humans to provide them with nesting sites.

Siltation from mining has also been implicated in the declines of the flattened musk turtle in Alabama (Dodd, 1990) and the Fitzroy River tortoise in Australia (Cann, 1993, 1998). It is also cited as a potential danger for the pig-nosed turtle in Kakadu National Park in northern Australia (Georges & Rose, 1993).

If one had to choose the group of United States river turtles most vulnerable to siltation and pollution, which would it be? Ken Dodd, a biologist with the National Biological Service singled out the map turtles as a good candidate in a 1977 article on declining herps (see also Seigel & Dodd, 2000). With few exceptions, map turtles eschew small creeks and ponds while inhabiting clear, clean lotic habitats in large rivers where they feed on mollusks and aquatic insects. Probably it is their food more than themselves that require clear unpolluted water, but the effect is the same. Unlike the adaptable false map turtle studied by D. Moll (1976b), most map turtles are too specialized to switch to other foods when the mollusks and insects decline. As their favorite foods disappear, so do the turtles. It is not surprising then that seven species of *Graptemys* are presently listed on state protected lists and two (yellow-blotched and ringed map turtles) are endangered species on the Federal List (Levell, 1995; Baillie & Groombridge, 1996–2001).

Wetlands

Particularly apropos to turtle declines are human alterations of wetlands (Mitchell & Klemens, 2000). Inasmuch as a strong majority of riverine organisms begin their lives in backwaters and other wetland habitats associated with rivers, these are critical habitats controlling riverine diversity. In this regard, turtles are no exception. Early life history stages of some and probably most freshwater turtles inhabit areas of slow current and shallow water, often with abundant aquatic vegetation. In the case of river turtles, this could mean embayments of the river proper, backwaters, and associated swamps and marshes. However, due to their small size and secretive nature, the habitats of juveniles are poorly documented. Congdon et al. (1992) reported a positive correlation between water depth and turtle size up to sexual maturity in painted and snapping turtles in a Michigan marsh. Riverine backwaters and/or associated wetlands are key habitats reported for juvenile Indian tent terrapins (E. Moll, 1987a), black knobbed map turtles (Lahanas, 1982), common sliders (Moll & Legler, 1971; Hart, 1983; D. Moll, 1990; Bodie & Semlitsch, 2000a), African forest turtles (Loveridge, 1941), smooth softshells (Plummer, 1977), and false map turtles (Bodie & Semlitsch, 2000a).

It is particularly significant to the conservation of turtle biodiversity then, that these critical habitats are being rapidly destroyed (see also Bodie & Semlitsch, 2000a; Bodie et al., 2000; Bodie, 2001; and chapter 2, this volume). Riverine backwaters are being annihilated by various types of river development projects discussed below. Wetlands have long been drained for agriculture and are one of the favorite targets of land developers. In short, these critical aquatic habitats are disappearing at a particularly alarming rate. Exemplifying this trend, in the contiguous United States, some 120 million acres of the original 215 million acres of wetland have already been destroyed and what remains is still being eliminated at a rate of 80,000 acres per year (Williams, 1996). In Peninsular Malaysia some 510,000 ha of former wetlands have been converted to rice production, representing 90–95% of the original freshwater swamp forest that existed on the Peninsula (Malaysian Wetland Working Group [MWWG], 1986). The Pantanal, the world's largest wetland at 77,200 square miles, is being threatened with a massive development project called Hidrovia that has considered dredging and channelizing the Paraguay-Paraná River to allow ocean-going vessels to travel to the center of the South American continent (Eckstrom, 1996).

Water Pollution

Water pollution is often cited as a factor leading to the decline of turtles in aquatic systems. The term "water pollution" has many connotations, but commonly it refers to a build up of human and animal wastes and/or industrial waste in aquatic systems. In many cases it is impossible to separate the effects of sewage from industrial waste because municipal waste is often a mixture of both. In the United States, over half of the 5.8 million kilometers of flowing water is significantly polluted (Palmer, 1994). Pollution can act directly, killing turtles outright, or indirectly by destroying the organisms on which a turtle depends for food or some other critical aspect of the habitat required by the turtle. Other than pesticide research, turtles have been little studied regarding pollution effects. As such, most reports of turtle declines are anecdotal or are based on circumstantial evidence rather than on rigorous study of the situation. A typical report simply notes the absence or decline of a species or a higher incidence of disease in turtles on a part of a river which has become more polluted. The following are representative:

1. Pope (1939) reported a decrease in the numbers of the common cooter (*Pseudemys floridana*) around Raleigh, North Carolina, following pollution of waters by sewage and dye works refuse.
2. Based on discussions with commercial fishermen on the Mississippi and Missouri Rivers, Anderson (1965) reported that false map turtles were abundant in the rivers up to 1940 but then became uncommon. They attributed the decline to river pollution and reported that increasing discharge of chemical pollutants and industrial waste had almost eradicated turtles for many miles below Kansas City and St. Louis.
3. Harding and Bloomer (1979) recorded declines in wood turtle populations in New Jersey after spraying for gypsy moths occurred in the 1950s and 1960s, although habitats remained unchanged otherwise.
4. Declines of populations of the ringed map turtle, a federally endangered species, from the Pearl River have resulted from habitat modification and water quality degradation (Stewart, 1988). In a status study of the species, McCoy and Vogt (1980a) reported that no ringed map turtles were seen on the river from Jackson downstream to Terry, Mississippi, a portion severely polluted by Jackson sewage and industrial waste.
5. Ernst (1985) listed water pollution along with industrial expansion, demand for property, drainage of wetlands, and chemical control of mosquitoes as factors reducing the chances that the red-bellied turtle (*Pseudemys rubriventris*) can persist in suitable habitat along the Delaware River in Pennsylvania.
6. Lovich (in Ernst et al., 1994) noted that populations of the yellow-blotched map turtle (*Graptemys flavimaculata*) and the Pascagoula map turtle were common on the Leaf River above a pulp-processing plant, but absent below the point of discharge. He suggested that the highly tannic effluent from the plant may have eliminated the insect and mollusk populations upon which the turtles feed.
7. Buhlmann and Gibbons (1997) reported pollution as a factor in the declines of at least ten of the twenty imperiled river turtle species that they list for southeastern United States.
8. Mascort (1997) reported declines in European pond turtles (*Emys orbicularis*) associated with pesticide use in northeastern Spain.
9. Cann (1993) reported the absence of Fitzroy River turtles and three other species previously encountered in a Fitzroy River tributary receiving mine

tailings. (Cann also encountered a dying northern Australian snapping turtle just upstream from this site.)

10. Dodd (1988) noted a disease causing emaciation and lesions on the plastrons of flattened musk turtles, and Lovich et al. (1996) described the prevalence and histopathology of an erosive shell disease in turtles (mainly sliders and river cooters) of Lake Blackshear, an impoundment of the Flint River in Georgia. These diseases are suspected of being associated with poor water quality.

River Development

Humans alter riverine habitat for a variety of reasons, and in a variety ways (Mitchell & Klemens, 2000). They alter rivers to provide such benefits as commercial navigation, raw materials for construction, flood control, water supplies for cities, for irrigation, for hydroelectric power, and even recreation. Some 19,300 km of United States rivers have been modified for example, by the construction of 75 reservoirs, 276 navigation locks, and 13,670 km of federally maintained levees (U.S. Army Corps of Engineers, 2000, in Bodie, 2001). They achieve their goals by an assortment of techniques including channelization (ditching), dam building (impounding), lock and levee construction, and sand/gravel mining. The fates of river turtles are inseparably intertwined with that of the rivers they occupy. All of these alterations can and do impact riverine turtle populations. In an evaluation of the conservation status of turtles in southeastern United States, Buhlmann and Gibbons (1997) reported that declines of map turtles, musk turtles, the alligator snapping turtle, and several cooters have been the result of pollution, channelization, damming, shooting, harvesting, and disease. The Madagascan big-headed side-neck and Hoge's side-necked turtle in southeastern Brazil have also been negatively impacted by a combination of habitat alteration and pollution (IUCN/SSC, 1989). The stream-dwelling flattened musk turtle has declined as a result of a similar potpourri of destructive pollution and habitat alterations (Dodd et al., 1988).

Damming of Rivers

Damming rivers has become a very widespread human activity, and few rivers in industrialized countries have escaped impoundment or channelization. Of the original 5,760,000 km of rivers in the United States, 360,000 km have been channelized and 75,000 sizable dams block almost every river in the contiguous forty-eight states, leaving only 2% of the rivers free flowing (Palmer, 1994). The situation is probably worse in the developing counties. Most, with the support of multilateral development banks such as the World Bank, are also seeking to extirpate any free-flowing river within their political boundaries. After a several-decade binge of building huge impoundments throughout Africa and South America, the dam-building obsession has now shifted to Asia. India continues

Figure 6.3
Srinagrind Dam, shown here, and the Srinagrind Reservoir formed by it have completely changed the character of the Khwae Yai branch of the Mae Klong River system—home of the critically endangered Thai narrow-headed softshell (*Chitra chitra*) in Thailand.

with the controversial Sardar Sarovar Dam on the Narmada River, even after losing support from the World Bank. In 1991, it resumed building the 250 meter-high Tehri Dam in the headwaters of the Ganges. Malaysia has begun a 230 meter-high dam in the rainforest of Borneo, while Thailand has constructed dams to regulate flow on the Mae Klong river system, the principal home of the critically endangered Thai narrow-headed softshell and in other major rivers as well (fig. 6.3). Vietnam, Cambodia, Laos, and Thailand have recently announced the formation of an organization to plan dam building on the Mekong River. Now, China is building the largest and most expensive hydroelectric dam yet on the Yangtze River. When completed, 113,000 acres of fertile river valley will be inundated, 1.4 million people will be transposed, several rare species including the Chinese alligator will potentially be extirpated, and a host of archeological treasures submerged (Pearce, 1995). After the Yangtze is dammed, only two of the world's thirty largest rivers (which do not flow into the Arctic Ocean) will be free of dams. The main stems of the Zaire and Amazon Rivers are the last of the free-flowing giants, and no doubt their time is limited.

The building and operating of dams essentially turns lotic habitats into lentic ones. Riverine habitat is inundated along with associated nesting, nursery, and feeding areas. Large natural river-connected lakes (lagoons) provide suitable habitats for some large riverine species such as the Central American river turtle and the Nile softshell (D. Moll, 1986; Loveridge & Williams, 1957). Large, artificial impoundments which fragment, and greatly change the environments of the fluviatile habitats they replace (e.g., Thirakhupt & van Dijk, 1995 "1994"; Vandewalle & Christiansen, 1993) affect "trapped" riverine species in different ways. Nile softshells again seem to do well in large African reservoirs, and turtle diversity remains high in impoundments such as Kentucky Lake (Lindeman, 2000). Species such as ringed map turtles, flattened musk turtles, and western pond turtles have declined as a result of being isolated in and by impoundments, however (U.S. Fish and Wildlife Service, 1988; Dodd, 1990, Reese & Welsh, 1997). Again, it is the river specialists that suffer most from the impoundment of free-flowing rivers. In addition to the habitat loss, these specialists can be affected as follows:

1. In lentic impoundments, lotic forms may be thrown into competition with eurytopic species. Along the Mississippi River Drainage, impoundments promote the numbers of common sliders and/or painted turtles, while the relative importance of map turtles and smooth softshells in the community declines. Vandewalle and Christiansen (1993) found only three generalized species of turtles (common snappers, spiny softshells, and painted turtles) in Red Rock Reservoir, an impoundment of the Des Moines River in eastern Iowa, whereas rivers here were found to contain between five and eleven species.
2. Dams prevent upstream sediment from replacing that being lost downstream through erosion; hence sand beaches and sand islands—which are crucial nesting sites to a variety of riverine species—eventually disappear. The Kenyir Dam completed on the Terengganu River in Malaysia in the early 1980s has contributed to the loss of favored sand island nesting sites of the river terrapin. Sand islands in the river are being eroded away, and the dam blocks replacement sediments. One island that has been closely monitored had an average of sixty-one river terrapin nestings from 1983 through 1987. Then, in 1988 and 1989 only seven nested. The reduced nesting appeared correlated with the height of the beach which, by 1988, was barely above water level.
3. In these artificial lentic waters biological and physical phenomena often contribute to the loss of food species of particular turtles. Mollusks, which are important foods of *Graptemys*, are particularly affected. Impoundments tend to have reduced dissolved oxygen and increased CO_2 and siltation that lead to the declines in most mollusk populations (Harman, 1974). Silt reduces light penetration, thereby reducing mollusk food stuffs. It also chokes their gills with sediment that inhibits proper respiration (Fuller, 1974). Impoundments also disturb the fish fauna, driving out glochidial hosts (Fuller, 1974) and cover up riffle areas important to a variety of aquatic arthropods (Dodd, 1977).

4. Productivity below the dam is also affected negatively by alterations in flood cycles and temperature effects. The reservoir action of hydroelectric dams prevents adequate discharges for water to exceed the river banks and flood the adjacent flood plain. Fish production in rivers that annually inundate the flood plain can greatly exceed (ca. 1000%) that in equivalent rivers lacking flood plains (Bernacsek, 1984). Thus, damming of rivers containing fish-eating specialists such as narrow-headed softshells and Asian giant softshells could significantly affect their populations by reducing their chief food source. Thermal affects may also affect productivity below the dam. Cold water released from the hypolimnion of large impoundments can have a devastating effect on the invertebrate fauna for considerable distances downstream from the dam. Harman (1974) reported that a few snails of the genus *Physa* were the only mollusks found in the cold streams (15°C below normal) draining the hypolimnion of two reservoirs in southeastern New York. Although concerns for resident turtles were expressed by Anderson (1965), Lake Taneycomo, a dammed hypothermic remnant of the White River in southern Missouri probably supports as many turtle species as it did before it was altered (eight) (although there are no data available concerning how their relative abundances may have changed) (D. Moll, unpub.). The modern resident species cope with the colder waters in the reservoir by clustering in the shallower bays and normothermic tributaries of the lower warmer portion of the reservoir where food supplies remain available (Thomas, 1993; Turner, 1995; Davis, 1996; Wilson, 1999).
5. Dams interfere with movements of migratory river turtles. Netting (1944) noted that spiny softshells could reinvade streams that had been cleaned up following periods of pollution, but that upstream migration by smooth softshells was prevented by the navigational and flood-control dams that block most rivers. The decline of the Thai narrow-headed softshell was attributed to the damming of Thailand's Mae Klong River, the turtle's only remaining habitat in Thailand (Thirakupt & van Dijk, 1995). They reported that dams blocked movements, fragmented populations, and reduced habitat quality. In addition, the periodic release of water from upstream dams inundates nesting areas during this turtle's nesting season, thereby preventing or delaying nesting, and drowning nests and egg clutches which have already been laid (Kitimasak, pers. com.) (fig. 6.4). The aforementioned example of the Kedah River in northeastern Malaysia is a particularly striking case where a down-river tidal barrage blocks river terrapins from migrating upstream to nest on river beaches, and also prevents painted terrapins from traveling downstream to nest on sea beaches.
6. Dams, by dividing rivers into lentic and lotic stretches, fragment river turtle populations into a series of small isolated units that are vulnerable to inbreeding depression and environmental stochasticity (Soulé & Wilcox, 1980; Gilpin & Soulé, 1986) as well as to pollution and disease. Dodd (1990) has looked at the effects of habitat fragmentation for the flattened

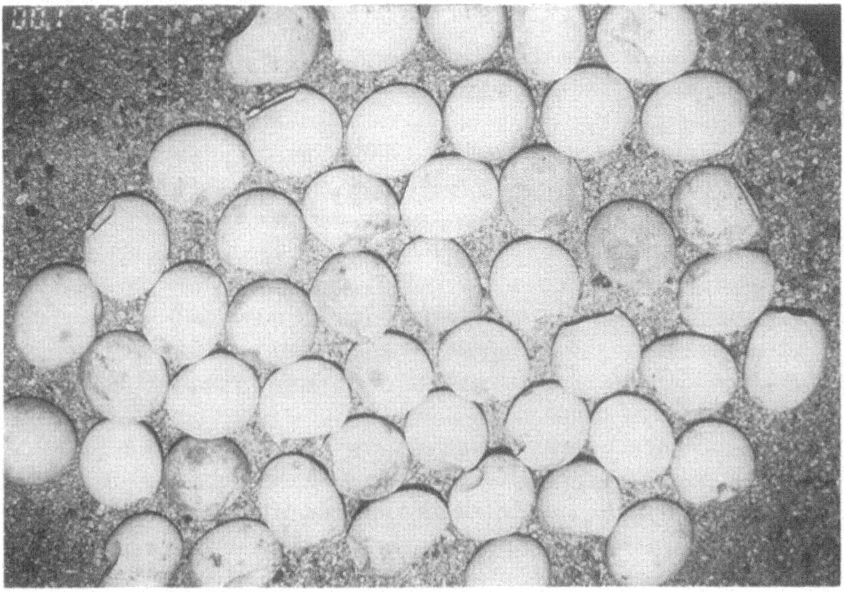

Figure 6.4
A clutch of Thai narrow-headed softshell eggs excavated from a nest in degraded nesting habitat subject to flooding in Thailand's Mae Klong River system. Most of the eggs are indented and impaired, and none from this clutch hatched. (Photo by Wachira Kitimasak)

musk turtle (*Sternotherus depressus*) on the Warrior River in Alabama. The former range of the turtle has been fragmented by at least three causes: impoundments, sedimentation, and pollution. Dodd estimates that only 6–9% of the historic range of the turtle is occupied by healthy populations. Most populations are isolated by extensive stretches of unsuitable habitat. In 1985, a contagious disease was suspected to be a key factor in the reduction one of the larger isolated populations by 50% in one season (Dodd, 1988).

Channelization

Channelization, another flood control technique, deepens, widens, and straightens sections of rivers. With regard to river turtles, the following effects can be particularly significant. By straightening the river or stream the numbers of riverine miles are reduced, along with the amount of aquatic habitat. Deepening the channel carries off water faster and helps to lower the adjacent water table. Stream banks may be raised with levees or dikes to further reduce flooding. The combination of lowered water table and reduced floods leads to the elimination of backwaters and other associated wetland habitats whose productivity and shelter from the current are important to the juveniles and/or adults of a number of chelonians. The sides of the channel are often lined with riprap or concrete, thereby eliminating riparian flora and thus reducing bank-side productivity. Dredging may eliminate sand bars (important to nesting and basking) and riffles, and destabilize the bottom, thus reducing key habitats for aquatic arthropods and mollusks (significant chelonian foods). In an exceptional situation, however, the production and deposition of sandy spoil mounds and beaches via dredging in the lower Appalachicola River of the Florida panhandle has provided good nesting habitat for alligator snapping turtles and other turtle species, and has allowed some species (e.g., spiny softshell, Barbour's map turtle) which bask and/or nest on sand banks to establish themselves in the lower reaches of the river (Ewert & Jackson, 1994). Snags and logs removed from the channel to facilitate navigation reduce or eliminate basking sites for the turtles. Finally, down river from the channeled areas, rivers become more prone to flooding due to the greater and more rapid flow of water.

Riverine specialists such as softshells and map turtles, with their narrow habitat and food requirements, are predictably vulnerable to channelization. Doody (1995) and Buhlmann and Gibbons (1997) found increased nest mortality of Louisiana softshells due to flooding resulting from channelization. Christiansen (1981) and Christiansen and Bailey (1988) attributed the declines of softshell turtles in Iowa to the straightening of rivers. Similarly, Johnson (1987) included channelization as one of the causes of decline for smooth softshells and false map turtles in Missouri. McCoy and Vogt (1980b) noted that black-knobbed map turtles were denser on winding areas of the Alabama and Tombigbee Rivers having moderate current and basking logs than on channelized areas with heavy barge traffic. Stewart (1988) reported that ringed map turtles had declined in areas of the Pearl River near Bogalusa, Louisiana following channel and floodplain modifications. McCoy and Lovich (in press a,b) have included channelization as a potential threat to populations of Escambia and Pascagoula map turtles in the southeastern United States.

Other species which reportedly have suffered from channel modification include wood turtles, Blandings' turtles, and European pond turtles (e.g., Bellrose et al., 1977; Harding, 1990; Ballasina, 1995).

Sand Mining

Sand mining is often associated with river and land development as sand is a basic constituent in many aspects of construction, particularly as a key ingredient in mortar and concrete. Sand banks along rivers offer an accessible source of this commodity. Herein lays the problem, as sand banks are also needed by many of the world's river turtles for nesting. Until recently, most sand removal in developing countries was done by hand-loading of small boats. With such methods, it took a long time to bring about appreciable damage. Moreover, as long as no dams were built upstream, the sand would be replaced whenever the river floods.

More recently, modern technology has taken over, and with bulldozers and backhoes to remove the sand and trucks, and tractors to haul it away, extensive tracts of beach can be razed in a matter of days (E. Moll, 1997) (fig. 6.5). If dams have been constructed upstream, such sand banks disappear forever. In a 1989–90 survey to assess the status of the river terrapin, in Bangladesh, India, Indonesia, Malaysia, and Thailand, E. Moll observed sand mining on every major river system visited (E. Moll, 1990b). Damage ranged from slight to almost complete removal of sand beaches as in the Kedah, Muda, and Kelantan Rivers of Malaysia. The latter three rivers all supported viable populations of river terrapin at one time, but now—lacking the sand needed for nesting—these populations dwindle toward extinction.

The National Chambal River Gharial Sanctuary supports several large, rare sand-nesting turtles that have been exploited to near extinction in much of India (i.e., Indian narrow-headed softshells, red-crowned roofed river turtle and three-striped roofed river turtle) (fig. 6.6). The remoteness of much of the refuge has helped to protect these species, but unfortunately some of their best nesting sites are sand banks near the highway bridge connecting Rajasthan and Madhya Pradesh, making them easily accessible to exploitation. When the Sanctuary was created in 1978, the Indian government granted construction concerns permission to mine sand within a two-mile stretch of the bridge. In recent years, funds for enforcement of this order have disappeared and sand removal now occurs well outside of the set limits. Long lines of trucks and tractors can be seen every day hauling sand from these banks, and turtle nests are being destroyed at an ever-

Figure 6.5
Sand mining along the Chambal River, India.

Figure 6.6
The Chambal River, India—site of a sanctuary benefiting both crocodilians and river turtles.

increasing rate. Fortunately, some of the tributaries of the Chambal are still free-flowing, and with each monsoon the sand continues to come downstream, replacing that being hauled away. Should all tributaries become impounded, this major nesting area will soon disappear.

Introduced Species

Due to a variety of human activities, both accidental and purposeful, animal and plant species are regularly being introduced into areas where they do not occur naturally (Mitchell & Klemens, 2000).

The introduction of foreign species can negatively impact natural populations both directly and indirectly. Introduced species may be efficient predators that directly impact turtle populations. For example, Thompson (1983) and Cann (1993) suggested that introduced foxes could be causing declines in populations of the Murray River turtle and the Mary River turtle in Australia. In western North America, the introduced bull frog has become a serious predator of western pond turtles (Salzberg, 1998b) and Sonoran mud turtles (Ernst et al., 1994). Doody (1995) and Buhlmann and Gibbons (1997) found that 21% of all predated smooth softshell nests that were observed in Louisiana were destroyed by introduced fire ants. Conversely, foreign species can indirectly affect turtle survival by spreading new diseases, by damaging the native species habitat, or by competing with native species for resources. Water buffalo are perceived as a major threat to the pig-nosed turtle, because they are known to trample turtle nests and the riparian vegetation upon which the turtle depends for food in the dry season (Georges & Rose, 1993; Georges & Kennett, 1989). The white amur (*Ctenopharyngodon idella*), a fish used to control aquatic vegetation, competes with herbivorous turtles. Studies carried out at Lake Conway, Florida following introduction of the white amur found that two common herbivorous turtles (cooters and Florida red-bellied turtles) and an omnivore (the stink pot; *Sternotherus odoratus*) that feeds on mollusks using aquatic macrophytes for a substratum, declined in areas where the fish reduced the aquatic vegetation (McDiarmid et al., 1983).

An introduction which turned out to be beneficial for native North American molluskivorous river turtles was that of the pollution-tolerant Asiatic clam (*Corbicula fluminea*). This species spread widely through American rivers in the mid-twentieth century, just as many native mollusks were succumbing to the increasingly degraded water quality of these habitats (Starrett, 1971; Thompson & Sparks, 1977). The Asiatic clam became the staple food for many of these species (e.g., Shealy, 1976; D. Moll, 1980; Mitchell & Klemens, 2000). The zebra mussel (*Dreissena polymorpha*) is a more recent molluskan invader with a serious capacity to degrade river ecosystems, and force the extinction of more native mollusks (Cwynar, 1995). Its value as a turtle food resource is not yet established, although its utilization by some common map turtles populations is known (P. Lindeman, pers. com.; D. Moll, unpubl.). If the zebra mussel should also come to deplete populations of Asiatic clams in many river systems, then molluskivorous river turtles might also be seriously impacted (Mitchell & Klemens, 2000).

Clear-cut examples of competition between introduced and native chelonians are lacking. Competition is often presumed likely when two somewhat similar species occur sympatrically. Dancik (1974) recorded a diverse array of exotic species resident in Chicago's Des Plaines River, and American turtles such as the alligator snapper (pet-trade escapees or released animals) occasionally turn up in European rivers (e.g., the Main River of Germany). The common slider has become the species of most concern, however. This highly adaptable and successful species not only has the greatest natural range of any non-marine chelonian (northern Indiana in the United States of America to Venezuela or Argentina depending on whether *dorbigny* is considered a separate species or subspecies of *scripta*), but it has also been introduced in many other areas of the world through the pet trade and commercial aquaculture ventures (see table 6.2). A number of local herpetologists have expressed concern that this widespread exotic may have a negative impact on native fauna or habitat in their region: Israel (Bouskilla, 1986), Italy (Ferri & di Cerbo, 1995; Frizenda & Ballasina, 1990; Giugliano, 1988), Malaysia (Kiew, 1984), South Korea (Platt & Fontenot, 1992), Singapore (Lim & Lim, 1992), South Africa (Newbery, 1984; Boycott & Bourquin, 1988, 2000), Spain (da Silva & Blasco, 1995), Thailand (Thirakhupt & van Dijk, 1995 "1994"), Western United States (Bury & Holland, in press), and Taiwan (Chen & Lue, 1998).

Table 6.2
Reports of the red-eared slider (*Trachemys scripta*) locations outside of its natural range.

Extralimital location	Breeding	Source
United States:		
Arizona	B	Hulse, 1980
California	B	Mooney et al., 1986
District of Columbia	B	Ernst, 1990
Florida	B	Wilson & Porras, 1983
Hawaii	B	Ernst et al., 1994
Maryland	B	Ernst et al., 1994
Michigan	B	Edgren, 1943
New Jersey	B	Stein et al., 1980
New York	B	Behler, 1996 in litt.
Pennsylvania	B	Manchester, 1982
Australia	?	Mansergh et al., 1993
Canada (Ontario)	?	Wright, 1995
France	B	C. Arvy, 1996, in litt.
Germany	?	Ernst et al., 1994
Great Britain	?	Ernst, 1990; Daniels, 1994
Guadeloupe, West Indes	?	Schwartz & Henderson, 1991
Holland	?	Green, 1991
Israel	?	Bouskila, 1986
Italy (Lombardy)	?	Ferri & Di Cerbo, 1995
Japan	?	Ernst, 1990
Malaysia	?	Kiew, 1984; E. Moll, 1995
Panama	?	E. Moll, 1995
Saipan	B	Scott Vogt, 1996, in litt.
Seychelles (Mahe Island)	?	Justin Gerlach, 1997, in litt.
Singapore	B	Lim & Lim, 1992; Moll, 1995
South Africa	?	Newbery, 1984
South Korea	?	Platt & Fontenot, 1992
Spain	?	Da Silva & Blasco, 1995
Sri Lanka	?	Green, 1991
Taiwan	B	Chen & Lue, 1998
Thailand	?	Thirakhupt & van Dijk, 1995 "1994"

"B" indicates established or breeding populations; "?" indicates unknown status.

Despite its adaptability, the introduction of the common slider does not necessarily mean that native fauna will be severely damaged. In an example related by Wilson and Porras (1983), the native fauna appears to have gotten the upper hand in competition with the slider. They report that by the 1960s, the introduced common slider was breeding in large populations in South Florida, where they were coexisting with the native Florida cooter and Florida red-bellied turtles. As of

1983, however, slider populations there appeared to be on the decline. The authors attributed this to the broader adaptability of the Florida red-bellied turtle. In another example, in central Italy Luiselli et al. (1997) discounted the concerns of other biologists (e.g., Giugliano, 1988; Frizenda & Ballasina, 1990) that the common slider may be a serious competitor of the European Pond Turtle. Luiselli et al. (1997) found that the introduced slider populations remained small in that region due to low winter survivorship of hatchlings and low adult reproductive success. Due to these factors they were not considered a serious threat to the native pond turtle at that time.

Another species that survives and reproduces well in foreign environments is the Chinese softshell (see chapters 5 and 7, this volume), a species that is being widely distributed for commercial farming ventures. Like the red-eared slider, the Chinese softshell, is being raised in farms both for food and for the pet trade. Being more secretive in its habits than the slider, there are fewer reports of introductions for this softshell, but it is widespread in regions where it is raised commercially (particularly South East Asia) (van Dijk et al., 2000a). Individuals have been observed in southern peninsular Malaysia (E. Moll, unpub.), in ponds within the city of Bangkok (Thirakhupt & van Dijk, 1995), and elsewhere in rural Thailand (D. Moll, pers. obs.). Substantial feral populations have become established in disturbed areas of Singapore (Lim & Lim, 1992). Outside of Asia, the turtle has been introduced into Brazil (Anon, 1981), the Hawaiian Islands (Brock, 1947; McKeown, 1978; McKeown & Webb, 1982), the Bonin Islands, Timor, and Japan (Ernst et al., 1994).

Certainly, the potential for ecological damage accompanies each successful introduction of these foreign, highly adaptable species, and rigorous long-term studies of introduced populations should be encouraged. It is only through such study that the magnitude of the problem can be assessed.

Stream and Riparian Managment

In an important recent paper concerning stream and riparian management for turtles which summarize the problems river turtles face from indirect threats and from direct exploitation, Bodie (2001) listed the following ten practices which harm river turtles: (1) reduction of snags and log jams; (2) riparian corridor drainage; (3) channelization; (4) dam building and impoundment construction; (5) flow regulation; (6) reduction of sand bars and beaches; (7) human riparian use; (8) pollution and siltation; (9) management for monotypic conditions; and (10) unsustainable use. We have similarly addressed these issues in this and earlier chapters, and concur with Bodie's "top ten" list of river turtle problems. Bodie (2001) further considered general management plans which could provide a blueprint for managers charged with maintaining and/or recovering turtle stocks faced with the above-listed threats. In addition to the obvious recommendation to reduce direct exploitation as much as possible, important tenets of Bodie's plan include:

1. Maintaining the connection between stream and riparian habitats through groundwater and over-bank flow in order to maintain riparian habitat complexity and provide woody debris for basking species through both erosion and deposition of riparian vegetation. The timing of the connection (and river flow levels) should conform to the historical hydro-period pattern as closely as possible so that flooding of riparian habitats and subsequent recession of water levels allow resident turtles to use feeding, nesting, and hibernacula habitats as they traditionally have in the natural river.
2. Mechanical and other disturbance of riparian nesting and over-wintering habitats should also be prohibited throughout the year.
3. Based upon analyses of flood plan usage patterns by North American and Neotropical river turtles, a minimum 150 meter riparian zone (preferably buffered by a wider area extending beyond the 150-meter core) should be protected on either side of river's banks as critical river turtle habitat.
4. For the maintenance of maximum river turtle and other species' diversity, broad-scale management for conditions that maintain the lotic character of the river should be encouraged, while the opposite (e.g., dam and reservoir construction) should be discouraged.
5. Sources of pollutants and silt entering the river must be identified and eliminated, or ameliorated to the extent possible.

As pointed out by Galat et al. (1996), Sparks et al. (1998), and Bodie (2001), passive management (i.e., allowing rivers to behave as natural rivers do) may accomplish many of these objectives, but in modern managed rivers the need for policy changes which factor in the requirements of river wildlife as well as commerce, and agriculture must be addressed before this seemingly simple requirement can be met (Sparks et al., 1998). For example, a proposal by the U.S. Army Corps of Engineers to regulate water flow in the Missouri River in order to better mimic the river's natural seasonal patterns of ebb and flow—specifically to benefit the needs of the endangered pallid sturgeon, and provide more reliable nesting habitat for endangered least terns, and piping plovers—has raised the ire of numerous federal, state, and commercial interests which have concerns about the resulting flow pattern and its effects on the lower Missouri (Quaid, 2002). The issue remains unresolved at this writing, as the Bush Administration has postponed indefinitely the river flow alteration plans' implementation.

In a more positive development, Missouri and other states have taken advantage of some of the havoc wrought by the great Missouri/Mississippi River floods of 1993 to naturalize river flow and maintain wetlands in some areas. As the cost of reclaiming a large amount of flood damaged bottomland agricultural property and levees is prohibitive, federal and state agencies have bought up thousands of acres to be left as "floodways" (open areas where future floods can spread out and slow down) and river connected wetlands (Vance, 1994).

Conservation, Management, and Rehabilitation

7

Profile 7. *For ages it has been a practice of the Sultans of Perak to reserve certain waters for their own fishing, and certain jungle tracts (usually surrounding a hot spring of mineral water) for their own hunting. There they would resort, annually or oftener, and with their relatives, chiefs, and followers take their kingly pleasure, as it was duly chronicled had been the custom of their ancestors.*

In the lull after the first heavy rains, that is, about the month of December, when the river has been swollen to flood-height for a couple of months, the tuntong or river-turtles ascend the Perak River in considerable numbers and lay their eggs on certain convenient sand stretches in the neighbourhood of Bota, about 100 miles from the river's mouth.

The most frequented of these laying grounds is a place called Pasir Telor (egg-sand), just below Bota, and it is here that the ladies of the Court annually assemble to dig up the eggs, which the Malay considers one of the greatest delicacies known to him.

The river-turtle is a great deal smaller than the sea-turtle, but it lays a larger egg, and one much more valued by Malays.

As soon as the river rises, watchers are stationed on the sands, and the turtles are said to lay three times. The nests are dug between two and three feet under the sand, and contain from about fifteen to thirty-five eggs each. During the laying season boats are not allowed to stop at the sands for fear they should disturb the turtles.

When the first set of eggs has been laid and the turtles have returned to the river, the watchers open the nests and send the eggs up to the Sultan. The second set of nests is opened by the royal party, and the third is left to hatch, an operation that takes six months. There is no sitting, the young turtles simply emerge from the sand, walk down into the river and swim away.

It is said that if the first and second nests are left untouched, the turtles themselves open them and scatter and destroy the eggs; but that, after the third 'lay,' they take their departure, having accomplished their task.

Directly the watchers report that the turtles have made the second nests, the Sultan and his family, with the neighbouring chiefs and their families, take boat and paddle down the stream to Pasir Telor.

Fifteen or twenty large house-boats and several bamboo rafts containing about one hundred and fifty people make an imposing procession. The rafts are simply floating houses, with mat walls and a high thatched roof, and are manned by crews of from four to sixteen polers; but the boats are graceful and picuresque barges, of which the foundation is a long dug-out of hard wood drawing very little water, the freeboard is raised by the breadth of one or two planks, and over the stern half of the boat is built a palm-thatched covering on a slight wooden frame, while curtains secure privacy. Inside this house, the roof of which rises in a sharp curve towards the stern, sit and lie on mats and cushions the owner and his family or friends. The crew occupy the forward half of the boat, where they sit to paddle down stream or stand to pole up. The steersman has a high seat in the stern, from whence he is able to see clear of the cabin-roof.

The covered portion of the barge which carries the Sultan's principal wife is decorated with six scarlet-bordered white umbrellas. Two officers stand all day long, just outside the state-room, holding open black umbrellas with silver fringes, and two others are in the bows with long bamboo poles held close together and erect. The royal bugler sits on the extreme end of the prow, and from time to time blows a call on the antique silver trumpet of the regalia. Flags are flown, other boats carry gongs and drums, and altogether the pleasure-fleet makes a brave show and a considerable noise, attracting the attention of all the dwellers on the riverine.

The journey from the Sultan's palace at Kuala Kangsar occupies two days, and on the morning of the third all the ladies of the party, with all their attendants and children (a good many still in arms), disembark for the ceremony of digging out the turtle-eggs.

The ladies are in their smartest garments and wear their costliest jewels. It is a blaze of brilliant-coloured silks, of painted sarongs, cloth-of-gold scarves, and embroidered gauze veils; of bright sunshades, gold bracelets, necklaces, and bangles; of curious jewelled brooches, massive hair-pins, and rings flashing with the light of diamonds and rubies.

The men appear in jackets, trousers, and sarongs of hardly less striking hues; but the horror of Western dyes and Western schemes of colour has not yet demoralised the Malay's innate sense of beauty and fitness, and nothing offends the eye as all this wealth of bravery moves slowly across the strand.

A scorching sun shines down on the gaily-clad figures with their background of dark jungle, on the yellow sands and sparkling river, with its burden of picturesque boats, and gives light and shadow to a charming picture.

The watchers have marked with twigs the various nests, and each lady of rank, with her little crowd of attendants, makes for one of these, and with her hands begins to dig up the sand in search of the eggs. But the nest is deep down, and the sides of the hole have a way of falling in on the digger, so a man or boy is desired to remove the overburden and make things easy for the lady. The overlying sand is quickly scooped out until one or two of the white eggs are disclosed, and then the lady, sitting on the edge and stooping far down, can just manage to reach the nest, and the eggs are carefully handed up.

Besides the pleasure of actually removing the eggs with one's own hand, of displaying to admiring eyes a vision of taper fingers and rounded wrist, of showing how little it matters that the costliest garments should trail in the sand, there is the rivalry of whose nest yields the largest number of eggs. Anything over twenty-five is considered a satisfactory find.

By the time all the nests have been rifled, the sands are growing so hot under the rays of the fiery sun that bare feet can hardly endure what is little short of torture. There is an almost hurried return to the boats, the finery is exchanged for simpler garments, and all the men and many of the ladies take to the river, and there disport themselves in a manner that is refreshing to sun-scorched bodies and the eyes of the Western spectator who is fortunate enough to see how it is possible to be unconventionally natural and yet perfectly modest.

It is only on such occasions as this that a strange man can see these ladies unveiled and even so he is not expected to look at them or go very near them; but their bathing-costume differs hardly at all from that which they commonly wear, and they thoroughly enjoy this opportunity of reveling in the clear waters of the sand-bedded stream.

Then every one scrambles back into the boats, which are pushed off into deep water, the rowers seize their paddles and with beat of gong and the musical notes of the silver serunai, with jest and laughter, pennons waving, and bright eyes sparkling behind the rainbow-coloured blinds, the picturesque flotilla glides on its course down the long sunny reach, in and out amongst the islets, round a heavily-wooded, deeply-shadowed headland, past the riverside hamlets and the orchards, the stately palms, the clusters of bamboo that overhang the water like great plumes of pale green feathers, and so ever onward through sunlight and shadow till another bourne is reached.

Excerpt from *With a Casting Net*, by Sir Frank Swettenham [1895] reprinted in Roff [1967].

It is interesting to note in Swettenham's account that the Malays were already practicing conservation methods for the river terrapins in the nineteenth century. By protecting the beaches from poachers and leaving the third laying to hatch, they maintained moderate recruitment to offset the egg harvest. Similar conservation measures existed on the Amazon River in the nineteenth century to protect the heavily exploited giant South American river turtle, including the protection of females and leaving a third of the eggs laid on the beaches to hatch (Goeldi, 1906 in Parsons, 1962; see also a brief history of exploitation patterns of this species in chapter 5, this volume.).

Unfortunately, conservation actions such as these have been the exception rather than the rule in human interactions with river turtles. As emphasized in earlier chapters, populations of most riverine turtle species have seriously declined from nineteenth century levels, but unlike the general declines reported for amphibians, most of the causes are obvious. The burgeoning commercial exploitation for food, traditional medicines, and pets, expanding populations of traditional and introduced predators, and extensive habitat destruction exemplified by such practices as sand mining, damming, channelizing, and pollution of flowing waterways have been key factors. We know far better how we reached this state of decline than how we can reverse the trend.

Since the early beginnings of river turtle conservation chronicled by Swettenham and Goeldi, a variety of conservation techniques have been tried world-wide. As there are few proven methods, most actions have been based on what seemed logical or was simply cloned from the methods being used in sea turtle conservation programs. The latter may not have been the best approach for as Pritchard warned in a 1980 paper (Pritchard, 1980b) "sea turtle conservation remains without a theoretical framework, and almost all techniques that have been used remain unproven and riddled with paradox."

Conservation methodology is generally divisible into two divergent approaches, one emphasizing in-situ (on site) techniques and the other ex-situ (off site) techniques. The former aim at protecting animals and their habitat while minimizing intrusion into the life history of the target species. The latter are more manipulative, requiring removal of the animal and/or its eggs from the natural habitat for varying periods. Protected areas, habitat improvement, protective laws, and predator control are in-situ techniques, while hatcheries, head starting, captive breeding, translocations, and reintroductions exemplify ex-situ techniques.

For the most part, existing conservation programs have not been operating long enough, nor have their results been adequately monitored, to make definite conclusions on which combination of techniques work best. Among the difficulties in evaluating these programs is that large river turtles are typically long-lived and can require between one and three decades just to mature. The best models presently available for late-maturing turtles suggest that many decades will be required before substantial increases in populations can be observed (Congdon et al., 1993, 1994; Crowder et al., 1987; Frazer, 1997). Another difficulty is that most conservation programs, having limited funding and tend to concentrate available resources into the conservation effort rather than into evaluation.

This chapter looks at the current conservation philosophy for dealing with declining turtle populations and examines the methodology available for rebuilding ebbing populations along with potential benefits and drawbacks.

In-Situ (On Site) Techniques

In the absence of sure-fire techniques, in-situ approaches tend to be the conservative's choice. These techniques allow turtles to carry out their life histories in the fashion that has been dictated through eons of natural selection, and so human error can be largely eliminated. In-situ methods are strongly advocated by the IUCN/SSC Tortoise and Freshwater Turtle Group Action Plan (IUCN, 1989). Also, as turtles cannot exist in nature without suitable habitat, habitat protection, acquisition, and maintenance should be priority inclusions in any conservation plan, including those emphasizing the ex-situ approach.

Protected Areas

These can be one of the most effective means of conserving river turtles because habitat and community remain intact. Inasmuch as the target species and its ecosystem are preserved, a protected area is the quintessential in-situ technique. Ideally, everything remains in a natural state, and the organisms carry out their life histories in the time-tested manner. Also all age groups receive protection, a key criterion in the conservation of long-lived, slow- maturing species (see section on Conservation Philosophy, this chapter). Some positives and negatives associated with protected areas are summarized in table 7.1.

Questions that arise when planning for protected areas are typically ones that theorists have been debating for over two decades since conservation biologists first considered practical application of the equilibrium theory of island biogeography (MacArthur & Wilson, 1967) to the design of nature reserves: What shape? How big? How many? In a riverine reserve, the question of shape is dictated in large part by the river itself, but this becomes a more valid question should buffer zones be included. As a river is an open system, the health of the reserve is heavily dependent on the watershed lying upriver. Ideally, buffer zones could be set up protecting this watershed from such dangers as deforestation, and organic and chemical pollution. The security of the reserve would be further strengthened if upriver buffer zones were established that prevented damming and sand mining of the river.

The question of how large has at least two components. Of foremost importance are the autecological requirements of the species. The reserve needs to be

Table 7.1
Positives and negatives associated with protected areas.

Positives

1. The community and habitat are protected entire, thus maintaining conditions naturally. Emphasis is synecological rather than autecological.
2. All age groups receive protection.
3. Moderately high-profile action advertising that conservation is being done.

Negatives

1. May be expensive to purchase, protect, and maintain a long-term financial commitment that many poorer countries will be unwilling or unable to keep (see Gomez, 1982).
2. Protected areas become a predictable location for those interested in finding endangered species whether these be predators, poachers, scientists, or tourists.
3. Long-time residents may have to be evicted. Traditional hunting and fishing grounds may be lost.

large enough to include all of the habitats required by the target species during the course of its life history. Minimally, this includes areas required for nesting, feeding, and predator avoidance of all ontogenetic phases. Every attempt should be made to have this information available during the planning stages of the reserve. The second component is more difficult to obtain and concerns how large of a reserve is required to adequately house a viable population. There is little argument that small populations are more subject to extinction, catastrophes, inbreeding depression and loss of heterozygosity. But how small is too small in regard to population size?

Apropos to this subject are the concepts of minimum viable population (MVP) and population viability analysis (PVA). An MVP is an estimate of the lowest number of organisms that can maintain a viable population. Early attempts to answer this question, did so by deriving a few magic numbers for general application. The best known are the figures 50 and 500 offered by Franklin (1980) and Soulé (1980). The former number was touted as the minimal population size necessary, to (at least temporarily) avoid problems with inbreeding and the latter, the minimal number necessary to maintain long-term adaptability. Such numbers were readily embraced by many conservation organizations, anxious to have absolute goals for which to strive. However, it has become evident that using a universal population size for species as diverse as desert pup fish, fruit flies, Indian rhinoceros, and river terrapin is probably too simplistic. More recently, the goal of an MVP has shifted to trying to estimate the size at which a population's chances of extinction become likely and then attempt to keep populations well above that size (Meffe & Carroll, 1994; Meffe et al., 1997). Debate on criteria for calculating minimum viable population size continues. Today, the term "minimum viable population" is often replaced with less absolute terminology (e.g., genetic effective population) and the realization that a number of factors can alter the critical size of a viable population. For example, Hedrick and Gilpin (1997) point out that the genetic effective size of a metapopulation is influenced by the carrying capacity of a patch, extinction and colonization rates, abundance and source of founders, total local patches and rate of gene flow between patches. Shafer (1990) suggested that until minimum population size for species can be better estimated, other approaches should be used to estimate reserve size.

Population viability analysis (PVA) is a process that incorporates the MVP concept but does not attempt to estimate a minimum size required to keep a population viable. Rather, the PVA uses simulation models to examine a profusion of interacting factors that can put a species at risk of extinction in order to predict the probability that a particular population will survive for a certain period of time into the future (e.g., 100 years). Alternatively, it can be used to predict the size of a population (and reserve size) necessary to withstand predicted adversities for a set length of time into the future. PVA is still relatively new, and like any new technique a variety of problems need to be addressed. They are limited by the fact that available data are often insufficient and time is critical (particularly where threatened species are concerned). To date, the majority of PVAs have been based on genetic and stochastic demographic data, while a large percentage of the causes and threats of extinction are ecological in nature (e.g., habitat destruction). Certainly one of the reasons for this discrepancy is that ecological systems tend to be highly complex and can confound making accurate estimates of time to extinction, but if PVAs are to be made more cogent, this is a key area that needs to be addressed. PVAs are being prepared for a growing number of species, and even with the aforementioned shortcomings the process appears to have considerable potential for the future. A review of population viability analysis including its practical difficulties can be found in Boyce (1992).

Inasmuch as PVA may not be practical in all situations, alternative methods of determining the size of the reserve are needed. Here again, the autecological requirements of the group can continue to serve as a guide as to what needs preserving. A less precise but useful methodology is to determine the distribution of the target turtle(s) on a particular river while identifying the area or areas having the largest subpopulations. These areas are presumably closest to optimal on the river for maintaining the species, and so are logical sites to consider for reserves. Using a core or source-population design, a river turtle sanctuary should include a large breeding nucleus along with optimal nesting and feeding habitats (E. Moll, 1985c, 1990a). The core or source population(s) in these favorable habitats will serve to supply sink populations in less favorable habitats throughout the river. Conservation goals of this methodology would be best served if the target species is totally protected from exploitation over the entire river but allowing limited subsistence hunting outside of the sanctuary may also be workable. Ideally, the protected population grows to a level where population pressure forces excess individuals from the protected area where they can continue to be taken by subsistence hunters. So long as the breeding nucleus is maintained intact, the survival of the species is maintained. The larger the breeding nucleus that can be protected, the fewer the problems with inbreeding depression and loss of heterozygosity.

Wilcox (1984) suggested another strategy for determining the size of a reserve. First, identify the target species with the most comprehensive habitat needs, then consider the pathway:

species → minimum viable population → habitat
requirement → minimum area requirement

Again, this method hinges on adequately calculating the minimum viable population of a key species. If this is successful, the area selected will then be more than ample to also protect smaller species with less demanding requirements (once again the minimum viable population of these species must be determined). This umbrella approach protects a variety of species simultaneously and is exemplified by the crocodile reserves in India (see Case Study 1). In these reserves all species are protected and so river turtles reap the same benefits as its patron species, the crocodile (Moll, 1985c, 1990a).

Case Study 1: India's Crocodile Conservation Program. India's crocodile conservation program offers an instructive example of melding in-situ and ex-situ techniques to achieve a desired conservation result, and although this program was not aimed at river turtles *per se*, they have reaped many of the same benefits as the crocodilian target species.

The exploitation of Indian crocodiles mushroomed following World War II, mainly as a result of more sophisticated hunting methods and increasing demand for crocodilian hides to fuel the luxury leather industry. By the early 1970s, populations of the three species of crocodiles occurring in India, the mugger (*Crocodylus palustris*), the saltwater crocodile (*Crocodylus porosus*), and the Gharial (*Gavialus gangeticus*) had suffered calamitous declines. In addition to being heavily exploited for their hides, gharial and mugger populations also were being significantly impacted by indirect factors such as incidental capture in fishing nets, habitat destruction from dam building, and channelization (Saharia, 1982; Bustard, 1990).

In an initial effort to stem the decline, India placed the three crocodilians on Schedule I of the Wildlife (Protection) Act, 1972 giving them complete protection under the law. Legislation also passed prohibiting the export of their hides and products. Next, India sought the help of the U.N. Food and Agriculture Organization and U.N. Developmental Program to develop a crocodile farming industry in the country. In 1974, Dr. H. R. Bustard—a crocodile and sea turtle researcher—was brought in as a consultant to provide a status and feasibility study. Bustard conducted the study in 1974 and submitted a report that became the basis for a crocodile breeding and conservation project that was initiated in 1975.

The multi-pronged action plan of the project included:

1. Setting up captive rearing stations where eggs from wild nests and captive adult breeding stock are hatched and the young head-started to sizes deemed safe from a majority of predators.
2. Establishing eleven sanctuaries ranging in size from 20 square kilometers (Manjira in Andrea Pradesh) to 12,568 square kilometers (National Chambal River Gharial Sanctuary in Uttar Pradesh, Madhya Pradesh, and Rajasthan) to provide suitable habitat for the release of the captive-reared animals.
3. Creating programs to train Indians in the captive rearing and management techniques needed to operate the conservation program. In particular, the Central Crocodile Breeding and Management Training Institute was established at Hyderabad for the purpose.
4. Organizing publicity and awareness programs promoting public involvement in the management program.

Throughout the remainder of the 1970s and much of the 1980s, the crocodile conservation program successfully carried out its mission of re-establishing crocodilian populations in India. The wild Gharial population increased from an estimated 60–70 adult and sub-adult individuals in 1974 (Bustard, 1990) to 1500 individuals in 1993 (Anon., 1993). A restocking program for the saltwater crocodile in Bhitarkanika National Park in Orissa had increased the numbers from less than thirty in 1974 (Bustard, 1990) to 584 in 1994 (Kar, 1994). And while there were no accurate estimates of the numbers of the widespread mugger in 1974, its populations were greatly depleted from former times and they were considered rare through most of the range (Saharia, 1982). In 1993, muggers were reported from over fifty localities and the numbers were estimated at 3000 to 5000 (Anon., 1993). The success of this conservation program cannot be measured only in terms of increases within crocodile populations. By prohibiting hunting, fishnets, and egg collection, the sanctuary system benefited wildlife in general. Crocodile and religious sanctuaries were the only sites where E. Moll (1985b,c), regularly saw large numbers and a wide variety of riverine chelonians in India, including rare species such as the red-crowned roofed terrapin.

The Indian Crocodile Conservation Program exemplified how successful conservation could be achieved by combining in-situ and ex-situ techniques. From 1975 to 1988 it was a showpiece of conservation action and was touted as a model for other countries in the region to follow. However more recently, it better illustrates the problems of maintaining successful conservation programs once achieved. By 1988, the novelty of the project had begun to wear off and funding was beginning to dry up. During the glory years of the project, a number of Indian nationals had been trained as researchers to study and carry out crocodile conservation in the country. But now with less publicity and flagging government interest, a second-line of professionals trained to take their place was largely absent. High costs of captive-rearing and the protection and maintenance of the sanctuary system, along with a paucity of additional sites for release of captive bred crocodilians, have made increasing or even maintaining the program problematic. Initially the project was a joint venture of India, the FAO, and the UNDP of the United Nations, but eventually most of the costs of maintaining the program fell to the Indian Federal and State governments alone. The costs of operating the program spiraled far above those initially projected.

Consider the National Chambal River Gharial Sanctuary, one of the project's great success stories. The estimated budget for the Sanctuary was 10 lakhs per annum (ca. U.S. $23,500) in 1979, but the cost during the sixth five-year plan has been estimated at Rupees 1.3 Crores (ca. U.S. $306,000). By the early 1990s, with less money available for enforcement, the effects of the cutbacks were already apparent. Fishermen were openly using gill nets which had been outlawed in the Sanctuary since its inception. Sand miners, obtaining sand for construction, were destroying sand nesting banks within the Sanctuary well beyond the two-mile stretch adjacent to the Rajasthan/Madhya Pradesh highway that they had been allotted. Poaching was also on the rise.

Captive rearing centers have also suffered from reduced funding, but a bigger problem is that they have become too successful. More crocodiles and gharials have been produced than there are places to locate them, and so rearing facilities are badly overcrowded, with over 12,000 muggers presently stockpiled in the rearing centers. One of the obvious solutions would be to harvest the excess crocodiles for leather and other products. Sustainable use has become a key element in the conservation strategies for crocodile species worldwide (Ross 1998). However, the Indian government prohibits commercial use of captive-bred crocodiles, thus leaving no current solution to the burgeoning problem. Hence, the Indian government has instructed breeding centers to stop producing any new crocodilians and the project has ground to a halt.

Recently, future projections for crocodile conservation in India have become somewhat brighter. In 1998 the IUCN/SSC Crocodile Specialist Group (Western Asia Region subgroup) held a conference at Jiwaji University in Gwalior, Madhya Pradesh, India to discuss the plight of Indian crocodiles. The meeting resulted in a six-part action plan for dealing the present problems (Anon., 1998). Included are plans to: re-establish a liaison with the Ministry of Environment and Forests; identify research priorities that will further the crocodile conservation effort; establish publicity and awareness campaigns to change the crocodilian image among local populations; train a new generation of crocodile researchers and managers to carry out the conservation mission; develop regional interaction and coordination; and seek necessary funding to carry out the action plan. Hopefully this renewed interest in India's crocodilians will allow the conservation program to survive and return to its former status as a prototype that other countries can emulate.

In some cases, multiple reserves may be desirable on the same river. In certain instances habitat type becomes more important than refuge size. Consider the river terrapin of tropical Asia. River terrapins are river specialists that rarely utilize lentic habitats. They feed in estuaries near the sea, but migrate long distances upriver to nest on riverine sandbanks above tidal influence. Obviously, setting aside a large reservoir as a reserve for the rheophilic river terrapin is unlikely to benefit the turtle no matter how large it is. Further, a large reserve in the river that does not include both feeding and nesting areas will likely be less effective than two smaller reserves, protecting each of these critical habitats. This does not mean that size is unimportant. If a reserve can be gazetted in an unpolluted estuary lined with the mangrove tree (*Sonneratia*), the fruits of which are a staple food of the terrapin, then more is better. The larger the food resource, the larger the river terrapin population that can be supported and the fewer small population problems will be experienced. Similarly, a reserve upstream, among the extensive sand banks used for colonial nesting by the terrapin, should set aside as much of this crucial habitat as possible. The more sand banks that can be protected from destruction, the more widespread nesting can be. With more dispersed nesting, predators cannot concentrate their efforts, later cohorts of nesting females will dig up fewer nests of previous cohorts, and catastrophes such as unseasonable floods are less likely to destroy all of the annual reproductive output.

Multiple reserves on a river serve as insurance to catastrophes allowing for recolonization if one or more subpopulations are extirpated. The following situations indicate a need for multiple reserves:

- All habitats used in the course of the turtle's life history are not contiguous. Feeding areas, nesting areas, and/or nursery areas are separated by large areas of unused habitat, (except during movement between these habitats) (e.g., giant South American river turtles, river terrapins, and painted terrapins exemplify species that move moderate to long distances between feeding and nesting areas as discussed above).

- Local or sub-populations of a species occur in habitats separated by long stretches of unsuitable habitat (e.g., the Fitzroy River turtle of Australia inhabits riffles which may be some distance apart along the river [Cann, 1998]). Its nesting sites appear to be in the vicinity of the riffle habitats. If the Fitzroy River turtle is the only target species, several small reserves each protecting a set of riffles (feeding area) and nesting sites may be as effective and more economical than gazetting a large area of river containing relatively small amounts of critical habitat.

Perhaps the most compelling arguments against choosing a series of small nature preserves over a large contiguous one are rooted in genetics. Multiple small reserves may assume a classic metapopulation structure—a metapopulation being a series of local populations in scattered habitat patches with intermittent or regular gene flow that periodically become extinct and then recolonized from other patches. Using a simulation model, Gilpin (1991) has demonstrated that even though a population may be relatively large overall, a metapopulation structure will lead to rapid losses of genetic variability characteristic of small populations. Each time a local population becomes extinct and is recolonized from another patch, some heterozygosity is lost. He further states that since few species in the wild possess the low heterozygosity predicted from this process, they presumably do not have or at least maintain a metapopulation structure for extended periods. It is more likely that somewhere in the range of most, one or more large populations survive extinction and serve as a refugium protecting genetic variation. In a free-flowing river, the metapopulation problem becomes less significant. The river serves as a natural corridor permitting dispersal between populations thus increasing colonization and gene flow. Dams, however, exacerbate the problem by increasing fragmentation and isolation (Dodd, 1990; Gibbs & Amato, 2000; Mitchell & Klemens, 2000). In such human-created metapopulations, human intervention may be required to maintain heterozygosity, reduce inbreeding depression, and recolonize extirpated populations.

Genetic considerations should thus be given adequate consideration when weighing decisions concerning the size and number of nature reserves (see section on Genetic Considerations below). Economics will certainly have to be considered as well. More often than not, final decisions will be based more on economic and less on biological arguments. Reserves can be expensive to set up and maintain. Potential expenses include purchasing large tracts of real estate and then hiring a permanent staff to enforce rules and protect species from exploitation. What is best for the animals involved may be too expensive for some countries or agencies to consider. Furthermore, rivers tend to be multiple use areas serving a variety of human needs including transportation, recreation, and food supply, and governments may be unwilling to limit competing uses of a stretch of river solely for conservation purposes.

For wide-ranging or migratory species, establishing reserves in both feeding and nesting areas may prove impractical. If only one area can be protected, then usually the nesting area is the best choice. While it can be argued that turtles require adequate food resources to survive, generally river turtles appear to be considerably more flexible in their food habits than in their reproductive habits. Protecting the nest site is especially favored for large colonial nesting river turtles (e.g., giant South American river turtles, river terrapin) whose concentrations and predictable returns to specific beaches make them highly vulnerable to exploiters and predators. Nesting sites are typically characterized by a high mortality of females, hatchlings, and eggs. While eggs and hatchlings are to some extent expendable, there can be no more damaging loss to a population than the large, reproductive females. Each large female river turtle can potentially lay hundreds (even thousands) of eggs in a life time but on average only two per turtle have to survive and become adults if the population size is to remain stable. The importance of adult females will be a recurring theme in this chapter.

Existing protected areas are as varied as the turtles that they protect. Examples of existing programs include:

- Belize has designated a series of protected zones for Central American river turtles in major waterways of the northern part of the country, aimed at incorporating all major habitats of the turtle (Polisar & Horwich, 1994).
- In the United States, Mississippi has designated approximately 13 kilometers of the Pearl River as a Ringed Sawback (*Graptemys oculifera*) Sanctuary (George, 1990). Presently, the philosophy of the sanctuary is simply to protect the turtle and its habitat but not actively to manage the population.
- Brazil's strategy has been to protect nesting sites in order to conserve their giant South American river turtle populations. History and descriptions of the program can be found in Vogt (in press) and IBAMA (1989). As a result of efforts of a turtle biologist named Jose Alfinito, the Brazil government in 1975 established a program through the agency IBDF to protect the turtle and its nesting beaches. In 1990, IBDF had become IBAMA and established a national center for freshwater turtles (CENAQUA). From 1975 to 1992 the program reported releasing some 18 million hatchlings from the protected beaches. Today, there are 115 protected nesting areas, on fifteen rivers in nine states.
- In India, turtles reap the benefits of refuges established for other species, particularly crocodilians and tigers, and in religious sanctuaries associated with Hindu temples (see Case Studies 1 and 2).
- The Project Tuntung Sanctuary, which was established to protect river terrapin on the Perak River in Malaysia, exemplifies an extractive reserve. Adult river terrapins are protected here, but egg collectors are allowed to exploit major nesting beaches and sand mining is allowed along the entire nesting region (see Case Study 3).

Case Study 2: Indian Religious Sanctuaries and Refuges. It is difficult to imagine a setting more picturesque and peaceful than that of the Bateshwar riverfront on that afternoon in early April. In this small city of temples, the Yamuna flowed slowly along a lengthy line of immaculately white buildings of varied shapes and structure (fig. 7.1). The rays of the descending afternoon sun transformed the town into a glowing waterside jewel sharply in contrast with the otherwise stark landscape. Actually, our primary purpose in coming to Bateshwar had not been to witness the unusual beauty of this striking scene but to view an especially large assemblage of turtles which inhabited this site.

The Yamuna River has long been famous for its turtles. Whereas ancient statuary depicts Mother Ganga as standing on the back of a crocodile, the vahan or vehicle of Mother Yamuna has traditionally been the turtle. Today, few parts of this river can be said to teem with turtles but at Bateshwar it still does. Here one can see what India's 'river of turtles' must have been like in the past.

Scanning the river with binoculars we could see turtles basking, floating and swimming almost everywhere; peaked silhouettes of Indian tent turtles in ludicrous basking poses on logs and protruding sticks in the river. These sun-worshippers were balanced precariously on their perches, with neck, limbs and toes maximally outstretched to gather the warm, soothing rays of the sun. On the

Figure 7.1
The Yamuna, known as India's "river of turtles," flows past a Jain temple in the village of Bateshwar which provides food and sanctuary for river turtles in its vicinity.

riverside sandbanks, ponderous soft-shells mimicked the behaviour of their tiny, hard-shelled cousins.

From a vantage point on the steep steps of the bathing ghat at a Jain-Hindu temple, a score more turtle heads continually appeared and then disappeared in the water below me. I recognized at least five species including a number of rare, painted roof terrapins (Kachuga kachuga). Throughout the day villagers, pilgrims and other visitors came to bathe in the holy waters and to feed doughballs, puffed rice and vegetables to the turtles. Though aquatic burials also took place here, there were no funerals that day.

The remarkable number and variety of chelonians at Bateshwar is attributed to at least two factors. Foremost, this stretch of river along the town is a religious sanctuary where the turtles are neither exploited nor harmed. Secondly, an abundance of food is available from the aquatic burials and the offerings provided by the faithful from the steps of the temple. Sanctuaries such as these are quite important from a conservation standpoint. River turtles are being heavily exploited throughout much of India today, and as a result, populations of certain species have undergone a serious decline. Therefore to find areas such as Bateshwar, where a breeding nucleus of these important species is being protected and the turtles are living harmoniously along with one of their chief predators, is indeed encouraging (E. Moll 1985b).

Religious shrines or sanctuaries dedicated to turtles or including turtles are scattered throughout Asia where they may be associated with the Buddhist, Islamic, and Hindu religions (Annadale & Shastri, 1914). We have visited examples of these in India, Bangladesh, Thailand, and Malaysia and have concluded that only those found in India benefit the turtles. In most, turtles are kept captive in concrete or tiled ponds lacking nesting sites where they remain until they die.

The Indian sanctuaries are often truly that. The turtles are fed and protected, but frequently remain in contact with wild populations. Basically there are two variations. One common type is a tank (pond) associated with a Hindu Temple in which a population of large softshells are kept, fed, and protected. In at least some of these a river runs nearby and no barriers prevent the turtles or their offspring from returning to the wild. One such temple is located at Kotipalle, Andhra Pradesh near the mouth of the Godaveri River. In the temple pond, E. Moll observed a number of Leith's softshells eating puffed rice and hibiscus flowers provided by local villagers (E. Moll, 1990a). In the second type of sanctuary, a section of river near a temple or group of temples is given protection. Bateshwar (described above) is an excellent example of such a sanctuary. A similar sanctuary can be found at Mathura on the same river. The river sanctuaries are far better for conservation than the first type. All species of turtles are protected, not just the large softshells. Other than being fed, the turtles live under entirely natural conditions. The large numbers of the rare red-crowned roofed turtle at Bateshwar testify as to the effectiveness of these sanctuaries in conserving river turtles.

Habitat Improvement and Restoration

This can be beneficial to both protected and unprotected areas. It is important to remember that sanctuaries on rivers are open systems. Just protecting the area of a reserve does not prevent damage due to chemical pollution or siltation originating upstream. Management of a river must consider what is happening throughout the entire watershed, particularly upriver and immediately downriver from the site of interest. Hence, the most beneficial forms of habitat improvement involve improving water quality for the entire river, including the tributaries, throughout the drainage basin. River restoration projects are gaining in popularity in developed nations (National Research Council [NRC], 1992). Such projects attempt to remove sources of silt, sewage, and chemical pollution, while re-creating former river flow patterns and flood plains (see chapter 6, this volume). Natural materials are used to reinforce the channel and hold the banks (i.e., tree trunks, boulders, roots and vegetation) as opposed to the concrete and steel that has typically been used in the past. Some plans include the reversal of channelization and even the removal of dams. In regard to the latter, many hydropower dams in the United States will require re-licensing by the federal government in the coming years offering an important opportunity to restore natural hydrologic conditions to some rivers (Richter et al., 1997).

On a smaller scale, a variety of actions can be used to improve nesting sites. Many river turtles seem to be attracted to extensive stretches of open sand for nesting purposes. Females can thus be enticed to particular nesting beaches by removing debris and vegetation that covers the sand. Prior to the river turtle nesting season on the Perak River in Malaysia, the Department of Wildlife and National Parks and individual egg collectors remove tropical grasses from nesting beaches and break up the silt crust that forms following floods (E. Moll, 1984a). Part of the IBAMA conservation program for South American river turtles consists of clearing logs and debris from the nesting beaches (IBAMA, 1989).

The creation of artificial nesting beaches has been tried with some success. In one experiment begun in 1950, juvenile giant South American river turtle that had been headstarted for 10 months were translocated to Lake Valencia in the Andean highlands west of Caracas and north of the Orinoco River. Parsons (1962) reported that the turtles were growing well and had begun nesting on artificial beaches constructed on several of the islands in the lake. However, a later report by Smith (1975) stated that no significant nesting had occurred on the lake's artificial beaches. In a more recent attempt, artificial beaches were constructed below the dam at Balbina on the Rio Uauma to accommodate giant South American river turtles blocked from reaching their natural nesting areas. Turtles have begun nesting on these man-made sites (Vogt, 1997, in press).

Habitat improvement is not limited to nesting sites. Providing basking sites and improving feeding areas exemplify other approaches. Lindeman (1999) found that map turtle density is positively correlated with deadwood density of the habitat. He concluded that anthropogenic removal of deadwood from rivers is likely harmful to populations of map turtles. An innovative program on the Pearl River in Mississippi removes snags and deadfall trees from the river channel to accommodate boats and water skiers, but deposits them nearer the bank to serve as basking sites for the threatened ringed sawback (map turtle) and other basking turtles (George, 1990). Moderate logging of riverside forests in Michigan can open up clearings that are conducive to wood turtle food plants such as raspberries, strawberries, and willows (Harding, 1990). Planting fruit trees such as figs (*Ficus*) or the mangrove tree (*Sonneratia*) in riparian areas and estuaries respectively would be potentially beneficial in those tropical areas where many chelonians are herbivorous.

A sound approach for planning effective habitat improvement requires studying the ecology of the target organism and in particular the limiting factors. Comparing habitat features of areas where the target organism is abundant with areas where populations are depleted can provide clues as to what could be altered to improve the turtle's environment.

Genetic Considerations in River Turtle Conservation

> Genetics should be viewed as "fine tuning" to be done after the "coarse tuning" of habitat protection has been accomplished. (Meffe & Caroll, 1994)

Conservation biologists are increasingly faced with maintaining declining populations of rare species. To a large extent, their initial efforts must be focused toward problems concerning habitat degradation and loss as these typically are the ultimate causes leading to the decline. But while habitat protection and improvement are first-priority actions, it has become clear that long-term conservation efforts are likely to fail if the genetic health of the target species is not considered in the recovery program (Frankel & Soulé, 1981). Over the past two decades, the role of genetics in conservation decisions and planning has continually grown, and genetics and conservation biology have become so intertwined that a new subdiscipline, conservation genetics, has emerged. Simply put, conservation genetics applies the tools and concepts of protein electrophoresis, mitochondrial DNA analysis, and nuclear DNA analysis (e.g., DNA fingerprinting and microsatellites). This section examines the roles that genetics can play in river turtle conservation.

A primary concern of conservation biology today is the preservation of biodiversity. In a more analytical sense, genetic diversity is the source of biodiversity and thus becomes the ultimate concern for conservation biologists. Each extinction reduces the extant genetic diversity a little more. As observed by Foose (1983), gene pools are becoming gene puddles. The diversity is displayed at three levels, the genetic variation within individuals (heterozygosity); the genetic variation within a population; and the genetic variation between populations. Each level must be considered in conservation biology (Meffe & Caroll, 1994). At the individual level the genetic problems of interbreeding develop, and knowledge of variation within a population (the gene pool) will determine the population's ability to adapt to new or changing conditions, while the loss of variation between populations results in coalescense of populations and a narrower range of options for the species. Both within- and between-population variability must be maintained to allow the full potential of evolutionary change of threatened species.

Heterozygosity is a concern for both captive and small wild populations. High heterozygosity is typically (though not always) associated with fitness (Hedrick & Miller, 1992), but too much or too little can lower fitness. For example, outbreeding depression can result when individuals from differing geographic areas are brought together in captive breeding or translocation programs. Matings between genetically dissimilar individuals can break up adaptive gene combinations and lower rather than raise fitness (Templeton, 1990, 1996). If at all possible, organisms to be reintroduced or translocated should come from nearby populations occupying similar habitats and, ideally, ones that previously shared some gene flow. Homozygosity can also bring about a reduction in fitness. Along with genetic drift, inbreeding can lead to increased deleterious traits in the progeny ("inbreeding depression"). The latter phenomenon commonly affects those characters that are closely tied to an individual's ability to survive and reproduce (Frankel & Soulé, 1981).

Inbreeding and inbreeding depression are not synonomous. Some species seem able to cope with frequent inbreeding without serious consequences. For example, inbreeding depression appears less common in populations having a long history of inbreeding (Charlesworth & Charlesworth, 1987). In at least some of the cases, inbreeding may have come about slowly, allowing selection time to weed out many of the deleterious alleles (Frankel & Soulé, 1981). Many endangered species are outbred, however. Small population sizes with increased inbreeding are a recent occurrence that can be accompanied by high levels of inbreeding depression (Hedrick & Miller, 1992). Inbreeding depression likely is less of a threat to historically small populations (or to species that naturally exist in metapopulations; Simberloff, 1988; Hanski & Gilpin, 1997; Hedrick & Gilpin, 1997) than to historically large populations that have been rapidly transformed into small ones.

As mentioned previously, habitat destruction and dam building are examples of factors that have fragmented previous large wild river turtle populations into smaller ones. Even if protected, the genetic health of these populations may suffer. Each fragment likely represents a "bottleneck" (less diversity than the population as a whole), and genetic drift can be expected to further reduce the remaining genetic diversity. Genetic transfusions (introducing animals from compatible captive or wild populations) may be a useful management technique to maintain genetic diversity in cases where natural gene flow between fragments no longer exists. The initial step is to make a genetic status survey (heterozygosity assessment) of existing populations and determine if reduced heterozygosity is a problem and, if so, to determine the best approach to remedy the problem.

For example, Gray (1995) used DNA fingerprinting to analyze the genetic status of the western pond turtle (*Clemmys marmorata*) in Washington, Oregon, and California. She confirmed the genetic break between a northern and southern subspecies of the turtle. The northern populations, which are threatened with extinction, showed a high degree of genetic similarity within populations.

Between-populations variability was small in the remnant coastal populations, but significantly higher between inland populations. Genetic variability was also high within southern populations.

Gray recommends that recovery efforts focus on the low within-population variability of the northern populations, but fearing outbreeding depression, she cautions against genetic transfusions with the genetically dissimilar individuals from southern populations. She concluded that management should focus on preservation of local gene pools, and increasing population numbers through captive breeding and reintroduction, or translocation from compatible wild populations.

Genetic management is typically more important for captive populations than for wild populations, because captive populations are generally smaller than those in the wild, and problems related to small population size (i.e., interbreeding, genetic drift) are magnified. One of the tools being used in the genetic management of captive populations is pedigree analysis. By making detailed pedigrees of the captive population, breeding programs can be designed to reduce inbreeding and thus inbreeding depression. In addition, the distribution of founder genes within the population can be tracked. As a primary goal of captive breeding is to reintroduce a species into its original habitat, it is important to maintain the founder alleles that adapted that population to that habitat (Hedrick & Miller, 1992). Captive-raised individuals that have been inbred are likely to have poor survivorship when reintroduced into the wild (Jiménez et al., 1994).

Although the most common application of genetic information in conservation has involved maintenance of genetic variation in and between populations, the number of conservation uses for genetic information is growing (see Avise & Hamrick, 1996). Several additional uses and potential uses for the conservation biology of river turtles include:

- Recognition of cryptic species requiring conservation attention. Avise (1989) recognized two types of systematic errors that could misdirect conservation efforts, taxonomic recognition of groups with little evolutionary differentiation and lack of recognition of phylogenetically distinct forms. Lovich and Gibbons (1997) referred to the latter as "covert species" and emphasized their importance to the conservation effort. As an example of such covert forms, consider snapping turtles. Phillips et al. (1996) recognized three distinctive groups of *Chelydra* (common snapping turtles) rather than one widespread polytypic species. This may have conservation significance in that now attention will be focused on the poorly known and seemingly rare middle and South American species.
- Determining population sources of migratory individuals and natal homing. Molecular genetic approaches have been important in revealing migratory patterns and natal homing (i.e., return to the same beach [or beaches in that general area] where they hatched) among sea turtles (Bowen & Avise, 1996; Gibbs & Amato, 2000). These same techniques could be beneficial in illuminating similar information in river turtles, and they are beginning to

be used to answer such questions (e.g., Sites et al., 1999; Bock et al., 2001; see chapter 2, this volume). Large vagile turtles living in major rivers often nest colonially on beaches far from their feeding areas. For example, the giant South American river turtle nests on over a hundred known sand beaches throughout the Brazilian Amazon, but during the rest of the year they spread out into flooded forests and oxbows to feed. From a conservation standpoint, it would be helpful to know if the turtles from one nesting beach are also concentrated in a particular feeding area. For example, should the nesting numbers decline, the cause may be found in the feeding area. Similarly it is not known whether hatchlings show natal homing. Such information could be significant in attempting to reintroduce nesting populations on depleted beaches.

- Determining demographic features and natural history of populations. Molecular genetic methodology can give researchers a handle on the numbers of individuals actually mating, the females producing the most offspring, and multiple paternity of clutches. These data can all be significant in calculating the genetic effective size of a population, which in turn allows researchers to predict the rate at which genetic diversity will be lost in populations. Galbraith et al. (1997) have demonstrated the utility of DNA fingerprinting in determining paternity and other aspects of freshwater turtle mating systems.
- Identifying biological material confiscated in illegal trade. Products from protected species can lack the superficial diagnostic characteristics needed for positive identification, but with molecular genetic methods it is often possible to identify the material not only to species but also to river systems from which it came.

Antipredator Techniques and Other Methods to Improve Recruitment

Most readers are probably well aware of the tremendous mortality that takes place on the colonial nesting beaches of sea and river turtles. Even if they have not actually visited a nesting beach during the reproductive season, they may have been taken there through the magic of television, courtesy of Jacques Cousteau, Marlin Perkins, David Attenborough, the Discovery Channel, or the Public Broadcasting System. The typical documentary begins with a scene showing one or more large female turtles laboriously moving up the beach, digging their nests and laying their eggs. At this point predators do not play an important part, as the adult turtles are large enough to deter most predators (jaguars, tigers, and humans excepted). However, in subsequent scenes predators take the starring role. Depending on the locality, multitudes of eggs are shown being exhumed, then consumed by raccoons, monitor lizards, hyenas, wild pigs, vultures, and the like. Then, just as one thinks things cannot get much worse for the turtles, the scene switches to reveal hatchlings exploding out of the surviving nests and running for the water as frigate birds, jabirus, storks, caracara, crows, or gulls swoop out of the skies to snatch them as foxes, coyotes, jackals, dingos, wild cats, or ghost crabs chase them over the sand. And just as the viewers are ready to cheer the scant few that have reached the water, the scene cuts away once more to an underwater view of predaceous teleost fishes, sharks, caiman, and crocodiles completing the banquet begun by the terrestrial predators.

If a poll were taken of the viewers immediately following one of these documentaries, featuring the question, "What is the best way to conserve turtles?" a near unanimous answer would be—protect the eggs and baby turtles from predation. Actually as we shall see shortly, this probably is not the surest approach to conserving long-lived animals such as turtles, but certainly it would help. Several techniques have been devised to increase survival of the eggs, hatchlings, and reproductive females at nesting beaches.

Predator control is one obvious action. In North America the raccoon has become one of the most damaging predators of turtle eggs (Mitchell & Klemens, 2000; chapter 2, this volume). Changes in land use patterns, a decline in value of raccoon pelts, and elimination of raccoon predators during this century have resulted in a population boom for this proficient nest robber. Christiansen and Gallaway (1984) reported that, following raccoon removal in 1979, a yellow mud turtle population in Iowa experienced markedly reduced nest and hatchling predation over the next two years. Stancyk (1982) reviewed efforts to control raccoon predation on loggerhead sea turtle beaches in the United States. Although intensive removal of raccoons greatly reduced first-night predation of nests temporarily, if predator removal was not continued, predation soon returned to former levels. Stancyk concluded that removal of mammalian predators is most effective in situations where immigration of new recruits is impeded (such as an island situation) or where manpower and equipment is available to carry out regular intensive removal effort.

Goodrich and Buskirk (1995) discuss a number of problems with predator control programs. In sanctuaries or reserves, where all species are protected, predator removal may not be possible. Cost and manpower needs can be excessive for many programs. It is often the least socially acceptable alternative, and in such situations other options need to be considered.

Nest protection can be a simpler and less costly alternative to predator control. For example, nest survivorship can be increased simply by transplanting eggs a few meters from the original nest site (Stancyk, 1982). We have also found that using a broom to sweep away tracks and nesting signs can be effective in deterring sight-oriented predators, including human poachers. Covering the nest with metal or plastic cages is effective for a variety of predators. Addison (1997) described a type of wire cage used with sea turtles, while Graham (1997) and Ewert and Jackson (1994) provide directions for constructing predator excluder cages that were effective in protecting hatchlings of red-bellied turtles and alligator snapping turtles, respectively. Eggs can also be transplanted to a beach hatchery where a common fence protects all of the nests (see section on Hatcheries, this chapter).

Predation is not the only factor leading to mortality at nesting sites. Depending on the locality, flooding and automobile traffic exemplify other serious hazards (see also chapters 2 and 6, this volume). Páez (1995) found that flooding rather than predation was the greatest source for nest mortality of yellow-spotted Amazon river turtles on the Caquetá River in Colombia. Pezzuti & Vogt (1999) recorded that approximately 56% (111 nests) of the (mainly) six-tubercled Amazon river turtle nests that they monitored on a nesting beach on the Japurá River in Brazil were destroyed by flooding. Von Hildebrand et al. (1988) and Soini (1995) also recorded quite variable, but often high, levels of egg mortality (to 100%) due to flooding in South American *Podocnemis* river turtles in Colombian and Peruvian populations they studied. Transplanting eggs to higher ground or to beach hatcheries to protect them from flooding has proven beneficial to both giant South American river turtles (IBAMA, 1989; Soares, 1995; Vogt, 1997) and yellow-spotted Amazon river turtles (Páez, 1995; Páez & Bock, 1997).

Highways separating turtle nesting and feeding habitats are often killing grounds of the highest order (e.g., Gibbs & Shriver, 2002). One study of roadkills for nesting diamondback terrapins recorded 4020 casualties along a 35-km stretch of New Jersey coastline from 1989 to 1995 (Wood & Herlands, 1995, 1997). Perhaps the most reliable solution to such a problem is to stretch fencing between the road and the turtle habitat and, if necessary to construct artificial nesting sites along the fence. Alternately, fencing can be used to funnel nesting females to an underpass or culvert that would allow them to safely reach the other side. Preliminary studies on tortoises (Fusari, 1982; Ruby et al., 1994; Boarman, 1995; Boarman et al., 1997) suggest that fencing and culverts can be beneficial conservation techniques, but to date the method has not been tried with river turtles.

Due to the expense and a host of other factors, fencing and culverts may not always be feasible. Turtle crossing signs have been tried in certain areas such as Portage, Wisconsin (Jones, 1994), Wabasha Co., Minnesota (Levell, 1995) and Spring Lake, Illinois to alert motorists (fig. 7.2). Such signs—particularly those coupled with reduced speed zones—could potentially reduce mortality in critical areas, but to date this has not been confirmed. A major problem is vandalism and theft of signs. For example, the Spring Lake experiment failed because the signs were continually stolen. Indeed, in Minnesota and Wisconsin the signs are removed following the nesting season for this reason.

Finally, conservationists should be aware that the loss of a female does not have to include the loss of all of her eggs. Unbroken eggs can remain viable for one to several hours depending upon the environmental conditions. If roads are regularly patrolled, eggs from road kills can be salvaged and hatched artificially (Lardie, 1976; Tucker, 1995; Wood & Herlands, 1995, 1997). A few eggs hatched will not compensate for the loss of an adult female, but at least the death will not be a total loss.

Sometimes a simple technological innovation can play a major role in the conservation of a species. The addition of turtle excluder devices or "TEDs" to commercial shrimping nets (in conjunction with the rigorous enforcement of laws which require their use) has reduced sea turtle by-catch and drownings as a result

Figure 7.2
A turtle crossing sign at Portage, Wisconsin, designed to alert motorists to crossing Blanding's turtles (*Emydoidea blandingi*) during their nesting season.

(see Meylan & Ehrenfeld, 2000). Similarly, diamondback terrapins are killed in large numbers as a result of drowning in commercial crab traps (Wood, 1997). Wood (1997) proposed a crab trap design which would exclude most terrapins from entering the trap, but not impede the entry of crabs. Widespread adoption of such a modified trap by commercial crabbers could save the lives of tens of thousand of terrapin annually.

Protective Laws, Conventions, and Organizations

Various laws, conventions and organizations contribute to protecting river turtles or their habitats. Available protection ranges from local game laws up to international conventions. Some of the most important are reviewed here along with their strengths and weaknesses.

International Treaties, Rules, and Conventions

CITES, The Convention on International Trade in Endangered Species, is arguably the best known and most wide-reaching of the international conventions. Established in 1973, CITES has now been ratified by 164 nations. The Convention regulates trade in those species (and their products) perceived to be threatened with extinction by listing them on three appendices. Appendix I species are considered most threatened, and no trade is allowable except under special conditions of a non-commercial nature (e.g., scientific study). Appendix II species are not imminently threatened with extinction, but could soon become so; trade is allowed only with export permits from the country of origin. Appendix III can include any species that a member nation wants to protect no matter what its status. Numerous river turtles are currently listed on CITES appendices (see IUCN Action Plan, 1989 and table 2.1, this volume).

CITES has been an extremely important tool for protecting wildlife that is being traded internationally. However, it has a number of limitations, the most obvious being that it can only protect species being traded across foreign borders—it is powerless to protect declining species being traded within a country. Also, CITES cannot be completely effective until all nations become members and are able and willing to enforce the convention within their own country. Countries also can sign the convention but take an exception on certain species which they want to continue to trade.

The International Union for Conservation of Nature and Natural Resources (IUCN) or World Conservation Union is a union of sovereign states, government agencies, and non-governmental organizations. Although the Union cannot legislate statutes, it does play a key role in conserving animals and plants on the international scale. Founded in 1948 under the auspices of UNESCO, the objectives of the IUCN are to promote and assist cooperation between governments, and national and international organizations involved with the protection of nature and especially declining species. The World Wildlife Fund (now World Wide Fund for Nature) was initiated in 1961 to fund the programs of the IUCN. Among its many contributions to the global conservation effort, the IUCN publishes the Red Lists formerly (the Red Data Books) that categorize the conservation status of threatened species in selected animal and plant groups, a practice begun in the mid-1960s. The present system of classification uses eight categories: Extinct (EX); Extinct in the Wild (EW); Critically Endangered (CR); Endangered (EN); Vulnerable (VU); Lower Risk (LR); Data Deficient (DD); and Not Evaluated (NE) (Baillie & Groombridge, 1996–2001). The threatened categories (CR, EN, and VU) are determined by quantitative criteria involving population size, population structure, rate of decline, and area of distribution. Table 2.1 lists the species of river turtles included in the CR, EN, EW VU, DD, and LR categories. The latter is subdivided into conservation dependent (cd), near threatened (nt) and least concern (lc) (Baillie & Groombridge, 1996–2001).

The Species Survival Commission (SSC)—one of the volunteer commissions of the IUCN—has been particularly important relative to river turtle conservation. A key objective of the SSC is to conserve biodiversity by developing and executing programs to study, save, and wisely manage species and their habitats. To carry out this expansive mission, the SSC has organized an extensive network of knowledgeable volunteers into specific "specialist groups." The Tortoise and Freshwater Turtle Specialist Group advises the SSC on matters concerning the conservation of river turtles. Begun in 1981, the group presently contains 105 members representing 27 countries. In addition to their advisory role, the group has participated in the publication of two action plans to promote conservation action for tortoises and freshwater turtles (IUCN, 1989 and Turtle Conservation Fund [TCF], 2002), and has summarized the available conservation and status information concerning the world's tortoises in a book entitled *The Conservation Biology of Tortoises* (Swingland & Klemens, 1989). Two similar volumes summarizing the conservation biology of riverine species and other freshwater turtles are in preparation.

Most recently, a partnership between the IUCN/SSC Tortoise and Freshwater turtle specialist Group, the IUCN/SSC Turtle Survival Alliance (a newly established [2001] alliance between organizations and individuals that share an interest in building a structured network of captive management and breeding programs into turtle and tortoise conservation efforts), and Conservation International's Center for Applied Biodiversity Science has been forged to provide a coordinated effort focusing on the conservation of the world's endangered chelonians (TCF, 2002). Briefly, the Turtle Conservation Fund, as the partnership is called, has produced an action plan (TCF, 2002) (see also the section concerning education in this chapter), which proposes funding support and conservation action in prioritized phases (see also Georges, 1993 for guidelines for setting conservation priorities for freshwater turtles). Phase 1 targets the requirements of the most critically endangered species, and then gradually expands in Phases 2 and 3 to address problems faced by less endangered forms and to ensure the survival in the wild of all extant chelonians. The phases incorporate coordinated *Action Components* which tie in in-range and out of range "assurance colonies" to provide breeding stock for those species where such is deemed warranted, field surveys and ecological research concerning targeted species, and the development of range country capabilities for enhancing the survivorship likelihood of wild populations of targeted species within their borders. While we continue to urge restraint in over-reliance on captive breeding methodologies where in-situ approaches still offer hope for recovery of species, this coordinated effort by dedicated chelonian biologists and conservationists seems to us to be the best chance yet for reversing the tide of chelonian destruction over the past two decades, and of maintaining current levels of chelonian diversity for the foreseeable future.

Another noteworthy international convention in regard to turtles, The Convention on Wetlands of International Importance (or Ramsar Covention), is exclusively concerned with protection of habitat. The convention obligates member nations to select at least one wetland that they will manage so that human activities will not destroy the ecological characteristics.

An example on a smaller scale is the 1997 ruling by the EC Council of Ministers (the European Parliament) prohibiting the importation of red-eared sliders throughout Europe because of the serious threat which they pose to endemic wildlife such as European pond turtles (Salzberg, 1998a).

National Legislation

A growing number of countries are establishing laws to protect declining species within their own borders. The U.S. Endangered Species Act of 1973 (ESA) is a good example of such legislation. The Act lists both species from the United States and abroad and prohibits the import or trade in any endangered species or its product. Listed species cannot be hunted, killed, injured or collected in the United States. The ESA prohibits federal agencies from authorizing projects that would jeopardize endangered species or their habitat. It provides for setting aside critical habitats (areas essential to the organisms survival) and authorizes the Fish and Wildlife Service to prepare recovery plans for each.

The Indian Wildlife Protection Act (1972) is another example of national legislation that protects some river turtles. The Act contains a number of schedules offering various levels of protection. River turtles are listed in Schedule I (rare and endangered species) and Schedule IV (small game species). The former are completely protected, while species in the latter category can only be captured or killed with a permit. Schedule I species include river terrapin, red-crowned roofed river turtle, Indian roofed turtle, Indian softshell, peacock softshell, Indian flap-shell, and Asian giant softshell. Two additional softshells, Leith's softshell and Indian narrow-headed softshell, currently reside on Schedule IV (Choudhury & Bhupathy, 1993).

Most Southeast Asian countries have similar acts, but the number of river turtles included varies considerably (Jenkins, 1995; van Dijk et al., 2000a). For example, the Asian giant softshell is the only river turtle protected on China's 1989 amended "List of Important Protected Wild Animals" (Zhao & Adler, 1993) and by Vietnams' Rare and Precious Forest Flora and Fauna Decree (Le Dien Duc & Broad, 1995). Laos protects the Asiatic softshell, black swamp turtle and Malayan box turtle. Indonesia's Conservation of Living Resources and Ecosystems Act of 1990 protects river terrapin, Malaysian giant turtle, narrow-headed softshells and Asian and New Guinea giant softshells.

Thailand's 1992 Wildlife Reservation and Protection Act offers one of the most comprehensive types of legislation for the protection of turtles in the region. This Act makes it illegal to capture, possess, or trade any chelonian except the Asiatic softshell. With any of the laws and conventions, the provisions are only as good as the enforcement. According to Thirakhupt and van Dijk (1995 "1994"), this Act is not being fully enforced in the countryside and in fact no one they interviewed even seemed aware of any turtles' protected status. Similar problems plague the efficiency of laws on the books in many other Asian countries and in other geographic locations also (e.g., see van Dijk et al., 2000a for specific regulations and recommended legislation and enforcement in these and other Asian nations they considered).

Mexico's official list has the impressive title of "NORMA Oficial Mexicana NOM-56-ECOL-1994. Que determina las especies y subespecies de flora y fauna silvestres terrestres y acuáticas en peligro de extinción, amenazadas, raras y las sujetas a protectión especial, y que establece especificaciones para su protección especial, y que establece especificaciones para su protección." Listed species are designated P for endangered, A for threatened, R for rare, and Pr for special protection. In regard to river turtles, the Central American river turtle is listed as P and *Pseudemys gorzugi* as R. A number of other species, that may utilize riverine habitats are listed Pr, including *Apalone ater* and *spinfera*, *Trachemys scripta*, *Staurotypus salvinii*, and *triporcatus* and nine species of *Kinosternon*.

State Legislation

In Malaysia, turtle exploitation has traditionally been the prerogative of the ruling Sultan of each state, and hence the federal Wildlife Protection Act does not include turtles. Seven states in peninsular Malaysia now have laws that protect some river turtles and regulate the taking of the eggs. Usually these laws are aimed at river terrapin, painted terrapin, and Malaysian giant turtles. They typically protect adult turtles and regulate the taking of turtle eggs (e.g., licensing). However, typically no limits are set on the numbers of eggs that can be taken. In East Malaysia, non-marine turtles are not protected in Sabah, but painted terrapins and Malaysian giant turtles are totally protected in Sarawak under the Fauna Conservation Ordinance of 1963 (Jenkins, 1995).

The United States offers both federal and state level protection to certain river turtles. Many states have their own list of endangered, threatened, and special concern species. The river turtles, *Graptemys flavimaculata*, *G. oculifera*, *Pseudemys alabamenisis*, and *Sternotherus depressus* are examples of species with both federal and state protection as endangered or threatened species. However, most protection of river turtle populations comes at the state level. Levell (1995) lists thirty-four states with legislation protecting river turtles or at least regulating their exploitation.

Game Laws

Other than the endangered/threatened lists, most national and state protection of river turtles comes in the form of game laws that dictate closed seasons, require licenses, and set possession and size limits.

Closed seasons usually aim at protecting species during periods when they are especially vulnerable such as the time of reproduction. In the United States, eight states have closed seasons for certain river turtles (Levell, 1995). For example, the open season for Missouri softshells is July 1 to December 31. As softshells nest from April to July in Missouri, this law exemplifies the use of a closed season to protect nesting females. In Venezuela, *Podocnemis unifilis* and *P. vogli* can be hunted from February 1 to March 31 (Pritchard & Trebbau, 1984). *P. vogli* nests from November to January, and *P. unifilis* from January to March. In this case, only females of *P. vogli* will be protected during their nesting season.

E. Moll (1990a) suggested a bimodal system of closed seasons for turtles in India that would offer protection to commercially exploited species without putting undue hardship on market vendors. Softshells and hard-shells in India tend to have distinctly different nesting peaks. Most softshells nest from June to December and most hard-shells from December to May. Setting a closed season for each group during their nesting period would still allow market vendors to sell one or the other group throughout the year.

Licenses help to regulate exploitation, but benefit conservation only when some of the money generated is invested in conservation action or when the license includes limitations on take and/or areas to be hunted. In the United States, twenty-four states require some type of license (usually a fishing license) to collect turtles. In most Malaysian states, the collecting of turtle eggs requires a license, as discussed previously. In the State of Perak, the license stipulates that a third of the eggs collected must be given to the Department of Wildlife for their hatchery program. Unfortunately, this requirement has not been rigidly enforced and eggs obtained for conservation purposes must be purchased at market price. The common licensing system in Malaysia actually encourages over-exploitation rather than conservation of turtles. On the East Coast of Malaysia, sea turtle/painted terrapin nesting beaches are licensed out to the highest bidder. As considerable money is expended for the license (ca. U.S. $1000–2000), the egg collector, in order to make a good profit, attempts to excavate as many nests as possible (fig. 7.3), and consequently nearly all of the nests are found and excavated. Recruitment from such beaches is virtually nil. This heavy exploitation to some extent will be self-regulating, in that eventually nesting populations will decline to the point that it will no longer be profitable to expend time and effort collecting the eggs. However, since painted terrapins nest on the same beaches as the far more abundant marine turtles, it will be the crash of the latter populations—not painted terrapins—that will determine when a beach is abandoned. Painted terrapin populations will be extirpated long before the sea turtle egg supply gives out.

Figure 7.3
Harvesting painted terrapin (*Callagur borneoensis*) eggs on a Malaysian sea beach nesting site.

Reasonable possession limits for river turtles can at least slow the rate at which they are lost from the environment. The aforementioned example of the Venezuelan hunting season for *Podocnemis unifilis* and *vogli* includes a daily limit of one turtle and a possession limit of two. Possession limits for river turtles also are included in the game laws of thirteen states in the United States (Levell, 1995). However, when the daily bag limit is a hundred turtles (as in West Virginia) or the possession limit is 12 dozen (as in South Dakota), the conservation value of the law is negligible.

Size limits for river turtles exist in seven U.S. states (Levell, 1995). The typical philosophy of such laws is to protect juveniles and harvest adults. Such a philosophy is significantly flawed, however, as larger, older individuals are most important in maintaining populations of long-lived species such as turtles (Frazer, 1992, 1997; Congdon et al., 1993, 1994; Galbraith et al., 1997; Tucker & D. Moll, 1997). From the conservation perspective, the best laws provide total protection to *all* life history stages (see section on Conservation Philosophy).

Ex-Situ (Off Site) Techniques

Ex-situ techniques involve removing one or more life history stages of the turtle from the wild in order to propagate the species. Although in-situ techniques are generally preferable, they may not always be viable options. For example, if protected areas are not feasible, if remnant populations are too small to persist on their own, or if individuals are too scattered for the sexes to find one another, then ex-situ techniques may be the best hope for survival. Ex-situ and in-situ strategies need not be considered mutually exclusive but can be complementary approaches to the problem. The major ex-situ techniques for turtles; hatcheries, headstarting, and captive breeding, should be regarded as temporary measures aimed at increasing species numbers in the wild to the point that the population becomes self-supporting without them. As such, in-situ actions such as setting up refuges, habitat restoration, reduction of pollution, etc., must be part of the total program providing suitable places for the turtles to go when they are released. Seigel and Dodd (2000) discussed the use of a "gradient approach" in the management program for endangered yellow-blotched map turtles of Mississippi's Pascagoula River, where the costs and benefits of non-manipulative (i.e., in-situ) and manipulative (i.e., ex-situ) approaches are considered and weighed carefully in the determination of an appropriate course of conservation action. They concluded that the more non-manipulative approaches were more likely to be successful and should be tried first in this, and almost every other case. They recommend that a similar decision-making procedure should be applied in the formulation of all conservation planning for threatened species. The need for incorporating the most current knowledge of conservation strategy and cost-benefit analyses into management and recovery plans was further stressed by Klemens (2000a).

Unfortunately, ex-situ techniques are not always chosen for the right reasons. They are often selected because of their high visibility and not because they are necessarily the most effective method for preserving the species. High visibility in this case means that hatcheries, headstart and captive breeding programs, by keeping large numbers of eggs, young or adults available at the project site, can impress the public, the media, and visiting officials that conservation is indeed going on. This, in turn, helps to assure that additional funding will be forthcoming and the program will be perpetuated. If this same money were being funneled into in-situ techniques, the results might be less visible to the public but more beneficial to the species.

Hatcheries

Hatcheries are one of the most widely used and least controversial of the ex-situ techniques. They have long been a mainstay of sea turtle conservation programs, and have been especially popular in river turtle conservation programs in Asia. Examples of such large-scale hatchery programs are the river terrapin and painted terrapin programs in Malaysia and Thailand. The oldest of these is the river terrapin hatchery on the Perak River that has operated continuously since 1967 (see Case Study 3).

Case Study 3: Last Curtain Call for the Drummer.

> With the coming of December, the drumming of the tuntong will again resound along the banks of the Perak. The trumpets, brightly decorated boats and bejeweled ladies with their fancy silks will be gone. The drumming will be softer now, for the turtles number in the hundreds rather than in the thousands. Still the drumming persists—and with it, the hope that the herds will someday regain their former magnificence so that future generations will have the opportunity to marvel at the tuntong's annual nesting spectacular. (E. Moll [1978a] "Drumming along the Perak.")

As described previously in chapter 5, the drummer, a.k.a. river terrapin, tuntong, and *Batagur baska*, has had better times. Once river terrapins ranged from East India to Cambodia, but now Malaysia, Cambodia, and Sumatra appear to be the last theaters that annually replay the river terrapin nesting spectacular, albeit on a much more modest scale than formerly. In India, Bhupathy (1997) could find only three sets of tracks and no natural nests in an extensive 1994 survey of river terrapin nesting sites of the Indian Sunderbans, the last remaining stronghold for the species in India. Similarly, a survey of reported nesting sites in the Bangladesh Sunderbans (December 1989 to April 1990) by S. M. A. Rashid (in E. Moll, 1990b) found no wild river terrapin. Direct evidence of river terrapins was obtained only from four markets along the periphery of the Sunderbans where four live males and three shells were located (Moll, 1990b). This despite Whitaker's (1983) report of a turtle collector in the region that was taking 200 river terrapin annually on hooks baited with mangrove fruit during the early 1980s.

The huge nesting aggregations described by Maxwell (1911) in Burma have also disappeared. Recent surveys of former nesting areas in the Ayeyarwady basin and delta region of Myanmar in 1993 by van Dijk (1994) sponsored by the Turtle Recovery Program and in 1999 by Thorbjarnarson (personal communication) for the Wild Life Conservation Society found no evidence that the population still exists. Heavy egg exploitation along with siltation of nesting beaches, fishing for adults with hooks baited with mangrove fruit and extensive drift netting appear to have been key factors in the demise of the Ayeyarwady population. Wild river terrapin are also thought to be extinct in Thailand (Thirakhupt & van Dijk, 1995 "1994").

This leaves Malaysia, Cambodia, and perhaps, Sumatra as the last harbors for viable river terrapin populations (E. Moll, 1990b). This circumstance is at least partly due to the predominant Islamic religion of these countries that discourages the eating of amphibious animals such as turtles. While this ban does not apply to the turtle eggs, the protection of adults appears to have been sufficient to stave off extinction of their terrapin populations, at least for the time being. Another positive factor has been Malaysia's pioneering efforts in establishing conservation programs for the river terrapin. Hatchery and head start programs for the turtle were initiated on the Perak River as early as 1967 and have since been expanded to include populations on the Kedah, Terengganu, and Dungun Rivers.

Malaysia's Conservation Program for the tuntong was begun by Mohamed Khan bin Momin Khan. As chief of the Perak Game Department, he began a small hatchery on the Perak River in 1967. Mohamed Khan went on to become Director General of the Department of Wildlife and National Parks in Malaysia. Although this position made him responsible for the impressive array of large game animals that occupy the Malay Peninsula (e.g., tigers, Sumatran rhinos and elephants, etc.), the chief (as he is known) never lost his interest in terrapins. The first Perak hatchery was just a fenced portion of the nesting beach which proved vulnerable to flooding. It was first moved to Game Department headquarters at Batu Gajah, but the distance from the nesting beaches (>15 miles over bumpy roads) made egg transport difficult. In the early 1980s, the Perak conservation program became known as Project Tuntung and was relocated near the river at Bota Kanan, Perak (fig. 7.4). The facilities comprised a concrete enclosure with a shallow pool shaded by a roof where adult river terrapin (62 females, 6 males in 1990) are kept for captive breeding and an adjoining large open area of sand divided by a fence into two parts. One part serves as a nesting beach for the captives. The other is used as a hatchery where terrapin eggs, obtained from captives, from a government nesting beach, or by purchases from local egg collectors, are incubated under semi-natural conditions. Next to the combination captive breeding enclosure and hatchery is a roofed nursery enclosure that includes four shallow concrete pools where different age groups of river terrapin juveniles are raised prior to release.

Egg collectors in the region must buy a license that assigns them a beach where they have exclusive collecting rights. In exchange for the license, collectors are required to provide a certain percentage of their eggs to the hatchery. Early on in the program, licensees were allowed to keep only a third of the eggs they collected, one third was to be given to the hatchery, and another third to the Sultan of Perak. However, the hatchery (and presumably the Sultan as well) found they

Figure 7.4
The "Project Tuntong" hatchery, headstarting, and captive breeding facility near Bota Kanan, Malaysia (sign in Malay).

were consistently shortchanged by this system. Even with the eggs obtained from the government beach and from captive stock, the numbers were inadequate to meet hatchery needs. Eventually, the hatchery resorted to paying market prices to the egg collectors in order to obtain adequate numbers of eggs.

When the eggs hatch, the young are moved to a concrete pool in the nursery facility. Here the young may be raised for anywhere from 1 to 3 years or more prior to release. Releases are usually made at the government beach and the young are allowed to run down the beach into the water.

Attempts to evaluate the success of the program have been hampered by various problems. The young are not marked in anyway prior to release, so there is no way of telling when and if they return to the nesting beaches. In addition, egg collectors not wanting the government and competitors to know how many eggs they are taking annually, often do not provide accurate information to interviewers. Nevertheless, attempts have periodically been made to census tuntung nesting. As suggested by the data presented in chapter 5 concerning the history of exploitation of river terrapins, the number of Perak River nesting females and nests showed no signs of increasing and a nesting census of Bota Kanan nesting beaches on the Perak River in 1986/1987 recorded only 190 nests and 4137 eggs (E. Moll, 1987b, 1989, 1990b). However, E. Moll had to rely on reports from the egg collectors who would not allow him to observe on the nesting beaches for fear of disturbing the females, and he thought the actual number of nests might be underestimated. Mohamed Khan (in Moll, 1989) suggested that as egg collectors must pay the State a royalty based on the numbers of eggs that they collect, they are likely to underestimate egg numbers. To rectify the problem, he pledged an effort by the Department of Wildlife and National Parks to station rangers at all of the nesting beaches in 1989/90 and 1990/91 to obtain an accurate estimate of nesting.

In these two censuses (believed to be the most accurate to date), 923 nests and 20,768 eggs were reported in 1989/90 and 849 nests and 20,608 eggs in 1990/91. Depending on whether the tuntong average one, two, or three clutches per year (this has not been established as discussed previously, but at least some lay a second clutch), the number of nesting females in 1989/90 would range from 308 to 923 females. If we compare that range with the early 1970s, when the maximum nests were 1204 (based upon Game Department records), and an estimate of numbers of females would range between 401 and 1204, it is evident that little progress has been made in increasing river terrapin numbers on the Perak River. If anything, the numbers have declined.

What is the reason for the lack of success? There are several possibilities: On the positive side, there is some information suggesting that wild river terrapin grow very slowly to maturity. Therefore the conservation program, despite being one of the longest running for any freshwater turtle, may not have been in operation long enough to have hatchery-produced females returning to nest in any number.

On the other hand, there are forces operating against population survival and certain questionable practices connected with the conservation program that could be responsible for the lack of success. Habitat alteration is one of the most important negative forces. River terrapin are heavily dependent on the productivity of estuaries. On the lower Perak, estates (plantations) have cleared riverside mangroves on which the turtles feed replacing them with brick brack (large rocks) to control erosion. They have also built tidal barrages that block entry into many of the small tributary streams where adult river terrapin feed during high tide (Moll, 1984a) and where juveniles seek shelter from the hazards of the river's main stem. Dams built upriver from the nesting areas have shut off the supply of sand that replenishes nesting banks as they erode. The problem is exacerbated by continued sand mining in the nesting area which removes sand from the river proper and hastens erosion. Although excessive sand loss is not an immediate problem, the Perak may someday experience the loss of high-quality tuntung nesting sites that has already happened on the Terengganu River following construction of the Kenyir Dam on Malaysia's east coast.

Regarding hatchery practices, headstarting of young prior to release is a major aspect of the program (fig. 7.5a,b). Yet the success of this practice remains to be proven in both sea turtle and river turtle conservation programs (see section on Headstarting in this chapter). By hatching and rearing the young under unnatural conditions, it is conceivable that the instinctive processes necessary for survival or return to the ancestral nesting sites could be affected. As the conservation program has seldom marked juveniles prior to release, there is no sure way of determining survivorship and recruitment of the headstarted turtles into the wild population.

(a)

(b)

Figure 7.5
(a) A box full of headstarted yearling river terrapin (*Batagur baska*) from the Bota Kanan facility; (b) their release into the Perak River, Malaysia.

Similarly, captive breeding has had little success in the Perak program. The captive population of 62 females produced only 113 eggs in 1989/90 and 370 eggs in 1990/91. These eggs were unnaturally brittle-shelled and have had a low hatching success according to hatchery personnel (25% versus 45% from eggs collected from the wild population). Part of the problem may be the small number of males (six), the diet (chiefly plants whereas nearly a third of the diet of wild individuals is mollusks), and the shallow tank (0.67 meter depth) where they are kept. E. Moll (1987b) recommended increasing the number of males, the protein/calcium in the diet, and the building of a larger/deeper aquatic enclosure to house the captive population. Sharma (1999) reported that an increased ration of fish in the captive's diet has helped to rectify some of these reproductive problems. A large circular enclosure was built in 1988 with a nesting area in the middle. However, an initial problem was that the enclosure was not sheltered from the sun and the water became too hot. The solution to this problem has been to provide a netting roof for the facility. As of 1990, approximately 50,000 eggs had been incubated at the Perak Hatchery and 21,530 juveniles had been released into the river. Sharma (1999) noted that the juveniles in captivity in 1994 were dying in large numbers from unknown bacterial and fungal infections, and we are obviously concerned that the release of infected juveniles could spread the disease(s) to the wild population.

Two additional programs have been established in the states of Kedah and Terengganu. These programs are smaller, comprising only a hatchery (fig. 7.6) and

Figure 7.6
Hatchery facilities for painted terrapins (*Callagur borneoensis*) near Terengganu, Malaysia.

headstarting facility. In the Kedah program all eggs are obtained from egg collectors, and juveniles may be raised for nine years or more prior to release. From its inception in 1978 to 1989, the Kedah hatchery purchased 6505 eggs and hatched 3881 (60% hatching success rate). However, the eggs are incubated in styrofoam boxes kept in a building which could skew sex ratios in TSD species, and if the river terrapin is found to have temperature-dependent sex determination this could prove to be a problem. No information on sex ratios of hatchlings is available. Over this 11-year period 1629 turtles have been released into the wild.

The Terengganu Program, initiated in 1976, differs in that most of its eggs have been obtained from a government nesting beach rather than from egg collectors. Until recently this was an island in the Terengganu River, but following the construction of the Kenyir Dam the island has eroded away to the point that most turtles have abandoned it for nesting. The Terengganu hatchery at Kuala Brang has shifted its egg collection to a new government beach on the Dungun River. Like the Kedah hatchery, eggs are incubated in styrofoam boxes (fig. 7.7). Between 1976 and 1989, 7403 eggs were collected, 6038 were hatched (81.6% hatching success rate), and 5217 turtles were released into the wild (see also Sharma, 1999). In 1991 with the shift to the Dungun River, 1120 eggs were obtained (267 from the government beach and 853 from egg collectors). Only 391 eggs were obtained from the old government beach on the Terengganu River.

Figure 7.7
Styrofoam boxes, plastic tubs, and natural, in-substrate sites are used for incubating river terrapin (*Batagur baska*), and painted terrapin (*Callagur borneoensis*) eggs in Malaysian hatcheries.

Outside of Malaysia, only Thailand and India have initiated token conservation efforts for river terrapin. Thailand maintains captive breeding populations for both river terrapin and painted terrapin at the Freshwater Wildlife Preservation Station on Klong Langu River in Satun Province of peninsular Thailand. In 1990, the river terrapin population comprised 52 adults (29 females, 23 males). Breeding records from the 1987/88 and 1988/89 seasons indicated 210 eggs laid and 133 hatched (63% hatching success) and 394 eggs laid and 143 hatched (36% hatching success), respectively. The population was still being maintained as of 1994 (Thirakhupt & van Dijk, 1995 "1994"), but apparently no reintroductions had occurred.

In 1988, following the discovery of a nesting beach at Mechua in a Sunderban Tiger Reserve, the West Bengal Forest Department of India established a captive breeding program for river terrapins at Saznakhali and River terrapin Rearing Centres in Pakhiralaya and Jinghahali (Ghosh & Mandal, 1990). Over the period 1988–1991, twenty-one nests were found in the wild and 645 eggs were collected and transported to Saznakhali for artificial incubation and captive rearing. The hatching success rate was 48%. In 1992, the project was head-starting 175 juveniles in several size classes and maintained twenty-five adults as breeders (Bhupathy et al., 1994; Bhupathy, 1995). By 1994, the Forest Department reported that forty captive raised terrapins had been released in the Harinbanga River, in the Sunderban Tiger Reserve (Bhupathy, 1997). As of 1994, Bhupathy could find only eleven individuals at the captive rearing facilities, and ten were being kept at the Alipur Zoo in Calcutta. The fate of the other head-started turtles is unknown. Bhupathy (1997) also reported that twenty-five river terrapin remained at the captive breeding center. Currently interest and funding in the project appears to have dwindled however.

Hatcheries can be as simple as moving the eggs to a nearby protected area of beach, or as complex as rearing eggs in elaborate, environmentally controlled facilities. The technique reduces egg mortality at nesting sites where a high percentage of nests are destroyed each reproductive season. This mortality may result from egg collection by humans, as in the case of Burmese or Malaysian River terrapin (Maxwell, 1911; E. Moll, 1976a,b,c, 1978a, 1984a), by predators as in Indian *Kachuga* (Rao, 1986, 1991), or by flooding as in Colombian river turtles (Páez, 1995; Páez & Bock, 1997). Simple instructions for setting up a beach hatchery can be found in Pritchard et al. (1983) the *Manual of Sea Turtle and Research and Conservation Techniques*.

Hatcheries should not be considered a cure-all, however. There is always the danger that establishing a hatchery will be perceived as solving the problem of a turtle's decline when in fact it more often than not is a band-aid type approach that palliates the condition without addressing the underlying causes of the decline (i.e., exploitation of adults, pollution, habitat destruction, etc.). For most species there is little accurate information available concerning how many eggs must hatch in order to maintain constant population levels. A convenient figure of 10% of the eggs laid is sometimes used, but this has little scientific basis. Nevertheless, if 10%

translates into hundreds or thousands of young turtles being released into the environment, the sheer numbers might convince conservationists that an ample conservation effort is being made. However, presuming that simply releasing large numbers of hatchlings into the wild will solve all conservation problems can be a dangerous supposition. The demise of the leatherback sea turtle (*Dermochelys coriacea*) in Malaysia is a sad reminder of how a hatchery program can lull conservationists into a false sense of security (see Case Study 4).

Case Study 4: Malaysian *Dermochelys coriacea* – A Conservation Effort That Failed. In the 1950s, the leatherback sea turtle rookery at Rantau Abang on the east coast of Terengganu, Malaysia boasted over 10,000 nests annually and ranked as one of the five largest in the world (Hendrickson & Alfred, 1961; Mortimer, 1989). This claim to fame was short-lived, however. By the mid-1990s the number of nests had plummeted to fewer than 100 per year or approximately 1% of the population levels in the 1950s (Chan & Liew, 1996). This catastrophic decline occurred despite a Fisheries Department hatchery for the species which had been in place since 1961. The fall of the Malaysian leatherback population from world class to endangered is examined herein.

Realization that the population was seriously declining developed slowly. Following an estimate of 10,000+ nests for 1956 by Hendrickson and Alfred (1961), few records were kept of nesting until the Fisheries Department began a ten-year tagging program in 1967. An analysis of these data was not published until the late 1980s when Chua (1988) calculated an average 3765 visits (nests) per annum between 1967 and 1976. Prior to this, a significant decline had been suggested by the findings of Siow and E. Moll (1982), who estimated 3500 nests in 1978 based on records of licensed egg collectors. Subsequent nesting surveys (1979 to 1984) were commissioned by Esso Oil Company to determine the environmental impact of off-shore oil development in the South China Sea. Since 1984 the Fisheries Department of Terengganu has collected nesting statistics for the leatherback. Chan and Liew (1996) assembled the data from all of the above sources and calculated that the decline, from 1956 to 1995 averaged 260 nests per year over the 39-year period. In graphing the decline, they observed two periods of particularly steep decline between 1972 and 1974 and again between 1978 and 1980. The first period coincided with the advent of trawlers in Malaysian waters and the rapid rise of the fishing industry in the region. The second period coincided with the development of the Japanese squid driftnet fishery in the North Pacific.

Certain causes for the leatherback's decline are apparent. The close coincidence between major periods of decline and the use of new and more efficient fishing techniques strongly suggests incidental capture and drowning by fishing gear as a major contributor to adult mortality. At the other end of the life history scale, over-harvesting of eggs has likely been the most significant mortality factor. Excavating of turtle nests has been a time-honored tradition among the coastal Malays. The great popularity of turtle eggs extends beyond mere taste and nutritional value, as they are reputed to have aphrodisiacal properties. Demand for the eggs has long exceeded supply, and egg harvests have typically approached 100% on Terengganu beaches. In 1951, the Terengganu State Government moved to control egg collecting by passing the "Turtles Enactment of 1951," which authorized the division of nesting beaches into egg-collecting tracts that could be auctioned off annually to the highest bidder. Although the law protected adult turtles, no limits were placed on the number of eggs that could be collected and near-total egg harvests continued. With adults suffering high mortality at sea and negligible recruitment, a population crash became inevitable.

Concern for the future of the leatherback prompted Malaysia to begin its first turtle conservation effort in West Malaysia. Proposed by the Malayan Nature Society in 1960, a hatchery was set up at Rantau Abang, Terengganu in 1961 and remained the only conservation measure for the turtle over the next 25 years. It is now obvious that the measures taken were inadequate to meet the task at hand. However, this knowledge was late in coming. At the time the hatchery was beginning, resource managers argued that almost all of the eggs laid would be naturally taken by predators, and so protecting 2–3% of the eggs laid was deemed adequate to maintain a stable population (Hendrickson & Alfred, 1961). Recognizing that adult mortality was increasing due to intensified fishing pressure, in 1978 an effort was made to incubate at least 15% of the eggs being laid (Siow & Moll, 1982). This percentage continued to increase until 1988 when all consumption of leatherback eggs was banned in Terengganu and the hatchery began to incubate all (or nearly all) of the eggs being laid (Chan & Liew, 1996). By this time, however, the population was in free-fall, and total protection of the relatively small numbers of eggs still being laid was inadequate to halt the decline. A high degree of infertility among eggs laid at the rookery due to a sex ratio imbalance (over-production of females) that resulted from hatchery techniques further exacerbated the problem. Chan calculated that hatch rates at Rantau Abang could not exceed 56.4% (Chan & Liew, 1996). Adult mortality from incidental capture and other factors had become too great to be replaced by such limited recruitment.

The rate of survival from egg to adult is unknown, but respected turtle scholars occasionally make educated guesses on the subject. Mortimer (1989) quotes two estimates based on a 75–85% hatch rate made by sea turtle biologists Nat Frazer from the United States and Colin Limpus from Australia. Frazer suggested that one adult female might survive out of 1000 eggs, whereas a more pessimistic Limpus thought one adult female might be produced out of 10,000 eggs laid. Mortimer reported that between 1961 and 1986 the hatchery at Rantau Abang incubated an average of 33,263 eggs annually. After adjusting for the 50% hatch rate typical at Rantau Abang, she concluded that as no more than twenty, or as few as two, new females might have been produced each year over this 25-year stretch if either Frazer's or Limpus' estimate was accurate. Such low recruitment would help explain the crash of the rookery over this time.

In *So Excellent A Fishe: A Natural History of Sea Turtles*, Archie Carr (1967) asserted that sea turtles could not be saved by controlling only one phase of the life cycle. *Dermochelys* conservation problems in Malaysia affirm this conclusion. A growing number of computer models simulating turtle population dynamics have reached similar conclusions: increasing the survival of early life history stages of long-lived vertebrates is unlikely to be effective in rebuilding declining populations unless the survivorship of older individuals is first returned to high levels (see discussion under Conservation Philosophy, this chapter). Rescuing a declining population of a long-lived species necessitates protection of all stages of the life history (Congdon et al., 1993).

The plight of the leatherback in Malaysia is bleak but not hopeless, as at least one other leatherback population recovered from even lower numbers. Nesting leatherbacks on the Tongaland Coast of Natal South Africa reached a low point of five nesting females in 1966, but with vigorous protection these numbers have increased to 124 in 1995 (Hughes, 1996).

A number of additional conservation measures are now in place to augment the hatchery effort in Malaysia. There has been a moratorium on the use of highseas drift nets since 1992. Large-mesh driftnets are also banned in Malaysian coastal waters, and a 14-km stretch of nesting beach at Rantau Abang has been gazetted as a sanctuary to prevent further development. A complete ban on egg consumption remains in effect, and hatchery procedures have been improved to increase hatching success (including incubating a certain percentage of the eggs at male-producing temperatures). Whether these measures will suffice to rebuild the rookery at Rantau Abang remains to be seen.

Unfortunately, in the case of long-lived species such as sea and river turtles, the effects of over-harvesting eggs or of hatching too few eggs in a hatchery are not immediately obvious. Rather, the population is covertly undermined from the bottom up. Recruitment is too low, but the young have cryptic habits and are seldom seen so their decline is not obvious. Adult females, being long-lived, continue to return to the nesting beaches each year and superficially the population may appear healthy for many seasons. When large cohorts of females finally begin to die off from old age, the nesting population, lacking recruitment of young adults, then declines rapidly. By then additional conservation action may be too late. Mortimer (1995a) has clearly described how overharvesting of eggs can destroy a green turtle population.

A variety of other problems can plague effective hatchery operation, among the more significant being reductions in egg viability from transplantation and negative effects resulting from altered hydric and thermal conditions of artificial nests. For these reasons, particular care should be taken in transporting the eggs and with duplicating the natural temperature and moisture environment of the wild nests as closely as possible. van Dijk (1990) noted that special care may be required during the incubation of hard-shelled eggs in particular.

Transferring eggs typically results in some reduction in viability. Data from sea turtles indicates that moving eggs lowers hatchling success from 20 to 70% (Mortimer, 1988). Obviously, a technique is only warranted if the survival of transplanted eggs exceeds that of natural nests. Careful packing of the eggs to be transported can greatly enhance their survivability. River terrapin eggs range from brittle to moderately pliable. In the latter type, thumb pressure produces indentation but some fracturing of the mineral layer of the shell usually results. Both types are easily damaged and require particular care during transportation. E. Moll (1978b) reported having packed 425 semi-pliable eggs of painted terrapins in styrofoam boxes, separated by layers of sand and transporting them over five miles of unpaved roads strapped to a bicycle. Individual clutches were then stored in sand-filled plastic garbage cans or styrofoam boxes until hatching. The hatching success was 83%.

Unlike chicken eggs, which should be turned as they incubate, turtle eggs should be kept in the same position throughout incubation. Early in development, the embryo adheres to the upper portion of the egg shell, and should the egg be turned over after that time then the weight of the eggs contents may rest on the embryo, leading to unnatural development or death. When transporting eggs, it is a good idea to move them as soon as possible after laying and to mark the top of the egg with a soft pencil to maintain the orientation.

Other potential problems can occur if the hydric and thermal characteristics of the natural nests are not duplicated. Hydric conditions can be particularly significant for species with flexible-shelled eggs. Embryos developing from eggs with a favorable water environment are able to metabolize more of their energy reserves, better utilize calcium from the egg shell, grow larger, and have greater hatching success than embryos developing in an unfavorable environment (Morris et al., 1983; Packard et al., 1981a,b, 1983; Packard & Packard, 1986; see also chapter 2, this volume). The benefits of developing in a favorable hydric environment may continue beyond hatching. For example, experiments by Miller et al. (1987) indicated that common snapping turtles hatching from eggs incubated on wetter substrates were faster swimmers and runners than those hatching from eggs incubated on drier substrates (see also Filoramo & Janzen, 2002).

Another potential problem of hatcheries is that sex ratios and hatchling sizes can be altered if characteristics of natural nests are not duplicated. Souza and Vogt (1994) determined that hatchlings of the yellow-spotted Amazon River turtle were larger from nests with eggs incubated at higher temperatures, even when hydric conditions among all nest was similar. Many, but not all, river turtles have temperature-dependent sex determination (see chapter 2, this volume). In other words, the temperature at which the eggs are incubated determines the sex of the hatchlings in many species. Typically, in most river turtles if incubation temperatures are too cool, the resulting hatchlings will usually be male-biased; and if too warm, female-biased. Thus, if artificial nests are dug deeper than normal, they could produce all males, and if too shallow all females. The leatherback hatchery in Terengganu, Malaysia may have produced predominantly female hatchlings for many years, leading to male impoverishment and a high level of infertility in nests (Chan & Liew, 1996). Incubating eggs in containers or in buildings can similarly alter the natural sex ratios of hatchlings. Many sea turtle conservation programs probably suffered significant set backs in the 1970s and early 1980s when eggs were routinely incubated in styrofoam boxes stored in shaded buildings—a process now known to promote masculization of hatchlings (e.g., Morreale et al., 1982; Mrosovsky, 1982, 1983; Spotila et al., 1983; Dutton et al., 1985). Although biasing sex ratios can be damaging to a population, it is also being debated as a possible tool in turtle conservation (see Case Study 5).

Case Study 5: **Debate: Manipulating Sex Ratios of Turtles.** Biologists have the ability to control sex ratios of hatchling turtles being produced in hatcheries. Over the past two decades, the problems and potential problems of altering sex ratios through hatchery practices have received most of the attention (see Hatcheries, this chapter). More recently, the discussion has shifted to whether sex ratio manipulation can have positive applications as a tool for conservation. Before looking at opposing views on this subject, a brief review of the technology may be useful.

Two methods of environmental sex determination (ESD) are available to manipulate sex ratios of turtles. Temperature-dependent sex determination (TSD) was reported for reptiles by Charnier in 1966 based on a study of an agamid lizard. Pieau (1971) first reported the phenomenon in turtles, though not all turtles have TSD, as discussed previously in chapter 2. TSD has been demonstrated in many hard-shelled species, but is not known among soft-shelled species. In those turtles having TSD, the sex of the embryo is determined by incubation temperature during the middle third of development (critical period). The threshold temperature (the incubation temperature that will produce equal numbers of each sex) varies with species and geographic locality. Generally, however, temperatures above the threshold produce females, and those below produce males. In some turtles there are two pivotal temperatures whereby females are produced at high and low temperatures but males are produced at intermediate temperatures (see chapter 2, this volume).

Hormonally induced sex determination is an alternative method of manipulating the sex ratio of turtles, that has been reported by a number of authors including Pieau (1974), Gutzke and Bull (1986), Gutzke and Chymiy (1988), Bull et al. (1988), Crews et al., (1989), and Vogt (1991, 1994). With this method, estradiol injected into the egg or applied topically during the middle third of the developmental period, produces all females.

The manipulation of sex ratios is already receiving some commercial application. For example, on an Iowa turtle ranch snapping turtle eggs are being incubated at male-producing temperatures because males attain market size more rapidly than females (Horine, 1996). Can manipulation of sex ratios also benefit conservation? Mrosovsky (1981) first suggested that feminizing of sea turtle populations was worthy of study. More recently, Vogt (1994) discussed the advantages of increased artificial incubation in turtle conservation programs. Under the assumption that female turtles are more valuable than males in the population, he argued that feminizing sex ratios would augment declining populations.

Vogt's advocacy of sex ratio manipulation elicited a rapid response. Mrosovsky and Godfrey (1995) cautioned that risks of feminizing populations be carefully considered, and that informative experiments should be carried out prior to wide-scale adoption of this technique. They drew parallels with headstarting—another "superficially attractive" conservation tool that was undertaken without sufficient thought as to how it should be evaluated. Girondot et al. (1998) used a theoretical model to reveal a potential problem with feminizing turtle populations (i.e., a selection for maculinizing alleles that can lead to a long-term increase in the male proportion of the population). They also found that the exogenous estrogens used in the hormonal method to feminize can also cause arrest of lengthening in the Müllerian ducts and in the opening of the tube's caudal end in the Wolffian ducts. In either case, females would be unable to expel eggs from the oviducts. Hence, they recommend actions that protect adults and in-situ techniques such as protecting nesting sites and improving natural conditions of incubation instead of feminization.

Lovich (1996) argued that many turtle populations have naturally unbalanced sex ratios which may be adaptive to the species, and that alteration of these ratios could have negative results. Lovich included a number of potential problems that could result from fewer males: a diminished effectiveness of multiple paternity and sperm competition in promoting fitness; reduced genetic diversity due to lowering the genetically effective population size; and decreased fertility of eggs. He also suggested that ecological consequences of excess females such as upsetting the balance of inter-sexual resource partitioning, need to be considered. Lovich concluded that there was currently inadequate evidence to evaluate fully the impact of sex ratio manipulation upon chelonian populations, but that available information argued against it.

Vogt's paper has already served as an important stimulus eliciting discussion on the use of manipulating sex ratios and more opinions likely will follow. On the one hand, ex-situ techniques such as these tend to be criticized as short-term technological fixes, treating symptoms rather than causes and replete with hidden problems. On the other hand, for many turtles these are desperate times and, if used judiciously, ex-situ techniques may be able to buy the time needed to establish in-situ methods of conservation. While agreeing with the aforementioned pleas for caution concerning the use of these techniques, we also feel that each deserves its day in court, and that ex-situ methods should not be eliminated from consideration just because they are controversial.

Headstarting

Whereas hatcheries attempt to improve survivorship of the eggs by protecting them from natural predation, headstart programs attempt to improve the survivorship of the hatchlings by growing them in captivity to a size that theoretically protects them from the high rates of predation they presumably face in the wild. If all this sounds innocuous enough, you will probably be surprised to learn that "headstarting" has become one of the more controversial methods in turtle conservation biology. Much of the debate has been waged in the sea turtle arena, and at the core of the debate has been the lack of evidence that headstarted turtles will return to their nesting beaches. In 1996, two Kemp's Ridley turtles became the first documented headstarted sea turtles to return to an experimental imprinting site when they nested at Padre Island in Texas (Shaver, 1996). In 1998, three additional headstarted turtles returned. Editorials in the *Marine Turtle News Letter* (Allen, 1990, 1992; Woody, 1990, 1991) and an article in *Science* (Taubes, 1992a) with letters of response (Wibbels, 1992; Shaver & Fletcher, 1992; Taubes, 1992b) provide the essentials of the headstarting debate. Proponents argue that these programs should be promoted because they increase the survival of juveniles and because the high visibility of headstart programs enlists public concern and interest. Critics argue that headstarting is expensive, the benefits remain to be demonstrated, and that it treats symptoms of the problem rather than the basic problem(s) causing the decline. They maintain that the method may even be harmful to turtles citing a number of biological questions including:

- Will headstarted turtles be able to properly imprint to nesting sites if they are released a year or so beyond hatching?
- Will an extended period of captivity with limited exercise and unnatural foods affect the turtles' ability to survive in the wild?
- Are turtles raised under crowded conditions likely to spread captive-borne diseases to wild populations?
- How will the absence of hatchlings affect natural ecosystems?

For reviews of headstarting as a tool in sea turtle management, see Mortimer (1995b) and Frazer (1994).

Headstarting is not strictly a technique of sea turtle biologists, however. Examples of headstarting programs and experiments among freshwater turtles include giant South American river turtles in Brazil (Vogt, 1995), Indian softshelled turtles on the Ganges River (Sinha, 1995); painted terrapins in Malaysia (Moll, 1990b); river terrapins in India (Bhupathy, 1995) and Malaysia (Moll, 1984a, 1985c); diamondback terrapins (Wood & Herlands, 1995, 1997) and American red-bellied turtles in the United States (Haskell et al., 1996); and Thai narrow-headed softshells in Thailand (D. Moll, pers. obs.). At certain beaches in Brazil, giant South American river turtle hatchlings are collected upon emergence from natural nests and kept for only one or two fortnights until the shell hardens and conditions are right for release (IBAMA, 1989). At the other extreme, in Malaysian programs some cohorts of juveniles may be kept for several years before release (fig. 7.8a,b). One year seems to be the most common period of headstarting.

The debate about headstarting is also not confined to the sea turtle biologists. In a review of turtle survivorship patterns, Iverson (1991) reported that turtles typically show type III survivorship in which most of the mortality occurs in the egg or juvenile life history stages. From this he deduced that reductions in mortality at these stages should have a greater impact on population increases than

(a)

(b)

Figure 7.8
(a) Tanks of headstarted young river terrapin at Bota Kanan; (b) a mixture of older juvenile river terrapin and painted terrapin at the Kedah facility, Malaysia.

reductions in adult mortality. In conclusion, Iverson recommended that "headstart programs" be applied more aggressively to terrestrial and freshwater turtles. Iverson's conclusions on the importance of early life history stages and the promotion of head starting were later challenged by several studies utilizing life-tables and computer models (Congdon et al., 1993, 1994; Galbraith et al., 1997; Heppell et al., 1996). The basic conclusions of these studies are that survival of adults and sub-adults are most critical to population stability in the long-lived turtles, and which conservation programs (such as hatcheries and headstarting) which concentrate on protecting a species only through the first year of life are unlikely to be successful until the causes of adult mortality are addressed. Heppell et al. (1996) found that headstarting can be helpful in rebuilding turtle populations when the survival of adults is maintained at high levels (with the proviso that headstarted juveniles are as viable as wild turtles).

Critics of headstarting programs (e.g., Pritchard, 1979; Dodd, 1982; Mrosovsky, 1983; E. Moll, 1984a, 1985c, 1989; Mortimer, 1988, 1995b) argue that headstarting remains an unproven/experimental procedure and thus should not be used as the sole basis for any conservation program involving long-lived species that may require many decades to prove effective. Despite the first documented returns of headstarted sea turtles to a nesting beach, headstarting remains a controversial procedure. In regard to freshwater turtles, there is some evidence that headstarting can improve juvenile survivorship (Haskell et al., 1996), but little evidence exists to indicate that it can be effective in rebuilding depleted populations of freshwater turtles. The Malaysian conservation effort—a 30-year-old ex-situ program—still shows no evidence of having reversed the decline of the river terrapin (E. Moll, 1989; see Case Study 3; Sharma, 1999). Until existing headstarting programs demonstrate some definite successes, the large sums of money required to run a headstarting program might better be spent on hatcheries, in-situ methods of conservation, and protecting more advanced life history stages.

Captive Breeding

Captive breeding for conservation purposes involves raising and reproducing species in captivity with the goal of releasing captive-bred specimens back into the wild to restock depleted populations (fig. 7.9). It is another high-visibility, ex-situ technique currently being used in some river turtle conservation programs and being considered for others. Captive breeding programs are part of the river terrapin conservation program for river terrapins in Malaysia and for Thai narrow-headed softshells (Kitimasak, 2002), river terrapins and painted terrapins in Thailand (see Case Study 3). In Thailand, the Thai narrow-headed softshell is critically endangered, and both terrapin species are thought to be extinct in the wild (Thirakhupt & van Dijk, 1995 "1994"). If the latter is correct, then all of the surviving members of both terrapin species are now in captivity at the Department of Fisheries facility in Satun Province. Case study 3, Sharma (1999), Moll and Moll (2000), and Kitimasak (2002) describe the Malaysian and/or Thai conservation

Figure 7.9
The river turtle captive breeding enclosure at Kuala Berang, near Terengganu, Malaysia.

programs which exemplify some of the problems of captive breeding of large river turtles.

Captive breeding and rearing are expensive, manpower-intensive, and suffer from many of the same problems discussed for headstarting (E. Moll, 1985c; Philippart, 1995; Rahbek, 1993). In addition to the expense and labor requirements, Snyder et al. (1996) list several additional problems of captive breeding. Herein, we look at several of these problems in context with river turtle conservation.

- Establishing self-sustaining captive populations. Large river turtles, because of their size and predilection for flowing waters, may not survive or breed well as captives in ponds or tanks, particularly when they are crowded (see Case Study 3).
- Genetic problems: using small captive populations to provide a high percentage of the recruitment for wild populations can produce a significant bottleneck in the gene pool (Philippart, 1995). There are also potential deleterious effects (outbreeding depression) of mixing gene pools through translocating captive-bred individuals to different river systems (Reinert, 1991). Long-term breeding within captive populations can lead to inbreeding depression. Techniques to alleviate inbreeding depression such as selective breeding from founding family lines are not being done in captive Asian river turtle populations, and may be impractical for many species that do not breed well in captivity (Snyder et al., 1996).

- Reintroduction: a recent review of programs to reintroduce captive bred animals into the wild reported low success rates (Beck et al., 1994). However, Snyder et al. (1996) suggested that animals whose behavior is largely instinctive (turtles?) may have the best chance of being successfully reintroduced (also see Case Study 1 concerning the Indian Crocodile Project). Whether reintroduction should be used as a conservation technique for reptiles has been debated by Burke (1991), Dodd and Seigel (1991), Reinert (1991), and Seigel and Dodd (2000), and will be discussed further below.
- Disease: due to the crowded conditions of many captive breeding facilities, contagious disease is a serious concern for captive stocks. The Cayman turtle farm has provided a wealth of information on diseases of captive-held sea turtles. Among the diseases recorded in green sea turtles at the farm were mycotic pneumonia (Jacobson et al., 1979), herpesvirus infections (Jacobson et al., 1985a,b; Kleese, 1980), chlamydiosis (Homer et al., 1994), and the coccidial parasite, *Caryospora cheloniae* (Jacobson, 1995). A profusion of viral, bacterial, mycotic, and protozoan, caused diseases are also known among captive freshwater species (Highfield, 1994; Izadjoo et al., 1987; Murphy & Collins, 1983). A serious consideration from the standpoint of conservation is infecting wild populations with such diseases by releasing asymptomatic captive individuals (Cunningham, 1996; Jacobson, 1993, 1996). A virulent upper-respiratory mycoplasma in the desert tortoise (*Gopherus agassizii*) is thought to have been introduced into wild populations of the Mojave Desert by the release of infected captive individuals (Jacobson, 1993). The presence of bacterial and fungal infections in headstarted captive *Batagur* juveniles noted by Sharma (1999) could be similarly introduced into wild populations.
- Preemption of other techniques: an expensive, high-profile technique such as captive breeding can divert attention away from the underlying problems causing the species decline and become a "technological fix" that delays solutions rather than remedy problems (Frazer, 1992; Rahbek, 1993; Phillipart, 1995).

Advocates of captive breeding programs can point to some notable successes (e.g., crocodiles [see Case Study 1]). Nevertheless, due to the expense, problems, and uncertainties in removing river turtles from the wild and then later reintroducing captive-bred animals, captive breeding is recommended here only as a last resort, when attempts to recover species or reduce threats to populations in the wild prove inadequate. When captive breeding is employed, it should be connected with recovery objectives for wild populations and should not be proposed as a long-term solution (Snyder et al., 1996). The IUCN/SSC Tortoise and Freshwater Turtle Specialist Group (1989) provides useful guidelines concerning captive breeding and the release of captive bred animals into the wild.

Where captive populations of river turtles already exist, we encourage breeding them, and the TCF Conservation plan described previously relies heavily upon coordinated captive breeding efforts to rehabilitate severely depleted species (TCF, 2002). The Centre for Herpetology/Madras Crocodile Bank (MCB) exemplifies a conservation oriented organization with turtle breeding programs. Since its inception in 1976, the MCB has maintained breeding groups of Indian turtles and tortoises, and they now have a total of 300 individuals of twenty-two species (Whitaker & Andrews, 1997). The results of such breeding can provide important knowledge of husbandry techniques, and can serve to supply zoos and other collections, thus reducing the drain on wild populations. Under carefully controlled conditions such programs may also be useful in restocking wild populations but, as stated above, such actions are fraught with potential dangers such as the spread of disease and outbreeding depression. The Herpetarium of the Zoologico Guadalajara—an organization practicing captive breeding and release of Mexican giant musk turtles and ornate sliders—has been careful to release most of the liberated offspring at the same sites where their parents were collected (Fanti, 1995).

Turtle Farming and Ranching

Turtle farming and ranching schemes usually involve captive breeding for commercial objectives. However, in some instances, they also are touted as conservation techniques under the claim that supplies of captive-bred turtles will reduce the demand on wild stocks and that hatchlings and/or headstarted juveniles from the project will be returned to the wild (e.g., Reichart, 1995; Vogt, 1995). Farms are distinguished from ranches by virtue of being self-sustaining. In a farm, the breeding stock, once acquired, is replaced by captive-bred individuals rather than wild-caught animals. In a ranch, eggs may be obtained in the wild and then hatchlings raised to marketable size. Wild females may be continually incorporated into the captive population to replace those that die.

Commercial farms/ranches for freshwater turtles either supply hatchlings for the pet trade or supply adults for food. Red-eared sliders and painted turtles from farms/ranches in the United States exemplify species being raised chiefly for the pet trade (Warwick, 1986; D. Moll et al., in press; as discussed in chapter 5). Central American sliders (*Trachemys scripta emolli*) were also being raised primarily for the San José pet trade in Costa Rica (Pritchard, 1993a). Pritchard saw some merit in this operation as local people were involved, and adult turtle hunting had ceased in the area (see Mora & Ugalde, 1991), a third of the hatchlings from the wild-harvested eggs were released, and the internal trade in *T. s. emolli* was thought to reduce the importation of exotic red-eared sliders to meet the demand for pets. The most successful of the commercial food species is the fast-growing Chinese softshell. Aqua-culture projects with this species can be found scattered throughout eastern and southeastern Asia (van Dijk et al., 2000a), and the culture of this species has even spread to Brazil (Anon., 1981). In the United States, snapping turtles are occasionally grown commercially in farm ponds. A farm/ranch in southeast Iowa, sells over 125,000 lb (56,000 kg) of snapping turtle meat annually at a wholesale price of $3–4.75 per pound (Horine, 1996).

The large South American river turtles (*P. expansa* and *P. unifilis*) are among the species often touted for such projects. Smith (1974, 1979) felt that both of these species had potential for domestication. Although the giant South American river turtle requires eight years to attain a marketable size of 23 kg, Smith (1975) calculated that a well-managed turtle pond could provide some 3960 kg of meat per hectare (22,000 pounds per acre) compared to 9 kg per hectare (50 pounds per acre) of cattle on the Amazon's nutrient poor pastures. Alho (1985) also estimated the market potential of raising the turtle, suggesting that a farmer starting with 5000 hatchlings could have 1500 mature turtles to sell in eight years. At the market prices of that time he predicted that the farmer would make $22,000 in gross income each cycle, or a net profit of $10,000 per cycle. Illegal farms/ranches have existed in Brazil for some time, but because the turtle was declared endangered in 1967 all trade in the species was clandestine. Now, conservation efforts have raised species numbers to the point that the Brazilian government has decided to reopen domestic trade and allow existing farms to trade turtles legally (Traffic International, 1997). In addition, IBAMA, the conservation arm of the Brazil government, has begun a program to formulate methodology for commercial farming of *P. expansa* and other species in order to slow the harvest of wild stock (Vogt, 1995). These farms can serve as experiments to determine if farming actually can benefit turtle conservation. Within some countries there is some evidence that such may be the case. In Taiwan, farmed Chinese softshells are the source of almost all turtle meat and eggs eaten, and some of the shell used for traditional Chinese medicine (although tremendous quantities of shell are still imported for the latter purpose [see chapter 5]). In addition, huge amounts of meat and live turtles from farms are exported abroad (Chen et al., 2000). The risks of the effects of establishing feral populations of native species remains to be determined, however.

Based on recent studies involving exploitation of long-lived turtles (Congdon et al., 1993, 1994; Galbraith et al., 1997), attempts at achieving an economically feasible and sustainable harvest of wild adults may be an unattainable goal. However, as the species that have been modeled to date are of omnivorous/carnivorous forms from northern populations, future studies should more closely examine tropical herbivorous species before completely dismissing the possibility. While we are not optimistic that conclusions will be different, we support a "leave no stone unturned" philosophy in this regard, and believe that this extremely important issue and conservation and management decisions related to it require a broader base of knowledge than currently exists. A question more apropos to this section of this chapter is whether farming/ranching has significant potential for conservation, and whether its development may alleviate or replace the need for harvesting wild river turtle stocks to meet market requirements.

As with headstarting, the verbal wars concerning the conservation value of farming were first fought within the framework of sea turtle management. Elements of the debate can be found in Reichart (1995) and Mrosovsky (1983) who supported the conservation benefits, and Ehrenfield (1974, 1980) and Dodd (1982) who found few conservation benefits in farming/ranching. Some of their arguments are reviewed in table 7.2. The Tortoise and Freshwater Turtle Specialist Group Action Plan (IUCN/SSC, 1989), while not expressing any opinion or judgment on commercial farms or ranches for crocodiles or sea turtles, felt that such operations offered little potential benefit for conservation of freshwater turtles at that time. Most recently, the pros and cons of commercial farming (especially of Chinese softshells in Asia have been discussed by various authors in van Dijk et al., 2000a).

Table 7.2
Selected arguments for (pros) and against (cons) using farming and ranching of sea turtles.

Pros

1. Farming/ranching provides economic incentives to protect wild populations (a strategy that is particularly important in less affluent countries).
2. They can reduce pressures on wild stocks by providing superior goods in large quantity that will satisfy the market demand.
3. Using doomed eggs and releasing a percentage of the hatch or headstarted turtles to the wild should mitigate the impact of the operation and ultimately benefit the wild population.
4. Useful research will result that can lead to closed-loop farming operations and improvement of conservation techniques.

Cons

1. Farming/ranching of endangered species to supply a foreign luxury trade undermines local conservation laws and regulations.
2. Commercial aquaculture will stimulate markets for turtle products, not reduce them. Exploitation of wild-caught individuals will simultaneously increase due to lower cost and difficulty of distinguishing between wild and captive-raised products.
3. "Doomed eggs" require a subjective determination, but no matter how determined could be more beneficial to conservation if they were removed to safe parts of the beach rather than to a commercial operation. Recent computer models indicate that the release of hatchlings and juveniles will be inadequate to rebuild depleted populations.
4. Although some research beneficial to conservation can result from commercial ranching, most will be directed at raising sea turtles for the market. Small operations may provide no beneficial contributions.
5. Farms/ranches may contribute ill-conceived yet well-publicized conservation schemes that detract attention away from better conservation methods.
6. Escapees which establish feral populations may compete with native species, spread disease, or reduce genetic fitness of wild conspecific populations.

Based on conservation methods proposed by Ehrenfield (1974, 1980); Dodd (1982); Mrosovsky (1983); Reichart (1982). All of these issues, and specifically con #6 (added by us) may also apply to farming river turtles.

Certainly, the type of ranching operations involved in supplying the pet trade or supplying turtle meat to local epicures, with their continued drain on local populations and potential to infect distant communities with introduced species, have few redeeming qualities for conservationists. However, certain types of closed-system farming operations, particularly those involving non-threatened species, may have more conservation potential.

For example, consider the situation in Bangladesh and West Bengal, India, the centers of turtle consumption in South Asia (E. Moll, 1990a; Choudhury & Bhupathy, 1993; Whittaker, 1997). The Bengali people of this region are notoriously fond of turtles, and animals from much of India and Bangladesh are shipped to markets in West Bengal to satisfy their chelonian cravings. However, the long-distance shipping of turtles by rail to West Bengal is expensive, and losses during shipment are considerable. Several species listed as protected on the Indian Wildlife Protection Act are included in the trade. E. Moll (1990a) and Whitaker (1997) have suggested that captive rearing of non-threatened species (either commercially or for home consumption) might prove to be more economical and reduce the demand for out-of-state turtles. Much of West Bengal and Bangladesh is peppered with ponds, and most villages have one to several within their confines. Some of these ponds are already used for raising fish, but it has been the practice to remove turtles from the ponds prior to stocking fish. However, some of the herbivorous Indian river terrapin could peacefully coexist with the fish (e.g., crowned river turtles, Indian roofed turtles). The feasibility of raising these turtles would depend upon a number of biological factors including age at maturity, growth, and fecundity of the species chosen. Whitaker (1997) has developed a pilot scheme to test the feasibility of rearing Indian flap-shells in village ponds.

E. Moll (1990a) and D. Moll (1995) have also suggested the possibility of raising a tried and proven turtle for farming projects in order to reduce demand on wild stocks. Considerable information is already available about the husbandry of the Chinese softshell, *Pelodiscus sinensis*, for example (e.g., Mitsukuri, 1905; Chen, 1976; Choo & Chou, 1983, 1984; Chou & Choo, 1983), and it is currently being raised profitably in many Asian locations (van Dijk, et al., 2000a). If non-endangered species can be raised profitably on a large enough scale to supply local turtle markets, then pressure on wild stocks theoretically can be reduced, and if these are truly closed-cycle farming operations, then humans and river turtles would both benefit. Chinese softshells have been used in Thai farms in place of the native softshell (*Amyda cartilaginea*) because of their greater fecundity and smaller space requirements (Thirakhupt & van Dijk, 1995 "1994"), and presumably this has benefited the native softshell. However, any benefits of importing Chinese softshells must be weighed against the problems that may result should they escape and become established in new environments. We feel that the situation in Asia is desperate enough that the risks associated with feral populations must be considered as the lesser of the evils, and that it is time to determine the full potential of closed-farming operations before wild stocks are completely gone.

Translocation

Moving organisms from one location to another is frequently proposed for conservation purposes. Templeton (1996) listed four uses of translocation to achieve conservation goals: to reintroduce a species into areas where they have become extirpated; to introduce an endangered species into satellite populations as insurance against catastrophes; to augment numbers in natural populations; and to increase the genetic heterogeneity of small populations. The latter usage may develop into a particularly important procedure as riverine populations become increasingly fragmented.

Results from the few attempts at translocating river turtles to date are inconclusive concerning the value of the technique. Medem (1969) reported that a viable population of *Podocnemis expansa* had been introduced above the rapids on the Caquetá River in the Colombian Amazon. Smith (1974) reported the following instances: *Podocnemis* (probably *expansa*) were introduced into Venezuela's Lake Valencia after being headstarted for 10 months. As of the time of Smith's report, no significant nesting had occurred on the lake's artificial nesting beaches. In 1965, "Operacion Tortuguillo" rescued some 80,000 hatchlings (apparently still in nests) from rising flood waters, relocating them in swamps and floodplain channels. The program was discontinued in part due to concerns that translocated turtles, deprived of normal nest emergence behavior, would be unable to imprint on their natal beach for future nesting.

Alfinito et al. (1976a) reported on an effort to translocate adult giant South American river turtles from the Rio Trombetas to the Rio Tapajós in Brazil. Pollution due to bauxite mining, along with unseasonal flooding, was destroying much of the reproduction on the Trombetas. To alleviate the situation, 130 adults were captured on the Trombetas, tagged and translocated to the Rio Tapajós. The authors reported that nesting increased on the Tapajós beaches in proportion to the numbers of females introduced, and some marked turtles had been identified on the Tapajós beaches. None of the introduced females was observed returning to the Trombetas beaches, but marked turtles did turn up in the cities of Obidos in Para and Itacoatira in Amazonas (distances from the release sites were not given).

The Thai Fisheries' Department has been relocating juvenile river terrapin and painted terrapin from their hatchery in Satun into reservoirs throughout the country. Thirakhupt and van Dijk (1995 "1994") criticized the effort on the grounds that these large, poorly vegetated reservoirs lacking suitable sand banks for nesting appear unlikely to provide suitable habitat for a species adapted to a lotic estuarine environment.

Arguments for and against translocation of species were discussed by Burke (1991), Dodd and Seigel (1991), Reinert (1991), McDougal (2000), and Seigel and Dodd (2000). The advantages of being able to reintroduce river turtles into formerly occupied habitats, or to translocate them from doomed to viable habitats, are obvious but must be weighed against possible consequences. Among the more serious problems are the low success rate in previous projects, the potential for spreading disease, social disruption, and genetic pollution. Still another potential problem of using translocation is that should a few cases prove successful, politicians will want to use this easy "solution" anytime a population of threatened or endangered species stands in the way of a favored project. Should translocation be necessary, the following actions can improve the chances of success:

- Causal factors of the original decline should be understood and rectified prior to release (IUCN/SSC, 1989).
- Releases should be timed to coincide with optimal environmental conditions (IUCN/SSC, 1989).
- Use wild-caught as opposed to captive-bred individuals (Griffith et al., 1989).
- Translocate from populations in the same general area (Templeton et al., 1986; Griffith et al., 1989; IUCN/SSC, 1989).
- A larger number of releases increases the probability of success (Griffith et al., 1989). However, simply using a large number of founders does not necessarily ensure that genetic diversity is being preserved. In order to design a sampling program that assures preservation of genetic diversity, genetic surveys of source populations should precede translocation (Templeton, 1990, 1996).

Conservation Philosophy

> We suggest that turtles may be especially vulnerable to population decline because they exhibit reproductive strategies incompatible with exploitation or significant loss of habitat. (Burke et al., 1994)

> Successful management and conservation programs for long-lived organisms will be those that recognize that protection of all life stages is necessary. (Congdon et al., 1993)

Technology and Halfway Technology

Thanks to Nat Frazer, one of our most eloquent chelonologists, "halfway technology" has become a household word among conservation biologists worldwide. The term actually originated with Lewis Thomas (1974), in his popular book, *The Lives of a Cell*, who used it to describe medical technology that treats symptoms but not the underlying causes of the disease. Frazer (1992, 1997) borrowed the term to describe conservation methods such as hatcheries, headstarting and captive breeding that also focus on symptoms of the problem (too few turtles) while ignoring basic causes of the decline (e.g., exploitation, habitat destruction). Frazer noted that *how* a problem is defined can be a major factor in arriving at an appropriate solution. If we perceive the major problem of today's threatened river turtle populations to be inadequate numbers, then we are likely to seek solutions that aim at increasing numbers of turtles. Hatcheries, headstarting, and captive breeding are methods that visibly can add numbers of turtles to the dwindling populations (hence their popularity). However, as Frazer observes, successful repatriation of these animals is highly unlikely unless the original causes of the population's decline have been ameliorated or solved. Besides the rap of being "half-way technology," ex-situ techniques also are criticized as short-term technological fixes and for their narrow focus on early life history phases. These shortcomings are analyzed below.

Hatcheries and headstart programs concentrate on protection of eggs and young juveniles respectively. However, a growing number of studies involving long-term capture release data, population models, and life-table analyses (e.g., Congdon et al., 1993, 1994; Crouse et al., 1987; Galbraith et al., 1997; Heppell et al., 1996; Tucker & Moll, 1997) indicate that such approaches are basically flawed. In long-lived species such as turtles, adult survivorship is the pivotal factor affecting population growth rate and recovery. Egg and juvenile survivorship are not. As turtles approach sexual maturity, their worth to the population increases because their "reproductive value" increases. An individual's likelihood of contributing to future reproduction correlates positively with its age (size). Findings from these models all imply that methods which increase survival of the early life history stages are unlikely to be effective in rebuilding declining populations unless the survival of older individuals is increased concomitantly.

Heppell (1998), in an attempt to identify management alternatives that can be applied across taxa, compared life-tables from several turtle populations using elasticity analysis, a method that calculates the proportional contribution of such factors as age-specific survival rate and fecundity to the annual population multiplication rate. Despite large variations in mean annual fecundity, annual survival, and age at maturity, she found considerable congruence in the elasticity patterns of the species of freshwater turtles examined. The high adult survival elasticity and low fecundity elasticity indicated further supports the contention that effective conservation efforts must aim at reducing adult mortality in order to reverse the decline of populations.

This does not mean that early life history stages should be ignored in conservation programs (Heppell, 1997, 1998). The first step in saving a declining freshwater turtle population should be to protect the older juveniles and adults, but once the survival rates of these older turtles are returned to high levels, then increasing the survivorship of nests and younger juveniles can effectively benefit population growth (Heppell, 1997, 1998; Heppell et al., 1996). If they are to be successful, conservation programs must protect all life stages of the turtle (Congdon et al., 1993, 1994; Crouse et al., 1987).

Another problem with the ex-situ approaches to turtle conservation problems is that they are chiefly technological solutions. Humans are presently living in an ever-increasing technology-dominated world, so it is little wonder that we have a penchant for technological solutions. However, as Conway (1988) asserts, technology is usually a palliative rather than a panacea for the problem. Because of their expense, ex-situ techniques will likely not be sustainable over the long term (Meffe, 1992), and their most important usage for conservation may be as stop-gap measures that can buy the time needed to put long-term, self-sustaining programs into place. Even here, some authors believe that technological solutions should be used only as a last resort in preserving biodiversity (Conway, 1988; Frazer, 1992). Proponents of ex-situ techniques argue that in-situ techniques are often ineffective in developing countries and may alienate local peoples to conservation practices in general (Gomez, 1982; Paez & Bock, 1997). Paez and Bock (1997) argue that in such situations ex-situ techniques (hatcheries, headstarting, and captive breeding) projects should be encouraged so long as they include mechanisms for evaluation. They point out that such programs can be a valuable source of biological information that can be helpful in future conservation efforts, whether or not the techniques prove successful.

Sustainability

Exploitation of river turtles becomes a conservation problem when it is unsustainable, but can it also be a conservation solution? Are there sustainable uses of river turtles that can promote their conservation, or is this concept best categorized as an oxymoron? The idea that economic incentives may be necessary to drive conservation efforts for turtles has been around for a long time. Most often it is applied to conservation in developing countries, the argument being that in poorer countries those people who are concerned about how they can adequately feed their own families will not be swayed to protect animal species unless there is something profitable in it for them. But show them that conserving species can be economically advantageous, and they will respond positively. Farming and ranching schemes have frequently been proposed as a means of improving local economies while concomitantly benefiting turtle populations (see section on Farming and Ranching above). Slogans such as "saving them through use" (Mittermeier, 1978) or "save them by eating them" (Vogt, 1995) are scattered throughout the conservation literature, but are they appropriate in light of modern conservation philosophy?

Over the past decade the notion of wildlife management programs combining economic development of human communities with the conservation of wildlife populations has grown in popularity (Klemens & Thorbjarnarson, 1995). This appears to correlate with the growth of the concept of "sustainable development"—terminology which was first presented in a 1987 United Nation's report, *Our Common Future* (World Commission on Environmental Development, 1987). It has since spread to become one of the most influential contemporary concepts dominating local, national, and world policies and actions including planning and funding of scientific, conservation, and management activities (Frazier, 1995). Initially, the term was defined as "development that meets the needs of the present without compromising the ability of future generations to meet their own needs" (World Commission on Environment and Development, 1987). However, Willers (1994) contends that the term is actually code for perpetual economic growth, a concept that is generally antithetical to conservation goals. Nevertheless, if the purse strings of the agencies involved in funding conservation projects are controlled by this philosophy, investigators working to conserve chelonians must deal with the concept.

With regard to turtles, a key question is, "Are they suitable subjects for sustainable use programs?" Based on information gleaned from computer models and the few long-term studies of turtles discussed above, chelonians appear to have reproductive strategies that are incompatible with exploitation or habitat loss. If these findings are applicable to river turtles at all latitudes, then most "sustainable" exploitation schemes involving regular removal of adult turtles from populations are destined for failure (fig. 7.10).

Sustained yield was one of the earlier scientific doctrines for managing resources over the long term, and it worked well as long as the resource was abundant relative to human demand (Salwasser, 1990). It will not work with depleted populations living in degraded environments. As stated by Frazier (1995): "Sustainable use of turtles (IF it is possible) can only be accomplished once the environments critical to them are managed in ways appropriate to long-lived species." Others argue that total protection of river turtles may not always be feasible. In developing nations where turtles may be important protein resources or an important part of the local economy, it may be politically more feasible to limit rather than to ban exploitation of a species entirely. The conservation plan for the Central American river turtles (see Case Study 6) exemplifies this philosophy. If harvesting cannot be prevented, then protection should be provided in the follow-

Figure 7.10
Harvest of large adult river terrapins (*Batagur baska*) and painted terrapins (*Callagur borneoensis*)—the age/size cohorts that river turtles can least afford to lose if viable populations are to be maintained.

ing order of priority based on the value of each group to the population: large adults, adults, large juveniles, small juveniles, eggs. The harvesting of eggs from nests which are constructed in locations which are likely to be flooded, or of those constructed late in the dry season nesting period and temporally close to the beginning of rising water levels would cause little problem for affected river turtles. This has been suggested as a type of sustainable exploitation which may be possible for several South American river turtle species (Fachín-Terán et al. (1996); Pezzuti & Vogt, 1999).

Case Study 6: Conservation of the Central American River Turtle. The Central American river turtle is a large (up to 65 cm carapace length), herbivorous, highly aquatic turtle inhabiting Atlantic drainages in Mexico, Belize, and Guatemala. The species is chiefly found in large, deep, lowland rivers and large, freshwater lagoons; however, they also occur in small tributary streams and brackish water estuaries (D. Moll, 1986). Chiefly a solitary nester, *Dermatemys* lays multiple clutches (1–4) of large, brittle-shelled eggs (10–13 per clutch) during periods of high water (Vogt & Flores-Villela, 1992b; Polisar, 1995, 1996, 1997). Nesting occurs from September to March in Chiapas, Mexico (Vogt & Flores-Villela, 1992b), and from late September to December in Northern Belize (Polisar, 1995, 1996).

Due to the diffuse nesting habits of the species, exploitation of eggs is insubstantial. Rather, most exploitation is directed at meat with adults and large juveniles emphasized (Vogt & Flores-Villela, 1992b; Polisar, 1995, 1997). Serious declines in Mexican populations were reported by Alvarez del Toro et al. (1979), and hunting has now virtually eliminated the species from much of its former range in southern Mexico (Vogt, in Polisar, 1994; Vogt et al., in press). In 1984, D. Moll observed that *Dermatemys* was still common to abundant in thinly populated areas of Belize, but had seriously declined in areas around certain human population centers (D. Moll, 1986) (fig. 7.11). Exploitation occurs year-round in Belize, peaking in April and May when rivers are low and turtles are easiest to catch. Demand is particularly great at this time as *Dermatemys* is a traditional Easter dinner in northern Belize. Currently, the species is listed as endangered by the United States (Levell, 1995), as an Appendix II species on CITES, APR 1 (highest priority for specific conservation action) by the IUCN/SSC (1989), and EN (endangered) on the IUCN's most recent Red List (1996).

Following studies of the status of *Dermatemys* in Belize, D. Moll (1986) recommended that the remaining intact populations of the species in the country be used to obtain the data needed to manage this unique turtle. In 1989, John Polisar was provided such an opportunity (Polisar, 1990a,b, 1991, 1994, 1995, 1996, 1997; Polisar & Horwich, 1994). Seven villages that comprise the Community Baboon Sanctuary, a grass-roots wildlands management project aimed at maintaining habitat for black howler monkeys, became concerned about over-harvesting of the turtle in their area, and invited him to study the population inhabiting a 32-km section of the Belize River which bisected the sanctuary. Over the next two years, Polisar studied this site as well as a variety of other areas in the country emphasizing reproductive biology, exploitation patterns, marketing

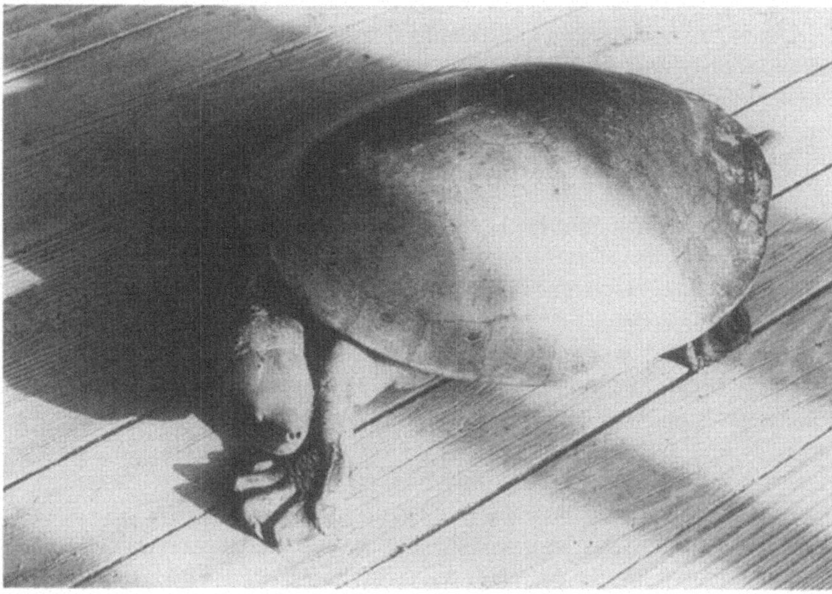

Figure 7.11
A Central American river turtle (*Dermatemys mawii*), purchased in the Belize City market, Belize, and destined for the dinner table of guests of the Hotel Mopan, Belize City.

patterns, and cultural traditions. At the conclusion of his study, Polisar sent a lengthy project report and a short illustrated summary to the government detailing his recommendations for conservation action that could function in a rural, undeveloped region with minimal wildlife enforcement. Additionally, Polisar further promoted his recommendations and educated legislators and the public with slide presentations (in village community centers as well as government and non-government offices), radio interviews, and multiple meetings and discussions.

Polisar felt that with the lack of enforcement capability, a complete ban on harvesting would be unfeasible and would merely lead to general non-compliance. Instead, he promoted allowing subsistence hunting to continue while restricting the activities of commercial fishermen who removed large numbers of the turtle for profit. This was important to maintain the continued support of the majority of the populace. He also promoted the establishment of sanctuary areas where no hunting would be allowed. The recommendations led to nationwide legislation in 1993 to protect and manage *Dermatemys*. Included within this legislation were the following key elements:

- Year-round possession limits that eliminate large-scale removals without affecting subsistence use.
- A short closed season (May 1–31).
- Establish several protected zones where selling and purchasing of *Dermatemys* is completely prohibited.

Although it is too early to determine whether this legislation will have the desired result, certainly it ranks as a positive step toward the conservation of this important chelonian.

Education and Research

Education is an important tool for shaping attitudes conducive to conservation projects as well as a key element in the success of any specific conservation program. Without widespread local support, few conservation programs can survive. For the most part such support can only be gained once the people of the region understand the necessity for the project and how they or their region may ultimately benefit. Education should occur at several levels categorized herein as "awareness," "understanding," and "training/application."

Awareness

Awareness simply alerts the target population to the existence of species, their needs, and their conservation problems. Such instruction is crucial for laying the ground work toward a pro-conservation attitude. A variety of techniques can be used to create awareness including T-shirts, posters, and bumper stickers bearing depictions of turtles and/or conservation slogans. Mass media approaches (e.g., radio and television programs) can also be effective in acquainting the public about turtles and their conservation problems. In particular, early education through books or classroom lessons and projects should be promoted. The earlier that such attitudes can be instilled in humans the better, but it is never too late to try.

In general, children seem to possess an affinity for turtles, and thus lessons involving turtle diversity, value, and problems are likely to be popular and useful in developing an appreciation of biodiversity and in creating a general conservation ethic. In the United States, preschool children are often introduced to turtles through cleverly written children's stories such as *Yertle the Turtle* (Seuss, 1958) and *Timothy Turtle* (Graham, 1946).

As the children grow older, somewhat more advanced books are available that provide biological information along with conservation messages. As one example, the juvenile level book, *Look Out For Turtles* by Berger (1992) notes that some people kill turtles for food, make ornaments from their shells, destroy their homes, and dump poisons into their water. It concludes by telling the readers that they have an important job to do for turtles and lists several things that they can do including: removing turtles from roads; helping to protect and clean the areas where turtles live; and asking for laws that will prevent sea turtles from being caught in fishing nets. *Turtles* by Baskin-Salzberg and Salzberg (1996) was written for children in the 7- to 12-year range. This attractive little book, which is liberally laced with excellent glossy photographs, introduces children to the fossil history of turtles, the present-day forms, and the conservation problems of the group. It also concludes with several things that everyone can do to help turtles including: leave them in the wild, don't buy them as pets, and do not buy products made from turtles. Addresses of organizations that can provide more information about helping turtles are listed at the end. A turtle classic best suited for the junior high level student and beyond, *Minn of the Mississippi* (Holling, 1951) traces the adventures of a three-legged snapping turtle from the cold waters of Minnesota to the bayous of Louisiana. Along the way the reader is treated to a wealth of information on natural history, rivers, boats, and rivermen.

An additional child-oriented approach is to prepare special coloring books promoting conservation themes. An excellent example of this approach is a coloring book created by Indraneil Das (1996) for children in India. On one page of the book is a drawing of a particular Indian turtle for coloring, and on the facing page is a short account of how to recognize the turtle, where it is found, why it is in danger, and how to color it.

For older children and adults, posters are another popular device to promote awareness, and to solicit information from the public concerning threatened species by conservation organizations (fig. 7.12). Some countries have posters depicting their turtles and tortoises (e.g., India and Thailand) and within the United States, posters of reptiles (including turtles) exist for several states (e.g., Illinois, Michigan). Posters can be especially effective in developing countries that have a high percentage of illiteracy. For the literate, field guides and popular books on reptiles and/or turtles specifically, have been very influential in creating public awareness. A sampling of popular turtle books with world-wide coverage include Pritchard (1967, 1979), Obst (1986), Ernst and Barbour (1989), and Rogner (1995, 1996). An updated version of Ernst and Barbour's *Turtles of the World* is now also available on CD-ROM (Ernst et al., 1998).

Regarding specific countries, more books have been written about the U.S. chelonian fauna than any other. The first, *Turtles of the United States and Canada*, published in 1939 by Clifford Pope was succeeded by *Handbook of Turtles* by Archie Carr (1952) which stood as the authority for the next two decades. In 1972, Carl Ernst and Roger Barbour took over with the publication of *Turtles of the United States*. The most recent book, *Turtles of the United States and Canada* (1994) by Ernst, Lovich, and Barbour has come full circle, being titled the same as Pope's original. Not only does this book create an awareness of the species and their habits but it also calls attention to specific conservation problems to a greater extent than its predecessors. *The Year of the Turtle* by Carroll (1991), a beautifully written and illustrated depiction of life of the spotted turtle (*Clemmys guttata*) and other northeastern U.S. aquatic turtles, exemplifies another form of educational turtle literature available to the public. A few countries other than the United States also have popular guides to their chelonian fauna. *The Turtles of Canada* (Froom, 1976), *Australian Freshwater Turtles* (Cann, 1998), Turtles of Thailand (Nutaphand, 1979), *The Turtles of Venezuela* (Pritchard & Trebbau, 1984), *The South African Tortoise Book* (Boycott & Bourquin, 1988, 2000), *Colour Guide to the Turtles and Tortoises of the Indian Subcontinent* and *Turtles and Tortoises of India* (Das, 1991, 1995), and *Tortues Continentales de Guyane Française* (Métrailler & Le Gratiet, 1996) are examples.

WANTED

Reports on Sightings of the ALLIGATOR SNAPPING TURTLE

The ALLIGATOR SNAPPING TURTLE is the world's largest fresh water turtle (record size is 219 lbs.). They average between 60-110 lbs. Missouri is the northern limit of the species range. Habitat loss and over-harvesting have caused a population decline of this species. The ALLIGATOR SNAPPING TURTLE is now listed as RARE in Missouri.

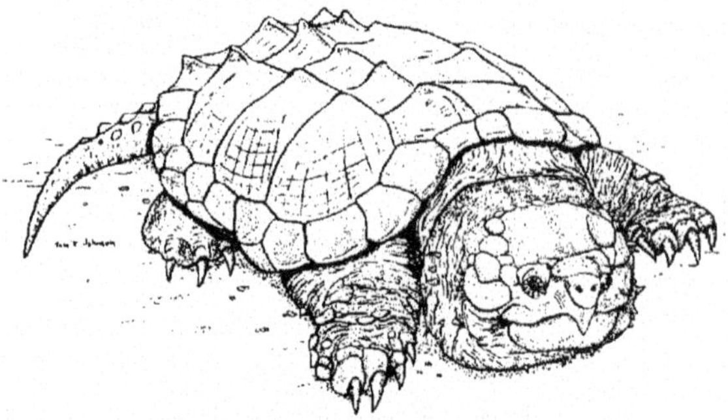

Anyone seeing an ALLIGATOR SNAPPING TURTLE in southern or southeastern Missouri is asked to contact their local Conservation Agent or the person listed below.

The common snapping turtle may be confused with the ALLIGATOR SNAPPING TURTLE. See illustrations and descriptions below.

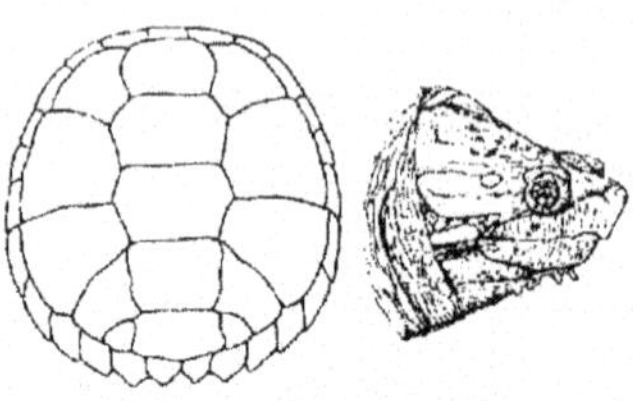

Common Snapping Turtles average 10-35 lbs. Head is pointed with no pronounced hooked beak. Upper shell without 3 prominent ridges and no "extra" scales.

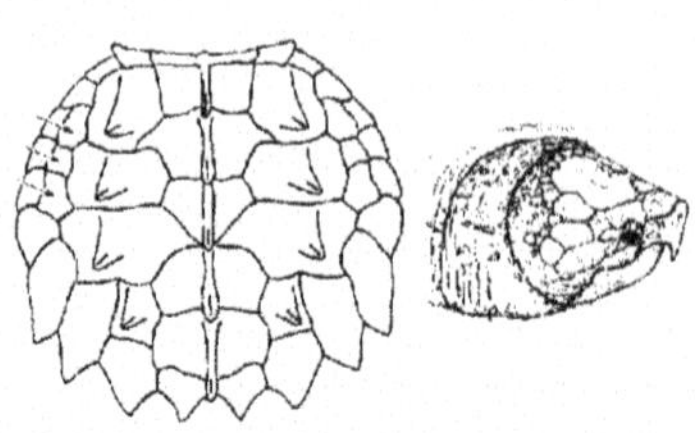

ALLIGATOR SNAPPING TURTLES average 60-110 lbs. Head is large and "chunky" with a pronounced hooked beak. Upper shell has 3 prominent ridges and a row of "extra" scales along each side of shell.

For reporting sightings of ALLIGATOR SNAPPING TURTLES, contact:

Tom R. Johnson, Herpetologist
Missouri Department of Conservation
P.O. Box 180
Jefferson City, Missouri 65102

Figure 7.12
A poster issued by the Missouri Department of Conservation to solicit information from the public concerning encounters with the alligator snapping turtle (*Macroclemys temminckii*).

More specific to conservation are the Red Lists of the IUCN (e.g., Baillie & Groombridge, 1996–2001) that publicize the threatened animals of the world, or more specifically lists for individual countries such as the four-volume *Chinese Red Data Book of Endangered Animals*. In this series, Volume 3 by Zhao (1998) concerns the threatened amphibians and reptiles. The recently published *Turtle Conservation* (Klemens, 2000b) provides a comprehensive overview of conservation issues and recommended conservation action for all the major chelonian groups.

Understanding

Understanding is the next step beyond just being aware of the existence of species or that they are threatened, or endangered. It is to comprehend why species are declining. In the United States, rain forest units are becoming commonplace in the early grades. Students learn that many species of animals and plants are becoming extinct because the rainforest is being destroyed. They are taught why rainforests are being cut, and the consequences that the cutting will have to all of life in the forest. Relative to river turtles, similar units could be designed around what is happening to rivers of the world and how this in turn affects organisms associated with the river.

Information in the scientific literature is frequently too complicated and abstract for the general public to understand or retain. This is where good teachers are needed to serve as intermediaries between the scientists and the general public. Innovative techniques are required to put the knowledge being accumulated by the conservation biologists into forms that allow general understanding. For example, Jeanne Mortimer (1995a) noted the difficulty of explaining to the public why there is such a long delay from when adult turtles and their eggs are removed from the populations until population declines are actually observed. She found that while the population dynamics involved were too difficult for the general public to grasp through words or mathematical formulas alone, they were receptive to diagrammatic representation of these concepts.

Still another way to promote public understanding is through analogy and word imagery. Frazer's (1992) association of the catchy term, "half-way technology," with the concept of treating symptoms rather than causes of a disease creates the sort of analogy that tends to stick in the human mind. Then, his labeling of headstarting and captive breeding as forms of half-way technology allows the public not only to understand but also to remember the conservation problems associated with these ex-situ techniques. More approaches such as these are badly needed in the race to educate the public to the conservation problems of river turtles.

Implementation Education

Implementation education represents the cumulative stage of the educational process, instructing how to apply the methods of conservation biology to the problem. Workshops and training courses made available by organizations concerned with conservation are useful in this regard. An excellent example of this type of program is the University of Michigan's innovative GREEN (Global Rivers Environmental Education Network) project (Stapp et al., 1995). In this program, teachers worldwide take their students to a local river and teach them techniques of studying water quality, watershed usage, and identifying socioeconomic factors leading to the deterioration of river quality. The students share their findings with local officials and with student groups in other countries. GREEN not only provides students with the opportunity to gain an in-depth understanding of their local environmental problems, but also encourages them to act on their findings and to share their information with other cultures on a global scale.

An example more specific to turtles is the recent Turtle and Tortoise Conservation Project in India, a three-year joint Indo-U.S. undertaking between the Wildlife Institute of India and the U.S. Fish and Wildlife Service (Bhupathy et al., 1994; Choudhury et al., 1997). Objectives of this multifaceted project were:

(1) To determine the conservation status of tortoises and freshwater turtles in India.
(2) To identify viable turtle populations and suitable habitats to establish protected areas.
(3) To set up captive breeding units for endangered chelonians aimed at reintroduction.
(4) To acquaint local wildlife officials with conservation problems of turtles in their region and train them in conservation methods suitable to solve these problems.

To satisfy the latter objective, at the end of three years of fact-finding involving field work, museum surveys, and sending out questionnaires to zoo and wildlife officers, a large workshop was held at the National Chambal Gharial Sanctuary in Madhya Pradesh. Federal and state wildlife officers, along with interested biologists from throughout India, were invited to the four-day workshop to disseminate the information collected during the study. Turtle conservation problems and how to deal with them were the main topics of discussion.

Recovery plans and action plans are publications that fall into the category of implementation education. Recovery plans include aspects of all three of the aforementioned levels. Such plans are required by the Endangered Species Acts of the United States and Australia and are also being prepared for species in other countries. Recovery plans typically contain three sections. The first usually provides information on the ecology of the species, the causes of its decline, and criteria to be met before downlisting a species from the endangered category. The second section outlines the actions proposed to achieve recovery. The last section details the implementation schedule and estimates the costs involved. Recovery plans are available for several river turtles including the river terrapin (E. Moll, 1984a), ringed map turtle (USFWS, 1988), Alabama red-bellied turtle, (USFWS, 1989), flattened musk turtle (USFWS, 1990), and yellow-blotched map turtle (USFWS, 1993).

Action plans typically concern groups of species and thus operate at a different scale from the species-specific recovery plans. The IUCN's Species Survival

Commission Specialist Groups have been instrumental in preparing Action Plans for a wide diversity of taxa. Typically, each Action Plan evaluates the status of species in the group based on input from the volunteer specialists, reviews current conservation action, and then outlines needed conservation projects. River turtles are included in the Action Plan prepared by the IUCN/SSC Tortoise and Freshwater Specialist Group (1989), and the more recent "A Global Action Plan for the Conservation of Tortoises and Freshwater Turtles: Strategy and Funding Prospectus for 2002–2007 by the Turtle Conservation Fund (2002). These two action plans provide some of the best overviews of the status of river turtles and their management on the global scale.

Conclusions

There is not single approach to the conservation of river turtles that can be applied to every species or every region. We have listed a number of the available tools which can be used. It is the task of the conservation biologist to mix and match these tools with others to fit each unique situation as suggested by Klemens (2000a) and by Seigel and Dodd (2000) with their "gradient" approach to designing conservation action. That which works in developed countries will not necessarily be successful for developing countries. The approach chosen will depend on intrinsic factors such as the status of the population and the biology of the species as well as extrinsic factors such as the predominant religion/culture of the region, local politics, and state of the economy.

Ideally, conservationists tend to favor in-situ approaches that provide complete protection of a declining species, its community, and its habitat. However, if the species is an important part of the local economy and/or part of a long-standing tradition (e.g., the traditional, Central American river turtle Easter dinner in Belize), it may be necessary to allow some exploitation while moving to protect significant portions of the population in order to marshal support from the local human population. Similarly, if species become so much in demand and their price so extreme (e.g., the Chinese three-striped box turtle [*Cuora trifasciata*]) that no wild population in a region is safe even in protected areas (such as the current exploitation of turtles from neighboring countries by the Chinese), then captive breeding of populations may be the only short-term approach available for saving them.

Again, every situation is different and it is impossible to provide a "one conservation program fits all" approach. Nevertheless, based on what we know about long-lived vertebrates and successful conservation programs, there are several basic premises that can serve as guidelines in considering conservation action for river turtles:

1. Education: the people involved or affected by the conservation program should be informed why the action is necessary and if or how this program will benefit them now or in the future. If possible, the local people should be given integral roles in the conservation program.

2. Ideally, all life history stages of a declining river turtle population should receive protection, but if this is not possible then adults and older juveniles should receive precedence.
3. In-situ techniques should be utilized where practical, but where ex-situ methods are warranted they should be adopted as temporary measures and affiliated with in-situ methods such as habitat acquisition, improvement, and maintenance.

Epilogue
Outlook for the Future

We are cautiously optimistic that in spite of the dire circumstances many river turtles find themselves in today, most—if not all—can be conserved to swim the waters of their home rivers into the foreseeable future. We do not feel that this optimism is false, and believe that there are a number of hopeful developments which we can build upon to attain this goal. First, while there is still much to learn, we are aware of the main problems river turtles and other chelonians face and we are rapidly coming to grips with the methodology that will be required to solve them. We certainly do not minimize the difficulties and challenges which lie ahead in implementing effective conservation action, but awareness and coordinated planning are the necessary first steps in the process of affecting change.

The idea of turtle biologists getting together to talk about turtles is not a particularly new concept. Symposia concerning turtle biology are occasionally held in conjunction with the major herpetological societies' annual meetings (e.g., at Williamsburg, Virginia in 1975, and Seattle, Washington in 1997), and Dick Vogt and the late Jack McCoy began the "Powdermill" series of turtle biologists' enclaves at the namesake site outside Pittsburgh in 1980—a practice which has continued at several-year intervals ever since. These symposia have been mainly concerned with chelonian biology, although conservation issues were often raised also. As awareness of the serious and widespread destruction of the world's turtle populations began to really sink in during the 1990s, the response of turtle biologists was to seek to better understand and find solutions to chelonian problems. The 1993 conference at Purchase, New York: Conservation, Restoration, and Management of Tortoises and Turtles—An International Conference (van Abbema, 1997) focused attention on these troubling issues, and in the same year a new chelonian society was founded, the Chelonian Research Foundation with a new journal, *Chelonian Conservation and Biology*, which concentrates on the issues suggested by its title. These important landmarks in chelonian conservation have been followed by the International Congress of Chelonian Conservation at Gonfaron, France in 1995 (Devaux, 1995) and the Workshop on Conservation and Trade of Freshwater Turtles and Tortoises in Asia held in Phnom Penh, Cambodia in 1999 (van Dijk et al., 2000a). As we write, the 2nd International Congress of Chelonian Conservation and the 5th Powdermill Conference—both focusing heavily on chelonian conservation issues—are scheduled for the summer of 2003 in Saly, Senegal and Manaus, Brazil respectively. These activities, the recent publication of *Turtle Conservation* (Klemens, 2000b), and the recent Turtle Conservation Fund's Action Plan (TCF, 2002) discussed in chapter 7 suggest that none of us with an interest in chelonian affairs is giving up in despair, and that there is an ever-growing groundswell of global interest in implementing conservation action to save the world's turtles. We believe that it is certainly time to employ the knowledge and awareness gained from these sources to implement coordinated conservation action in behalf of the world's chelonians—to shift from the planning and consideration mode of the 1990s into the action and implementation mode of the early 2000s (as is already underway in numerous cases around the world). Of course, we do not have all the information we need now, but we never will, and the situation is desperate enough to take bold action—to take risks if necessary, based upon the best information available now. We feel that the rapid implementation of the concepts and suggestions presented in this volume and elsewhere referred to above *will* save river turtles as well as chelonians of the sea, land, and wetlands of the earth.

The old adage—"if we are the problem, we are also the solution"—continues to come to mind. As human culture evolves we believe we will see a time before very long when direct exploitation of turtles for food and traditional medicine—arguably, the most serious problems facing river turtles—will decline. These changes may unfortunately be due in part to the reduction of wild stocks which formerly supplied demand, but will also occur as turtle-using societies modernize and shift away from dependence upon wild species for food and medicine. This process is occurring in Thailand for instance (van Dijk & Palasuwan, 2000), and there is encouraging evidence of a reduction of reliance upon traditional Chinese medicine in China and Hong Kong. A 1999 survey of medicinal practices in Hong Kong found that only 22% of outpatient medical consultations in the city were provided by traditional Chinese medicine practitioners, and officials believe the percentage elsewhere in mainland China may be even lower (Normille, 2003). Efforts are also underway in traditional Chinese medicine research to develop herbal alternatives for turtle shell in some medicinal potions (Asian Trade Working Group [ATWG], 2000). There is also recent encouraging news that China is beginning to take steps to better control its international trade in turtles (Xianlin et al., 2002). Taiwan is now able to satisfy its population's hunger for turtle meat and eggs (if not yet of shells) almost totally by Chinese softshell production in its closed-system turtle farms, whilst still also providing a huge export volume as well (Chen et al., 2000). If we can fight something of a holding action in the near future which will maintain core populations and habitats of suitable quality and size, we may once again see population increases and redistribution into habitats where populations are now extirpated as exploitation pressures diminish. The maintenance of sanctuaries and quality habitat, the promotion of public education, and conservation involvement, and the legislation and strict enforcement of laws prohibiting exploitation—particularly international trade in wild turtle products—are so vitally important in this regard, and must be promoted vigorously to maintain the necessary seed stock for better days. Magnusson's ideas (Magnusson, 2002) concerning the need to focus on regulations relating to the economics of trade in order to bring about "commercial extinction" at population sizes well above "biological extinction" levels are good ones, and deserve further attention in regard to river turtles in particular.

Finally, there is cause for optimism due to our allies in the battle—the turtles themselves. The modern nearly 300 species represent a tough and resilient order of animals which have been on the Earth for a very long time. They have survived to the present day through several of the most catastrophic periods in the Earth's history with, as Archie Carr (1952) put it, a "philosophy of meditation and passive resistance," that has served them well though their long history on Earth. Softshelled turtles and turtles of the same genus (*Podocnemis*) that swim in South America's rivers today watched the dinosaurs' demise (Carr, 1952). The giant South American river turtle—the largest and traditionally most heavily exploited species of that genus—is still present today, and is even common in some areas within its vast range. Even the critically endangered Asian river terrapin, despite ruthless, relentless exploitation for centuries, is hanging on in Malaysia, and is still turning up in viable numbers in previously unknown locations (e.g., Cambodia). Maybe the modern species are tougher and more resilient than we give them credit for. Their long history is not cause for complacency, but should provide hope that the grit and fiber they have inherited from their ancestors will, with a little help from ourselves, see them through their most current crisis and beyond.

Literature Cited

Acharji, M. N. 1955. A collection of chelonians and snakes from Chota Nagpur, Bihar. Indian Mus. 53:383–392.

Acuña-Mesén, R. A. 1993. Las Tortugas Continentales de Costa Rica. ICER Press, San Pedro de Montes de Oca, Costa Rica.

Acuña-Mesén, R. A., and D. Moll. In press. *Rhinoclemmys funerea* (Cope, 1876 "1875") Black Turtle. *In*: The Conservation Biology of Freshwater Turtles. P. C. H. Pritchard and A. G. J. Rhodin (eds.). Chelonian Research Foundation.

Addison, D. S. 1997. Galvanized wire cages can prevent nest depredation. Marine Turtle Newsletter 76:8–11.

Agassiz, L. 1868. A Journey in Brazil. 5th edition. Tricknor and Fields, Boston.

—. 1879. A Journey in Brazil, 9th edition. Houghton, Osgood and Company, Boston.

Akani, G. C., D. Capizzi, and L. Luiselli. 2001. Diet of the softshell turtle, *Trionyx triunguis*, in an Afrotropical forested region. Chelonian Conserv. Biol. 4:200–201.

Alfinito, J. 1975. A preservacao da tartaruga Amazonica. Brazil Florestal. 6:20–23.

Alfinito, J., C. M. Vianna, M. M. F. Da Silva, and H. Rodrigues. 1976. Transferencia de tartarugas do rio trombetas para o rio tapajos. Brasil Florestal. 7:49–53.

Alho, C. J. R. 1985. Conservation and management strategies for commonly exploited Amazonian turtles. Biol. Conserv. 32:291–298.

Alho, C. J. R., and L. F. M. Padua. 1982. Reproductive parameters and nesting behavior of the Amazon turtle *Podocnemis expansa* (Testudinata: Pelomedusidae) in Brazil. Can. J. Zool. 60:97–103.

Alho, C. J. R., T. M. S. Danni, and L. F. M. Padua. 1984. Influencia da temperatura de incubacao do sexo da tartaruga da Amazonia *Podocnemis expansa* (Testudinata: Pelomedusidae). Rev. Brasil. Biol. 44:305–311.

—. 1985. Influencia da temperatura de incubacao do sexo da tartaruga da Amazonia *Podocnemis expansa* (Testudinata: Pelomedusidae). Rev. Bras. Biol. 44:305–311.

Allen, C. H. 1990. Guest editorial: Give 'headstarting' a chance, pp. 12–16. Marine Turtle Newsletter, Vol. 51.

—. 1992. It's time to give Kemp's ridley headstarting a fair and scientific evaluation!, pp. 21–24. Marine Turtle Newsletter, Vol. 56.

Allen, E. R., and W. T. Neill. 1950. The alligator snapping turtle, *Macrochelys temminckii*, in Florida. Spl. Publ. Ross Allen's Rept. Inst. 4:1–15.

Altherr, S., and D. Freyer. 2000. Asian turtles are threatened by extinction. Turtle and Tortoise Newsletter 7–11.

Altman, J. C. 1987. Hunter-gatherers today: an Aboriginal-economy in northern Australia. Australian Institute of Aboriginal Studies, Canberra.

Alvarez Del Toro, M. 1960. Los Reptiles de Chiapas. Tuxtla Gutierrez, Chiapas, Mexico.

Alvarez Del Toro, M., R. A. Mittermeier, and J. B. Iverson. 1979. River turtle in danger. Oryx 15:170–173.

Anderson, J. 1878–79. Anatomical and zoological researches and zoological results of the Yunnan Expeditions. Reptilia and Amphibia. Calcutta.

—. 1898. Zoology of Egypt, 1: Reptilia and Batrachia, London.

Anderson, P. K. 1958. The photic responses and water approach behavior of hatchling turtles. Copeia 1958:211–215.

—. 1965. The Reptiles of Missouri. University of Missouri Press, Columbia, MO.

Anderson, R. V., M. L. Gutierrez, and M. A. Romano. 2002. Turtle habitat use in a reach of the upper Mississippi River. J. Freshwater Ecol. 17:171–177.

André, E. 1902. A Naturalist in the Guianas. Thomas Nelson and Sons, London.

Annandale, N., and Shastri, M. 1914. Relics of the worship of mud turtles (*Trionychidae*) in India and Burma. Proc. Asiatic Soc., Bengal 105:131–135.

Anon. 1981. Tartarugas, a nova experiencia comercial japonesa no Brasil. "Globo" Rio de Janeiro-Brasil.

—. 1991. Final listing rules approved for nine species, pp. 7–8. Endangered Species Technical Bulletin, Vol. XVI.

—. 1993. Crocodile conservation and management in India, pp. 1–27. Crocodile Specialist Group Workshop. J. Hutton (ed.), Madras, India.

—. 1998. Action plan: Biodiversity conservation in wetland habitats: Crocodiles, pp. 1–19. Juwaji University, Gwalior, M. P. India; Madras Crocodile Bank Trust, Mamallapuram, Tamil Nadu, India.

Arrhenius, O. 1921. Species and area. J Ecol 9:95–99.

Artner, H., and A. Hofer. 2001. Observations in the Qing Ping Free Market, Guangzhou, China, November 2000. Turtle and Tortoise Newsletter 14.

Ashton, R. E., Jr., and P. S. Ashton. 1985. Handbook of Reptiles and Amphibians of Florida. Part 2, Lizards, turtles, and crocodilians. Windward Publ., Inc., Miami, FL.

Atatür, M. K. 1979. Investigations on the morphology and osteology, biotope and distribution in Anatolia of *Trionyx triunguis* (Reptilia, Testudines) with some observations on its biology. Ege Universitesi, Fen Fakultesi Monografiler Ser. 18:1–75.

Atatür, M. K., P. Meyland, and E. Traskavak. In press. *Trionyx triunguis* (Forskal 1775) Nile softshell. *In*: The Conservation Biology of Freshwater Turtles. P. C. H. Pritchard and A. G. J. Rhodin (eds.). Chelonian Research Foundation.

ATWG. 2000. Conclusions and recommendations from the workshop on conservation and trade of freshwater turtles and tortoises in Asia. (Asian Turtle Trade Working Group), pp. 148–155. *In*: Asian Turtle Trade: Proceedings of Workshop on Conservation and Trade of Freshwater turtles and Tortoises in Asia. Vol. 2. P. P. van Dijk, B. L. Stuart, and A. G. J. Rhodin (eds.). Chelonian Research Foundation, Phnom Penh, Cambodia.

Auth, D. L. 1975. Behavioral ecology of basking in the yellow-bellied turtle, *Chrysemys scripta scripta* (Schoepff). Bull. Florida St. Mus., Biol. Sci. 20:1–45.

Avise, J. C. 1989. A role of molecular genetics in the recognition and conservation of endangered species. Trends Ecol. Evolution 4:279–281.

Avise, J. C., and J. L. Hamrick. 1996. Conservation Genetics: Case Histories from Nature. Chapman & Hall, New York.

Bailey, K. A., and C. Guyer. 1998. Demography and population status of the flattened musk turtle, *Sternotherus depressus*, in the Black Warrior River basin of Alabama. Chelonian Conserv. Biol. 3:77–83.

Baillie, J., and B. Groombridge. 1996–2001. IUCN Red List of Threatened Animals. Revisions from the web http://www.redlist.org/info/categories_criteria2001.html. IUCN, Gland, Switzerland.

Ballasina, D. 1995. Conservation and reproduction techniques at the Carapax Center, Italy, pp. 210–213. *In*: Proceedings of the International Congress of Chelonian Conservation. B. Devaux (ed.). SOPTOM, Gonfaron, France.

Barbour, T., and A. F. Carr. 1940. Antillean terrapins. Mem. Mus. Comp. Zool. 54:381–415.

Baskin-Salzber, A., and A. Salzberg. 1996. Turtles. Franklin Watts, New York.

Bates, H. W. 1863. The Naturalist on the River Amazon. John Murray, London.

Beck, B. B., L. G. Rapaport, M. S. Price, and A. Wilson. 1994. Reintroduction of captive-born animals. *In:* Creative Conservation: Interactive Management of Wild and Captive Animals. P. Olney, G. Mace, and A. Feistner (eds.). Chapman & Hall, London.

Belkin, D. A., and C. Gans. 1968. An unusual chelonian feeding niche. Ecology 49:768–769.

Bellrose, R. C., R. Sparks, F. L. Paveglio, Jr., D. Steffeck, R. Thomas, R. Weaver, and D. Moll. 1977. Fish and wildlife changes resulting from the construction of a nine foot navigation channel in the Illinois Waterway from LaGrange Lock and Dam upstream to Lockport Lock and Dam. Contract Report #DACW 23–67-C-0066). U.S. Army Corps of Engineers, Chicago District, Chicago.

Berger, M. 1992. Look Out for Turtles. HarperCollins Publishers, New York.

Bernacsek, G. M. 1984. Dam design and operation to optimize fish production in impounded river basins. CIFA Technical Paper, Rome.

Berry, J. F. 1975. The population effects of ecological sympatry on musk turtles in northern Florida. Copeia: 1975 692–701.

Berry, J. F., and J. M. Legler. 1980. A new turtle (genus *Kinosternon*) from Sonora, Mexico. Science Natural History Mus., Los Angeles Co. 325:1–12.

Berry, J. F., and R. Shine. 1980. Sexual size dimorphism and sexual selection in turtles (Order Testudines). Oecologica (Berlin) 44:185–191.

Bhadauria, R. S., A. Pai, and D. Basu. 1990. Habitat, nesting and reproductive adaptations in the narrow-headed softshell turtle *Chitra indica* (Gray) (Reptilia, Chelonia). J. Bombay Natural History Society. 87:364–367.

Bhaitt, K. 1989. New locality record for the Indian pink ringed terrapin. J. Bombay Natural History Soc. 86:252–253.

Bhupathy, S. 1995. Status and distribution of the river terrapin *Batagur baska*, in the Sunderbans of India, Sacon Technical Report 3. Salim Ali Center for Orinthology and Natural History, Coimbatore, India.

—. 1997. Conservation of the endangered river terrapin, *Batagur baska*, in the Sunderbans of West Bengal, India. J. Bombay Natural History Soc. 94:27–35.

Bhupathy, S., B. C. Choudhury, and E. O. Moll. 1994. Conservation and management of freshwater turtles and land tortoises of India. Final Report: turtle and tortoise conservation project. A Joint Indo-U.S. collaborative project of the Wildlife Institute of India and the U.S. Fish and Wildlife Service, Dehra Dun, India.

Bhupathy, S., B. C. Choudhury, F. Hanfee, Kalyar, S. M. Munjurul Hannan Khan, S. G. Platt, and S. M. A. Rashid. 2000. Turtle trade in South Asia: Regional summary (Bangladesh, India, and Myanmar), pp. 101–105. *In:* Asian Turtle Trade: Proceedings of a Workshop on Conservation and Trade of Freshwater Turtles and Tortoises in Asia. Vol. 2. P. P. van Dijk, B. L. Stuart, and A. G. J. Rhodin (eds.). Chelonian Research Foundation, Phnom Penh, Cambodia.

Bider, J. R., and W. Hoek. 1971. An efficient and apparently unbiased sampling technique for population studies of painted turtles. Herpetologica 27:481–484.

Bishop, S. C., and W. J. Schoonmacher. 1921. Turtle hunting in mid-winter. Copeia. 96:37–38.

Bjorndal, K. A., A. B. Bolten, C. J. Lagueux, and D. R. Jackson. 1997. Dietary overlap in three sympatric congeneric freshwater turtles (Pseudemys) in Florida. Chelonian Conserv. Biol. 2:430–433.

Bjorndal, K. A., and A. B. Bolten. 1990. Digestive processing in a herbivorous freshwater turtle: consequences of small intestine fermentation. Physiol. Zool. 63:1232–1247.

—. 1992. Body size and digestive efficiency in a herbivorous freshwater turtle: advantages of small bit size. Physiol. Zool. 65:1028–1039.

Bleakney, J. S. 1963. Notes on the distribution and life histories of turtles in Nova Scotia, Canada. Field-Nat. 77:67–76.

Boarman, W. 1995. Reduction in mortalities of desert tortoises and other vertebrates along a fenced highway, Abstract. International Congress of Chelonian Conservation. B. Devaux (ed.). SOPTOM, Gonfaron, France.

Boarman, W. J., M. Sazaki, and W. B. Jennings. 1997. The effects of roads, barrier fences, and culverts on desert tortoise populations in California, pp. 54–58. *In*: Proceedings: Conservation, Restoration, and Management of Tortoises and Turtles: An International Conference. J. van Abbema (ed.), State University of New York, Purchase, NY.

Bock, B. C., V. P. Paez, and N. F. Perez. 1998. Estudio preliminar con radiotelemetria sobre los desplazamientos de hembras de la tortuga *Podocnemis unifilis* en el Rio Caqueta, Amazonas, Colombia. Actual. Biol. 20:29–36.

Bock, B. C., V. P. Paez, and M. M. White. 2001. Genetic population structure of two threatened South American river turtle species *Podocnemis expansa* and *Podocnemis unifilis*. Chelonian Conserv. Biol. 4:47–52.

Bodie, J. R. 2001. Stream and riparian management for freshwater turtles. J. Environ. Management 62:443–455.

Bodie, J. R., and R. D. Semlitsch. 2000a. Spatial and temporal use of floodplain habitats by lentic and lotic species of aquatic turtles. Oecologia 122:138–146.

—. 2000b. Size-Specific Mortality and Natural Selection in Freshwater Turtles. Copeia 2000:732–739.

Bodie, J. R., R. D. Semlitsch, and R. B. Renken. 2000. Diversity and structure of turtle assemblages: Associations with wetland characters across a floodplain landscape. Ecography 23:444–456.

Bouskila, A. 1986. On the danger of red-eared terrapin *Chrysemys scripta* in natural habitats in Israel. Hardun 3:63.

Bowen, B. W., and J. C. Avise. 1996. Conservation of marine turtles, pp. 190–237. *In*: Conservation Genetics: Case Histories from Nature. J. C. Avise and J. L. Hamrick (eds.). Chapman & Hall, New York.

Bowen, B. W., and S. A. Karl. 1997. Population genetics, phylography, and molecular evolution, pp. 29–50. *In*: Biology of Sea Turtles. P. Lutz (ed.). CRC Press, Boca Raton, FL.

Boyce, M. S. 1992. Population viability analysis. Annu. Rev. Ecol. Syst. 23:481–506.

Boycott, R. C., and O. Bourquin. 1988. The South African Tortoise Book: A Guide to South African Tortoises, Terrapins and Turtles. Russel Friedman Books, Johannesburg, South Africa.

—. 2000. The Southern African Tortoise Book: A Guide to Southern African Tortoises, Terrapins and Turtles. Revised Edition. Southern Book Publishers (Pty) Ltd, Johannesburg, South Africa.

Boyer, D. R. 1965. Ecology of the basking habit in turtles. Ecology 46:99–118.

Braid, M. R. 1974. A bal-chatri trap for basking turtles. Copeia 1974:539–540.

Branch, D. M. 1988. Bill Branch's Field Guide to the Snakes and Other Reptiles of Southern Africa. Struik Publisher, Cape Town.

Brimley, C. S. 1942–1943. Reptiles and amphibians of North Carolina: the turtles, tortoises or terrapins. Carolina Tips. 5:22–23; 6:2–3, 6–7, 10–11, 14–15, 18–19.

Britson, C. A., and W. H. Gutzke. 1993. Antipredator mechanisms of hatchling freshwater turtles. Copeia 1993:1186–1188.

Broadley, D. G. 1981. A review of the genus Pelusios (Wagler) in southern Africa (*Pleurodira: Pelomedusidae*). Occ. Pap. Natl. Mus. Rhodesia, B. Natur. Sci. 69:633–686.

Brock, V. E. 1947. The establishment of *Trionys muticus* in Hawaii. Copeia 1947:142.

Brooks, R. J., D. A. Galbraith, E. G. Nancekivell, and C. A. Bishop. 1988. Developing management guidelines for snapping turtles, pp. 174–179. *In*: Symposium: Management of Amphibians, Reptiles, and Small Mammals in North America: 1988. R. C. Szaro, K. E. Severson, and D. R. Patton (eds.). USDA Tech. Serv. Gen. Tech. Rept., Flagstaff, AZ.

Brower, J. E., and J. H. Zar. 1984. Field and Laboratory Methods for General Ecology. William C. Brown Company Publishers, Dubuque, IA.

Brown, J. H., and B. A. Maurer. 1986. Body size, ecological dominance and Cope's rule. Nature 324:248–250.

Brown, W. L., Jr., and E. O. Wilson. 1956. Character displacement. Syst. Zool. 5:49–64.

Bryan, A. M., W. B. Stone, and P. G. Olafsson. 1987. Disposition of toxic PCB congenors in snapping turtle eggs: expressed as toxic equivalents of TCDD. Bull. Environ. Contam. Toxicol. 39:791–796.

Budhabhatti, J., and E. O. Moll. 1990. Feeding behavior of *Chelydra serpentina*: Life history notes. Herp. Rev. 21:19.

Buhlmann, K. A., and J. W. Gibbons. 1997. Imperiled aquatic reptiles of the southeastern United States: Historical review and current conservation status, pp. 201–231. *In*: Aquatic Fauna in Peril: The Southeastern Perspective. G. W. Benz and D. E. Colllins (eds.). Southeast Aquatic Research Institute, Decatur, GA.

Buhlmann, K. A., and M. R. Vaughn. 1991. Ecology of the turtle *Pseudemys concinna* in New River, West Virginia. J Herpetol. 25:72–78.

Bull, J. J. 1980. Sex determination in reptiles. Q. Rev. Biol. 55:3–21.

Bull, J. J., and E. L. Charnov. 1988. How fundamental are fisherian sex ratios? Oxford Serv. Evol. Biol. 5:96–135.

Bull, J. J., and R. C. Vogt. 1979. Temperature-dependent sex determination in turtles. Science 206:1186–1188.

Bull, J. J., W. H. N. Gutzke, and D. Crew. 1988. Sex reversal by estradiol in three reptilian orders. Gen. Comp. Endocrinol. 70:425–428.

Burger, J. 1976. Behavior of hatchling diamondback terrapins (*Malaclyemys terrapin*) in the field. Copeia 1976:742–748.

Burke, R. L. 1991. Relocations, repatriations, and translocations of amphibians and reptiles: taking a broader view. Herpetologica 47:350–357.

Burke, V. J., and J. W. Gibbons. 1995. Terrestrial buffer zones and wetland conservation: a case study of freshwater turtles in a Carolina bay. Conserv. Biol. 9:1365–1369.

Burke, V. J., J. W. Gibbons, and J. L. Greene. 1994. Prolonged nesting forays by common mud turtles (*Kinosternon subrubrum*). Am. Midland Naturalist 131:190–195.

Burke, V. J., J. L. Greene, and J. W. Gibbons. 1995. The effect of sample size and study duration on metapopulation estimates for slider turtles (*Trachemys scripta*). Herpetologica 51:451–456.

Burke, V. J., J. E. Lovich, and J. W. Gibbons. 2000. Conservation of freshwater turtles, pp. 156–179. *In*: Turtle Conservation. M. W. Klemens (ed.). Smithsonian Institution Press, Washington, D.C.

Bury, R. B. 1979. Population ecology of freshwater turtles, pp. 571–602. *In*: Turtles: Perspectives and Research. M. Harless and H. Morlock (eds.). Robert E. Krieger Publishing Company, Malabar.

Bury, R. B., and D. C. Holland. In press. *Clemmys marmorata* (Baird and Girard, 1852), Western Pond Turtle. *In*: The Conservation of Freshwater Turtles. P. C. H. Pritchard and A. G. J. Rhodin (eds.). Chelonian Research Foundation.

Busack, S. D., and S. B. Hedges. 1984. Is the peninsular effect of red herring? Am. Naturalist 123:266–275.

Bustard, H. R. 1990. The role of management in the conservation of endangered species with special reference to Indian crocodilians, pp. 463–483. *In*: Conservation in Developing Countries: Problems and Prospects. J. C. Daniel and J. S. Serrao (eds.). Oxford University Press, Oxford.

Butler, B. O., and T. E. Graham. 1995. Early post-emergent behavior and habitat selection in hatchling Blanding's turtles, *Emydoidea blandingii*, in Massachusetts. Chelonian Conserv. Biol. 1:187–196.

Cagle, F. R. 1942. Turtle populations in southern Illinois. Copeia 1942:155–162.

—. 1950. The life history of the slider turtle, *Pseudemys scripta troostii* (Holbrook). Ecol. Monogr. 20:31–54.

Cagle, F. R., and A. H. Chaney. 1950. Turtle populations in Louisiana. Am. Midland Naturalist 43:383–388.

Cahn, A. R. 1937. The turtles of Illinois. Ill. Biol. Monogr. XVI:1–218.

Cann, J. 1978. Tortoises of Australia. Angus and Robertson Publishers, Sydney.

—. 1993. Do not take our freshwater turtles for granted, pp. 363–365. *In*: Herpetology in Australia: A Diverse Discipline. D. Lunney and D. Ayers (eds.). Transactions of the Royal Zoological Society of New South Wales, Mosman.

—. 1998. Australian Freshwater Turtles. Beaumont, Singapore.

Cann, J., and J. M. Legler. 1994. The Mary River tortoise: a new genus and species of short-necked chelid from Queensland, Australia (Testudines; Pleurodira). Chelonian Conserv. Biol. 1:81–96.

Cantor, T. E. 1847. Catalogue of reptiles inhabiting the Malayan Peninsula and islands, pp. 607–620. *In*: Chelonia. J. Asiatic Soc. Bengal. Vol. XVI.

Carpenter, C. C. 1955. Sounding turtles: a field locating technique. Herpetologica 11:120.

Carr, A. F. 1952. Handbook of Turtles. The Turtles of the United States, Canada, and Baja California. Comstock Publ. Assoc., Cornell University Press, Ithaca, NY.

—. 1967. So Excellent a Fishe: A Natural History of Sea Turtles. National History Press, Garden City, NY.

Carr, A. F., and H. Hirth. 1961. Social facilitation in green turtle siblings. Anim. Behav. 9:68–70.

Carr, A. F., and L. T. Marchand. 1942. A new turtle from the Chipola River, Florida. Proc. N. Engl. Zool. Club 20:95–110.

Carr, J., and R. Mast. In press. Kinosternon herrerai Herrera's mud turtle. *In*: The Conservation Biology of Freshwater Turtles. P. C. H. Prichard and A .G. J. Rhodin (eds.). Chelonian Research Foundation.

Carr, T., and N. Carr. 1985. *Trionyx triunguis*: habitat. Herp. Rev. 16:30.

Carroll, D. M. 1991. The Year of the Turtle. Camden House Publishing, Inc., Charlotte, VT.

Chan, E. H., and H. C. Liew. 1996. Decline of the leatherback population in Terrengganu, Malaysia, 1956–1995. Chelonian Conserv. Biol. 2:196–203.

Chaney, A., and C. L. Smith. 1950. Methods for collecting map turtles. Copeia 1950:323–333.

Charlesworth, D., and B. Charlesworth. 1987. Inbreeding depression and its evolutionary consequences. Annu. Rev. Ecol. Systematics 18:237–268.

Charnier, M. 1966. Action de la temperature sur la sex-ratio chez l'embryon d' *Agama agama* (Agamidae, Lacertilien). Soc. Biol. Quest. Afric. 160:620–622.

Chaudhuri, B. L. 1912. Aquatic tortoises of the middle Ganga and Brahmaputra. Rec. Indian Mus. 7:212–214.

Chen, T. H., and K. Y. Lue. 1998. Ecological notes on feral populations of *Trachemys scripta elegans* in northern Taiwan. Chelonian Conserv. Biol. 3:87–90.

Chen, T. H., H. C. Lin, and H. C. Chang. 2000. Current status and utilization of chelonians in Taiwan, pp. 45–51. *In*: Asian Turtle Trade: Proceedings of a Workshop on Conservation and Trade of Freshwater Turtles and Tortoises in Asia. Vol. 2. P. P. van Dijk, B. L. Stuart, and A. G. J. Rhodin (eds.). Chelonian Research Foundation, Phnom Penh, Cambodia.

Chen, T. P. 1976. Aquaculture Practices in Taiwan. Page Bros. (Norwich) Ltd. (Distributed by Fishing News Books Ltd. Farnham, Great Britain): pp. 137–144.

Chessman, B. C. 1978. Ecological studies of freshwater turtles in southeastern Australia. Ph.D. Dissertation. Monash University, Australia.

—. 1983. Observations on the diet of the broad-shelled turtle, *Chelodina expansa*. Australian Wildlife Res. 10:169–172.

—. 1984. Food of the snake-necked turtle, *Chelodina longicollis* (Shaw) (Testudines: Chelidae), in the Murray Valley, Victoria and New South Wales, Australian Wildlife Res. 11:573–578.

—. 1986. Diet of the Murray turtle, *Emydura macquarii* (Gray) (Testudines: Chelidae). Australian Wildlife Res. 13:65–69.

—. 1988. Habitat preferences of freshwater turtles in the Murray Valley, Victoria and New South Wales. Australian Wildlife Res. 15:485–491.

Choo, B. L., and L. M. Chou. 1983. Effect of stocking density on the growth and survival of *Trionyx sinesis* hatchlings. *In*: Symposium on Research in Biology and Biotechnology in Developing Countries; November 2–4, National University of Singapore, Singapore.

—. 1984. Effect of a sand substrate on the growth and survival of hatchlings of the softshell turtle *Trionyx sinensis* (Weigmann). Aquaculture 40:325–331.

Chou, L. M., and V. L. Choo. 1986. The potential of soft-shell turtles culture in Singapore. *In*: Development and Management of Tropical Living Aquatic Resources: Proceedings of an International Conference held at Universiti Pertanian Malaysia, Serdang, Selangor, Malaysia from 2–5 August, 1983. Chan Hooi Har (ed.). Universiti Pertanian Malaysia, Serdang, Selangor, Malaysia.

Choudhury, B. C., and S. Bhupathy. 1993. Turtle trade in India: A study of tortoises and freshwater turtles. WWF-India (prepared by Traffic-India), New Delhi.

Choudhury, B. C., S. Bhupathy, and E. O. Moll. 1997. Conservation and management of freshwater turtles and land tortoises in India, pp. 301–306. *In*: Proceedings: Conservation, Restoration, and Management of Tortoises and Turtles: an International Conference. J. van Abbema (ed.). State University of New York, Purchase, NY.

Christiansen, J. L. 1981. Population trends among Iowa's amphibians and reptiles. Proc. Iowa Acad. Sci. 88:24–27.

Christiansen, J. L., and B. J. Gallaway. 1984. Raccoon removal, nesting success, and hatchling emergence in Iowa turtles with special reference to Kinosternon flavescens (Kinosternidae). Southwest. Naturalist 29:343–338.

Christiansen, J. L., and R. M. Bailey. 1988. The lizards and turtles of Iowa. Iowa Department of Natural Resources, Non Game Tech. Ser. 3:1–19.

Chua, T. H. 1988. Nesting population and frequency of visits in *Dermochelys coriacea* in Malaysia. J. Herpetol. 22:208–218.

CITES. 1973. Text of the Convention. *In*: Convention on International Trade in Endangered Species of Wild Fauna and Flora. C. Secretariat (ed.). Cites, Chatelaine-Geneve, Switzerland.

Clark, D. B., and J. W. Gibbons. 1969. Dietary shift in the turtle *Pseudemys scripta* (Schoepff) from youth to maturity. Copeia. 1969:704–706.

Clark, H. W., and J. B. Southall. 1920. Fresh water turtles: a source of meat supply. U.S. Bur. Fish. Doc. 889:3–20.

Clark, W. S. 1982. Turtles as a food source of nesting bald eagles in the Chesapeake Bay region. J. Field Ornithol. 53:49–51.

Clay, B. T. 1981. Observations on the breeding biology and behaviour of the long-necked tortoise *Chelodina oblonga*. J. R. Soc. West. Aust. 4:27–32.

Close, L. M., and R. A. Seigel. 1997. Differences in body size amoung populations of red-eared sliders (*Trachemys scripta elegans*) subjected to different levels of harvesting. Chelonian Conserv. Biol. 2:563–566.

Cochran, P. A., and D. R. McConville. 1983. Feeding by *Trionyx spiniferus* in backwaters of the Upper Mississippi River. J. Herpetol. 17:82–86.

Cole, G. A. 1994. Testbook of Limnology. 4th edition. Waveland Press, Inc., Prospect Heights, IL.

Compton, J. 2000. An overview of Asian turtle trade, pp. 24–29. *In*: Asian Turtle Trade: Proceedings of a Workshop on Conservation and Trade of Freshwater Turtles and Tortoises in Asia. Vol. 2. P. P. van Dijk, B. L. Stuart, and A. G. J. Rhodin (eds.). Chelonian Research Foundation, Phnom Penh, Cambodia.

Conant, R., and J. T. Collins. 1991. A Field Guide to Reptiles and Amphibians, Eastern and Central North America. Houghton Mifflin Co., Boston.

Congdon, J. D., J. L. Greene, and J. W. Gibbons. 1986. Biomass of freshwater turtles: a geographic comparison. Am. Midland Naturalist 115:165–173.

Congdon, J. D., G. L. Breitenbach, R. C. Van Loben Sels, and D. W. Tinkle. 1987. Reproduction and nesting ecology of snapping turtles (*Chelydra serpentina*) in southeastern Michigan. Herpetologica 43:39–54.

Congdon, J. D., and J. W. Gibbons. 1989. Biomass productivity of turtles in freshwater wetlands: A geographic comparison. *In*: Freshwater Wetlands and Wildlife, Conf 8608101, Doe Symposium Series No. 61. R. R. Sharitz and J. W. Gibbons (eds.). USDOE Office of Scientific and Technical Information, Oak Ridge, TN.

—. 1996. Structure and dynamics of a turtle community over two decades, pp. 137–159. *In*: Longterm Studies of Vertebrate Communities. M. L. Cody and J. A. Smallwood (eds.). Academic Press, Inc., San Diego.

Congdon, J. D., S. W. Gotte, and R. W. McDiarmid. 1992. Ontogenetic changes in habitat use by juvenile turtles, *Chelydra serpentina* and *Chrysemys picta*. Can. Field Naturalist 106:241–248.

Congdon, J. D., A. E. Dunham, and R. C. Van Loben Sels. 1993. Delayed sexual maturity and demographics of Blanding's turtles (*Emydoidea blandingii*): Implications for conservation and management of long-lived organisms. Conserv. Biol. 7:826–833.

—. 1994. Demographics of snapping turtles (*Chelydra serpentina*): Implications for conservation and management of long-lived organisms. Am. Zoologist 34:397–408.

Congdon, J. D., R. D. Nagle, A. E Dunham, C. W. Beck, and O. M. Kinney. 1999. The relationship of body size to survivorship of hatchling snapping turtles (*Chelydra serpentina*): An evaluation of the "bigger is better" hypothesis. Oecologia 121:224–235.

Conway, W. 1988. Can technology aid species preservation? pp. 263–268. *In*: Biodiversity. E. O. Wilson and F. M. Peter (eds.). National Academy Press, Washington, D.C.

Cox, J. 1984. Crocodile nesting ecology in Papua, New Guinea, p. 203. Livestock Dir., Dept. Prim. Indust. Fao, Un, Port Moresby, DP/PNG/74/029, Field Document No. 5.

Crews, D. T., T. Wibbels, and W. H. N. Gutzke. 1989. Action of sex steroid hormones on temperature-induced sex determination in the snapping (*Chelydra serpentiana*). Gen. Comp. Endocrinol. 76:159–166.

Crouse, D. T., L. B. Crowder, and H. Caswell. 1987. A stage based population model for loggerhead sea turtles and implications for conservation. Ecology 68:1412–1423.

Crowder, L. B., D. T. Crouse, S. S. Heppell, and T. H. Martin. 1987. Predicting the impact of turtle excluder devices on loggerhead sea turtle populations. Ecol. Appl. 4:437–445.

Cunningham, A. A. 1996. Disease risk of wildlife translocations. Conserv. Biol. 10:349–353.

Cwynar, T. 1995. Minimonsters "musseling" into Missouri. Missouri Conservationist 56:32–35.

Da Silva, E., and M. Blasco. 1995. *Trachemys scripta elegans* in southwestern Spain. Herp. Rev. 26:133–134.

Dalrymple, G. H. 1977. Intraspecific variation in the cranial feeding mechanism of turtles of the genus *Trionyx* (Reptilia, Testudines, Trionychidae). J. Herpetol. 11:255–285.

Dancik. T. 1974. A survey of the turtles of the Des Plaines River. Bull. Chicago Herp. Soc. 9:23–33.

Daniels, T. P. 1994. Red-eared terrapin at Loch Ardinning, Milngavie (V. C. 86). Glasgow Nat. 22:430.

Darlington, P. J., Jr. 1948. The geographical distribution of cold-blooded vertebrates. Q. Rev. Biol. 23:1–26, and 105–123.

—. 1957. Zoogeography: The Geographical Distribution of Animals. John Wiley & Sons, Inc., New York.

Das, I. 1990. The trade in freshwater turtles from Bangladesh. Oryx. 24:163–166.

—. 1991. Colour Guide to the Turtles and Tortoises of the Indian Subcontinent. R & A Publishing, Avon, UK.

—. 1995. Turtles and Tortoises of India. Oxford University Press, Bombay.

—. 1996. Freshwater Turtles: A Colouring Book for Children. People's Trust for Endangered Species, London.

Davenport, J., and E. A. Macedo. 1990. Behavioural osmotic control in the euryhaline diamondback terrapin *Malaclemys terrapin*: responses to low salinity and rainfall. J. Zool., London. 220:487–496.

Davenport, J., and T. M. Wong. 1986. Observations on the water economy of the estuarine turtles *Batagur baska* (Gray) and *Callagur borneoensis* (Schlegel and Muller). Comp. Biochem. Physiol. 84A:703–707.

—. 1992. Aerial and aquatic respiration in the black-rayed softshell turtle *Amyda cartilaginea*. Herp. J. 2:115–120.

Davenport, J., T. M. Wong, and J. East. 1992. Feeding and digestion in the omnivorous estuarine turtle *Batagur baska* (Gray). Herp Journ. 2:133–139.

Davis, S. D. 1996. Reproduction and diet of *Kinosternon odoratum* inhabiting a cold water reservoir in southwest Missouri. M.S. Thesis. Southwest Missouri State University, Springfield, MO.

De Lima, A. G., W. E. Magnusson, and V. L. Da Costa. 1997. Diet of the turtle *Phrynops rufipes* in Central Amazonia. Copeia 1997:216–219.

De Treville, D. 1975. A *Trionyx* carapace from an ancient Egyptian tomb. Chelonia 2:1–2.

Dean, R. H. 1980. Paper presented: Selected aspects of the ecology of the Central American mud turtle, *Staurotypus salvinii*. American Society of Ichthyologists and Herpetologists, Fort Worth, TX.

Delany, M. F., and C. L. Abercrombie. 1986. American alligator food habits in northcentral Florida. J. Wildlife Management 50:348–353.

Depari, J. A. 1996. Overwintering in the nest chamber by hatchling painted turtles, *Chrysemys picta*, in northern New Jersey. Chelonian Conserv. Biol. 2:5–12.

Devaux, B. 1995. Proceedings: International Congress of Chelonian Conservation, p. 344. SOPTOM, Gonfaron, France.

Dixon, J. R., and P. Soini. 1977. The reptiles of the Upper Amazon Basin, Iquitos Region, Peru, II: Crocodilians, turtles and snakes. Contr. Biol. Geol., Milwaukee Pub. Mus. 12:1–91.

Dobie, J. L. 1971. Reproduction and growth in the alligator snapping turtle, *Macrodermys temminckii* (Troost). Copeia 1971:645–658.

Dodd, C. K. 1977. Amphibians and reptiles: the declining species. Water Spectrum 10:24–32.

—. 1982. Does sea-turtle aquaculture benefit conservation? pp. 465–471. *In*: Biology and Conservation of Sea Turtles. K. A. Bjorndal (ed.). Smithsonian Institution Press, Washington, D.C.

—. 1988. Disease and population declines in the flattened musk turtle *Sternotherus depressus*. Am. Midland Naturalist 119:394–401.

—. 1990. Effects of habitat fragmentation on a stream-dwelling species, the flattened musk turtle *Sternotherus odoratus*. Biol. Conserv. 54:33–45.

—. In press. *Sternotherus depressus*, flattened musk turtle. *In*: The Conservation of Freshwater turtles. P. C. H. Pritchard and A. G. J. Rhodin (ed.). Chelonian Research Foundation.

Dodd, C. K., and R. A. Seigel. 1991. Relocation, repatriation, and translocation of amphibians and reptiles: Are they conservation strategies that work? Herpetologica 47:336–350.

Dodd, C. K., K. M. Enge, and J. N. Stuart. 1988. Aspects of the biology of the flattened musk turtle, *Sternotherus depressus* in northern Alabama. Bull. Florida St. Mus., Biol. Sci. 34:1–64.

Doody, J. S. 1995. A comparative study of two syntopic species of softshell turtles (*Apalone mutica and Apalone spinifera*) in southcentral Louisiana. M.S. Thesis. Southeastern State Louisiana University, Hammond, LA.

Doody, J. S., R. A. Sims, and A. Georges. 2001. Use of thermal springs for aquatic basking by the pig-nosed turtle, *Carettochelys insculpta*. Chelonian Conserv. Biol. 4:81–87.

Drummond, H., and E. R. Gordon. 1979. Luring in the neonate alligator snapping turtle (*Macroclemys temminckii*): Description and experimental analysis. Z. Tierpsychol. 50:136–152.

Dunham, A. E., and J. W. Gibbons. 1990. Growth of the slider turtle, pp. 135–145. *In*: Life History and Ecology of the Slider Turtle. J. W. Gibbons (ed.). Smithsonian Institution Press, Washington D.C.

Dunson, W. A. 1960. Aquatic respiration in *Trionyx spinifer* asper. Herpetologica 16:277–283.

—. 1967. Sodium fluxes in freshwater turtles. J. Exp. Zool. 165:171–182.

—. 1970. Some aspects of electrolyte and water balance in three estuarine reptiles, the diamondback terrapin, American and "salt water" crocodiles. Comp. Biochem. Physiol. 32:161–174.

—. 1976. Salt glands in reptiles. *In*: Biology of the Reptilia. Physiol. A. Vol. 5. C. Gans and W. R. Dawson (eds.). Academic Press, New York.

—. 1979a. Control mechanisms in reptiles, pp. 273–322. *In*: Mechanisms of Osmoregulation in Animals. R. Gilles (ed.). Wiley-Interscience, New York.

—. 1979b. Salinity tolerance and osmoregulation of the key mud turtle, *Kinosternon b. baurii*. Copeia 1979:548–552.

—. 1981. Behavioral osmoregulation in the key mud turtle, *Kinosternon b. baurii*. J. Herpetol. 15:163–173.

—. 1984. The contrasting roles of the salt glands, the integument and behavior in osmoregulation of marine and estuarine reptiles, pp. 107–129. *In*: Osmoregulation in Estuarine and Marine Animals. Vol. 9. R. Gilles and L. Bolis (eds.). Springer-Verlag, New York.

—. 1986. Estuarine populations of the snapping turtle (Chelydra) as a model for the evolution of marine adaptations in reptiles. Copeia 1986:741–756.

Dunson, W. A., and H. Heatwole. 1986. Effect of relative shell size in turtles on water and electrolyte composition. Am. J. Physiol. 250:R1113–R1137.

Dunson, W. A., and F. J. Mazzotti. 1989. Salinity as a limiting factor in the distribution of reptiles in Florida Bay: a theory for the estuarine origin of marine snakes and turtles. Bull. Marine Sci. 44:229–244.

Dunson, W. A., and E. O. Moll. 1980. Osmoregulation in sea water of hatchling emydid turtles, *Callagur borneoensis*, from a Malaysian sea beach. J. Herpetol. 14:31–36.

Dunson, W. A., and M. E. Seidel. 1986. Salinity tolerance of estuarine and insular emydid turtles (*Pseudemys nelsoni* and *Trachemys decussata*). J. Herpetol. 20:237–245.

Dupre, A. 1995. "Florida Freshwater Turtle" in Europa: The state of the situation, Abstract, p. 12. *In*: Proceedings: International Congress of Chelonian Conservation. B. Devaux (ed.). SOPTOM, Gonfaron, France.

Durrell, G. M. 1953. The Overloaded Ark. Faber and Faber, London.

Dutton, P. H., C. P. Whitmore, and N. Mrosovsky. 1985. Masculinisation of leatherback turtle, *Dermochelys coriacea*, hatchlings from eggs incubated in styrofoam boxes. Biol. Conserv. 31:249–264.

Eckstrom, C. K. 1996. A wilderness of water: Pantanal. Audubon 98:54–65.

Edgren, R. A. 1943. *Pseudemys scripta troostii* in Michigan. Copeia 1943:249.

Ehrenfeld, D. W. 1974. Conserving the edible turtle: can mariculture help? Am. Sci. 62:23–31.

—. 1980. Commercial breeding of captive sea turtles: Status and prospects, pp. 93–96. *In*: Reproductive Biology and Diseases of Captive Reptiles. J. B. Murphy and J. T. Collins (eds.). Meseraull Printing, Inc., Lawrence, KS.

Eldredge, N. 1979. Alternative approaches to evolutionary theory. Bull. Carnegie Mus. of Natural History 13:7–19.

—. 1985. Time Frames: The Rethinking of Darwinian Evolution and the Theory of Punctuated Equilibria. Simon and Schuster, New York.

Eldredge, N., and J. Cracraft. 1980. Phylogenetic Patterns and the Evolutionary Process: Method and Theory in Comparative Biology. Columbia University Press, New York.

Eldredge, N., and S. J. Gould. 1972. Punctuated equilibria: an alternative to phyletic gradualism, pp. 82–115. *In*: Models in Paleobiology. T. J. M. Schopf (ed.). Freeman, Cooper and Company, San Francisco.

Eley, T. 1984. Information on turtles for sales as food items in San Francisco and Oakland, Ca. In litt. to Dr. Curt Freese, Senior Wildlife Biologist. World Wildlife Fund, Washington D.C.

Elgar, M. A., and L. J. Heaphy. 1989. Covariation between clutch size, egg weight and shape: comparative evidence for chelonians. J. Zool. 219:137–152.

Emmons, L. H. 1989. Jaguar predation on chelonians. J. Herpetol. 23:311–314.

Enge, K. M. 1991. Herptile exploitation—Annual performance report. Plus Appendices A–D and Addendum to Table 2. Bureau of Nongame Wildlife, Division of Wildlife, Florida Game and Freshwater Fish Commission, Tallahassee, FL.

Ernst, C. H. 1985. Red-bellied turtle, pp. 267–270. *In*: Species of Special Concern in Pennsylvania. Vol. 11. H. H. Genoways and F. J. Brenner (eds.). Carnegie Museum of Natural History Special Publication, Pittsburgh, PA.

—. 1990. Systematics, taxonomy, variation, and geographic distribution of the slider turtle, pp. 57–67. *In*: Life History and Ecology of the Slider Turtle. J. W. Gibbons (ed.). Smithsonian Institution Press, Washington, D.C.

—. 1995. Freshwater and terrestrial turtles of the Unites States: Status and prognosis. Bull. Chicago Herp. Soc. 30:225–230.

Ernst, C. H., and R. W. Barbour. 1972. Turtles of the United States. The University Press of Kentucky, Lexington.

—. 1989. Turtles of the World. Smithsonian Institution Press, Washington, D.C.

Ernst, C. H., R. G. M. Altenburg, and R. W. Barbour. 1998. Turtles of the World. ETI, World Biodiversity Database.

Ernst, C. H., J. E. Lovich, and R. W. Barbour. 1994. Turtles of the United States and Canada. Smithsonian Institution Press, Washington, D.C.

Evermann, B. W., and H. W. Clark. 1920. Lake Maxinkuckee: a physical and biological survey, Vol. 1. The Department of Conservation, State of Indiana.

Ewert, M. A. 1976. Nest, nesting, and aerial basking of *Macroclemys* under natural conditions and comparisons with *Chelydra* (Testudines: Chelydridae). Herpetologica 32:150–156.

—. 1979. The embryo and its egg: development and natural history, pp. 33–416. *In*: Turtles: Perspectives and Research. M. Harless and H. Morlock (ed.). John Wiley and Son, New York.

—. 1985. Embryology of turtles, pp. 77–267. *In*: Biology of the Reptilia, Vol. 14. C. Gans, F. Billett, and P. F. A. Maderson (eds.). John Wiley and Son, New York.

—. 2000. *Chelydra serpentina osceola* (Florida snapping turtle) reproduction. Herpetol. Rev. 31:172.

Ewert, M. A., and D. R. Jackson. 1994. Nesting ecology of the alligator snapping turtle (*Macroclemys temminickii*) along the lower Apalachicola River. Florida Game and Fresh Water Fish Comm., Tallahassee, FL.

Ewert, M A., and C. E. Nelson. 1991. Sex determination in turtles: Diverse patterns and some possible adaptive values. Copeia 1991:50–69.

Ewert, M. A., and D. S. Wilson. 1996. Seasonal variation of embryonic diapause in the striped mud turtle (*Kinosternon baurii*) and general considerations for conservation planning. Chelonian Conserv. Biol. 2:43–54.

Fachín-Terán, A., M. Chumbe-Ayllon, and G. Taleixo-Torres. 1996. Consumo de tortugas de la reserva nacional Pacaya-Samiria, Loreto, Per°. Vida Silvestre Neotropical. 5:147–150.

Fachín-Terán, A., R. C. Vogt, and F. S. Gomez. 1990. Food preferences of turtles from a blackwater river community in the Rio Guapore near Costa Marques, Rondonia, Brasil. *In*: Abstracts: Annual Joint Meeting of the Herpetologists League and the Society for the Study of Amphibians and Reptiles, Tulane University, New Orleans, LA.

Fachín-Terán, A., R. C. Vogt, and M. D. S. Gomez. 1995. Food habits of an assemblage of five species of turtles in the Rio Guapore, Rondonia, Brazil. J. Herpetol. 29:536–547.

Fahey, K. M. 1987. Aspects of the life history of the river cooter, *Pseudemys concinna* (Le Conte), in the Talapoosa River, Tallapoosa County, Alabama. Ph.D. Dissertation. Auburn University, Auburn, AL.

Fanti, E. 1995. Reintroduction of reptiles born in captivity to their natural environment, p. 12. *In*: Proceedings: International Congress of Chelonian Conservation. B. Devaux (ed.). SOPTOM, Gonfaron, France.

Feehan, T. 1986. Turtle trade controversy re-ignited. TRAFFIC—USA. 7:4–5.

Ferri, V. 1995. *Emys orbicularis*: Situation and conservation projects in Lombardy. *In*: Proceedings: International Congress of Chelonian Conservation. B. Devaux (ed.). SOPTOM, Gonfaron, France.

Ferri, V., and A. R. Di Cerbo. 1995. Lombardy Arcadia Project: Initiatives and propositions for the control of the red-eared pond turtle (*Trachemys scripta*, Schoepff). pp. 298–300. *In*: Proceedings: International Congress of Chelonian Conservation. SOPTOM, Gonfaron, France.

Feuer, R. 1966. Variation in Snapping Turtles, *Chelydra serpentina* Linaeus: A Study in Quantitative Systematics. Ph.D. Dissertation. Salt Lake City: University of Utah, Salt Lake City, UT.

Fidenci, P. 2002. Chelonian notes along the Caura River, Venezuela, 2001. Turtle and Tortoise Newsletter 6–8.

Filoramo, N. I., and F. J. Janzen. 2002. An experimental study of the influence of embryonic water availability, body size, and clutch on survivorship of neonatal red-eared sliders, *Trachemys scripta elegans*. Herpetologica 58:67–74.

Finkle, R. G., D. L. Knickerbocker, and D. L. Claussen. 2000. Influence of hydric conditions during incubation and population on overland movement of neonatal snapping turtles. J. Herpetol. 34:452–455.

Fitch, H. S., and M. V. Plummer. 1975. A preliminary ecological study of the softshelled turtle *Trionyx muticus* in the Kansas River. Israel J. Zool. 24:28–42.

Flaherty, N., and J. R. Bider. 1984. Physical structures and the social factor as determinants of habitat use by *Graptemys geographica* in southwestern Quebec. Am. Midland Naturalist 111:259–266.

Foose, T. J. 1983. The relevance of captive populations to the conservation of biotic diversity, pp. 374–401. *In*: Genetics and Conservation: a Reference for Managing Wild Animal and Plant Populations. C. M. Schonewald-Cox, S. M. Chambers, B. MacBryde, and L. Thomas (ed.). Benjamin/Cummings, Menlo Park, CA.

Frankel, O. H., and M. E. Soulé. 1981. Conservation and Evolution. Cambridge University Press, Cambridge, UK.

Franklin, I. R. 1980. Evolutionary change in small populations, pp. 135–139. *In*: Conservation Biology: an evolutionary-ecological perspective. M. E. Soulé and B.A. Wilcox (ed.). Sinauer Associates, Sunderland, MA.

Frazer, N. B. 1992. Sea turtle conservation and halfway technology. Conserv. Biol. 6:179–184.

—. 1994. Sea turtle headstarting and hatchery programs., pp. 374–380. *In*: Principles of Conservation Biology. G. K. Meffe and C. R. Carroll (eds.). Sinauer Associates, New York.

—. 1997. Turtle conservation and halfway technology: what is the problem? pp. 422–425. *In*: Proceedings: Conservation, Restoration, and Management of Tortoises and Turtles:

an International Conference. J. van Abbema (ed.), State University of New York, Purchase, NY.

Frazer, N. B., J. W. Gibbons, and T. J. Owens. 1990. Turtle trapping: preliminary tests of conventional wisdom. Copeia 1990:1150–1152.

Frazier, J. 1995. Long term survival of contemporary anapsids in the face of hominid frivolity: the case of sustainable development, pp. 262–265. *In*: Proceedings: International Congress of Chelonian Conservation. B. Devaux (ed.). SOPTOM, Gonfaron, France.

Freiberg, M. 1981. Turtles of South America. T. F. H. Publications, Neptune, NJ.

Frizenda, S., and D. Ballasina. 1990. Le statut des chéloniens terrestres et d'eau douce en Italie. Bull. Soc. Herp. France 53:18–23.

Froom, B. 1976. The Turtles of Canada. McClelland and Stewart Limited, Toronto.

Fryer, G., and T. D. Iles. 1972. The Cichlid Fishes of the Great Lakes of Africa. T. F. H. Publications, Neptune, NJ.

Fugler, C. M. 1984. The commercially exploited Chelonia of Bangladesh: taxonomy, ecology, reproductive biology and ontogeny. Bangladesh Fish. Inf. Bull. 2:1–52.

Fuller, S. L. 1974. Clams and mussels, pp. 215–273. *In*: Pollution Ecology of Freshwater Invertebrates. C. W. Hart and S. L. H. Fuller (eds.). Academic Press, New York.

Fusari, M. 1982. A study of the reactions of desert tortoises in different types of fencing, pp. 125–132. Proc. Symp. Desert Tortoise Council.

Fuselier, L., and D. Edds. 1994. Habitat partitioning among three sympatric species of map turtles, genus *Graptemys*. J. Herpetol. 28:154–158.

Galat, D. L., J. W. Robinson, and L. W. Hesse. 1996. Restoring aquatic resources to the Lower Missouri River: issues and initiatives, pp. 49–71. *In*: Overview of River-Floodplain Ecology in the Upper Mississippi River Basin. Vol. 3. D. L. Galat and A. G. Frazer (eds.). U.S. Government Printing Office, Washington, D.C.

Galbraith, D. A., R. J. Brooks, and G. P. Brown. 1997. Can management intervention achieve sustainable exploitation of turtles? pp. 186–194. *In*: Proceedings: Conservation, Restoration, and Management of Tortoises and Turtles: An International Conference. J. van Abbema (ed.). State University of New York, Purchase, NY.

Galois, P., M. LéVeillé, L. Bouthillier, C. Daigle, and S. Parren. 2002. Movement patterns, activity, and home range of the eastern spiny softshell turtle (*Apolone spinifera*) in Northern Lake Champlain, Québec, Vermont. J. Herpetol. 36:402–411.

Gans, C. 1961. The feeding mechanism of snakes and its possible evolution. Am. Zoologist 1:217–227.

Garber, S. W. 1990. The ups and downs of the diamondback terrapin. The Conservationist/NYSDEC 44–47.

Garstka, W. R., W. E. Cooper, Jr., K. W. Wasmund, and J. E. Lovich. 1991. Male sex steroids and hormonal control of male courtship behavior in the yellow-bellied slider turtle, *Trachemys scripta*. Comp. Biochem. Physiol. 98A:271–280.

Gasith, A., and I. Sidis. 1984. Polluted water bodies, the main habitat of the Caspian terrapin (*Mauremys caspica rivulata*) in Isreal. Copeia 1984:216–219.

Gasperetti, J., A. F. Stimson, J. D. Miller, J. P. Ross, and P. R. Gasperetti. 1993. Turtles of Arabia. Fauna of Saudi Arabia. 13:170–360.

Gelatt, T. S., and J. D. Kelley. 1995. Western painted turtles, *Chrysemys picta bellii*, basking on a nesting common loon, Gavia immer. Can. Field Naturalist 109:456–458.

George, G. 1990. Status and conservation of *Graptemys barbouri*, *Graptemys flavimaculata*, *Graptemys oculifera*, and *Graptemys caglei*. pp. 24–30. *In*: Proceedings of the First

International Symposium on Turtles and Tortoises: Conservation and Captive Husbandry, Orange, CA.

Georges, A. 1982. Diet of the freshwater turtle *Emydura kreftii* (Chelonia, Chelidae), in an unproductive lentic environment. Copeia 1982:331–336.

—. 1987. The pig-nosed turtle Warradjan. Aust. Natural History 22:228–234.

—. 1993. Setting conservation priorities for Australian freshwater turtles in northern Australia, pp. 49–58. *In*: Herpetology in Australia: a diverse discipline. D. Lunney and D. Ayers (eds.). Transactions of the Royal Zoological Society of New South Wales, Mosman.

Georges, A., and M. Adams. 1992. A phylogeny for the Australian chelid turtles based on allozyme electrophoresis. Aust. J. Zool. 40:453–476.

—. 1996. Electrophoretic delineation of species boundaries within the short-necked freshwater turtles of Australia (Testudines: Chelidae). Zool. J. Linn. Soc. 118:241–260.

Georges, A., and R. Kennett. 1989. Dry-season distribution and ecology of Carettochelys insculpta (*Chelonia Carettochelyidae*) in Kakadu National Park, Northern Australia. Aust. Wildl. Res. 16:323–335.

Georges, A., and M. Rose. 1993. Conservation biology of the pig-nosed turtle, *Carettochelys insculpta*. Chelonian Conserv. Biol. 1:3–12.

Georges, A., J. Birrell, K. A. Saint, W. P. Mccord, and S. C. Donnellan. 1998. A phylogeny for side-necked turtles (Chelonia: Pleurodira) based on mitochondrial and nuclear gene sequence variation. Biol. J. Linn. Soc. 67:213–246.

Ghosh, A., and N. R. Mandal. 1990. Studies on nesting and artificial hatching of the endangered river terrapin *Batagur baska* (Gray) in the Sunderbans Tiger Reserve, West Bengal. J. Bombay Natural History Soc. 87:50–52.

Gibbons, J. W. 1967. Variation in growth rates in three populations of the painted turtle, *Chrysemys picta*. Herpetologica 23:296–303.

—. 1968. Growth rates of the common snapping turtle, *Chelydra serpentina* in a polluted river. Herpetologica 24:266–267.

—. 1970a. Sex ratios in turtles. Res. Popul. Ecol. 12:252–254.

—. 1970b. Reproductive dynamics of a turtle (*Pseudemys scripta*) population in a reservoir receiving heated effluent from a nuclear reactor. Can. J. Zool. 48:881–885.

—. 1990a. The slider turtle, pp. 3–18. *In*: Life History and Ecology of the Slider Turtle. J. W. Gibbons (ed.). Smithsonian Institution Press, Washington, D.C.

—. 1990b. Turtle studies at SREL: A research prospective, pp. 19–44. *In*: Life History and Ecology of the Slider Turtle. J. W. Gibbons (ed.). Smithsonian Institution Press, Washington D.C.

—. 1990c. Sex ratios and their significance among turtle populations, pp. 171–182. *In*: Life History and Ecology of the Slider Turtle. J. W. Gibbons (ed.). Smithsonian Institution Press, Washington, D.C.

Gibbons, J. W., and D. H. Nelson. 1978. The evolutionary significance of delayed emergence from the nest by hatchling turtles. Evolution 32:297–303.

Gibbons, J. W., and J. W. Coker. 1978. Ecological and life history aspects of the cooter, *Chrysemys floridana* (LeConte). Herpetologica 33:29–33.

Gibbons, J. W., and J. L. Greene. 1990. Reproduction in the slider and other species of turtles., pp. 124–134. *In*: Life History and Ecology of the Slider Turtle. J. W. Gibbons (ed.). Smithsonian Institution Press, Washington, D.C.

Gibbons, J. W., and J. R. Harrison, III. 1981. Reptiles and amphibians of Kiawah and Capers Islands, South Carolina. Brimleyana 5:145–162.

Gibbons, J. W., and J. E. Lovich. 1990. Sexual dimorphism in turtles with emphasis on the slider turtle (*Trachemys scripta*). Herpetol. Monogr. 4:1–28.

Gibbons, J. W., G. H. Keaton, J. P. Schubauer, J. L. Greene, D. H. Bennett, J. R. Mcauliffe, and R. R. Sharitz. 1979. Unusual population size structure in freshwater turtles on barrier islands. Georgia J. Sci. 155–159.

Gibbons, J. W., R. D. Semlitsch, J. L. Greene, and J. P. Schubauer. 1981. Variation in age and size at maturity of the slider turtle *Pseudemys scripta*. Am. Naturalist 117:841–845.

Gibbons, J. W., J. L. Greene, and J. D. Congdon. 1990. Temporal and spatial movement patterns of sliders and other turtles, pp. 201–215. *In:* Life History and Ecology of the Slider Turtle. J. W. Gibbons (ed.). Smithsonian Institution Press, Washington, D.C.

Gibbs, J. P., and G. D. Amato. 2000. Genetics and Demography in Turtle Conservation, pp. 207–217. *In:* Turtle Conservation. M. W. Klemens (ed.). Smithsonian Institution Press, Washington, D.C.

Gibbs, J. P., and W. G. Shriver. 2002. Estimating the effects of road mortality on turtle populations. Conserv. Biol. 16:1647–1652.

Gilbert, B. 1993. Snappers, those battering rams that walk. Smithsonian 24:93–101.

Gilles-Baillien, M. 1970. Urea and osmoregulation in the diamondback terrapin *Malaclemys centrata centrata* (Latreille). J Exp. Biol. 52:691–697.

—. 1973a. Isoosmotic regulation in various tissues of the diamond terrapin *Malaclemys centrata centrata* (Latreille). J. Exp. Biol. 59:39–43.

—. 1973b. Hibernation and osmoregulation in the diamondback terrapin *Malaclemys centrata centrata* (Latreille). J. Exp. Biol. 59:45–51.

Gilpin, M. 1991. The genetic effective size of a metapopulation. Biol. J. Linn. Soc. 42:165–175.

Gilpin, M. E., and M. E. Soulé. 1986. Minimum viable population: Processes of species extinction, pp. 19–34. *In:* Conservation Biology: The Science of Scarcity and Diversity. M. E. Soulé (ed.). Sinauer Associates, Inc., Sunderland, MA.

Girgis, S. 1961. Aquatic respiration in the common Nile turtle, *Trionyx triunguis* (Forskal). Comp. Biochem. Physiol. 3:206–217.

Girondot, M., H. Fouillet, and C. Pieau. 1998. Feminizing turtle embryos as a conservation tool. Conserv. Biol. 12:353–362.

Giugliano, E. 1988. Un caso di acclimatamento di tetuggine palustre (*Chrsemys scripta elegans*) in Piemonte. Bolletino Gruppo RANA Italia, 1:27–29.

Gleason, H. A. 1922. On the relationship between species and area. Ecology 3:158–162.

Goeldi, E. A. 1906. In part a condensation of Joa Martins da Silva Coutinho, Sobre as tartarugas do amazonas. Chelonias do Brazil. Bol. Mus. Goeldi de Hist. Nat. e Ethnographia (Para), Bul. Mensual de la Soc. Imperiale Zoologique d' Acclimatacion 4:699–756.

Goldman, C. R., and A. J. Horne. 1983. Limnology. McGraw-Hill, New York.

Gomez, E. D. 1982. Problems of enforcing sea turtle conservation laws in developing countries, pp. 537–539. *In:* Biology and Conservation of Sea Turtles. K. A. Bjorndal (ed.). Smithsonian Institution Press, Washington, D.C.

Goode, J. 1967. Freshwater tortoises of Australia and New Guinea (in the family Chelidae). Lansdowne Press, Melbourne.

Goodrich, J. M., and S. W. Buskirk. 1995. Control of abundant native vertebrates for conservation of endangered species. Conserv. Biol. 9:1357–1364.

Gordon, D. M., and R. D. MacCulloch. 1980. An investigation of the ecology of the map turtle, *Graptemys geographica* (Le Sueur), in the northern part of its range. Can. J. Zool. 58:2210–2219.

Graham, A. 1946. Timothy Turtle. Viking Press, New York.

Graham, T., and A. Georges. 1996. Struts for collapsible funnel traps. Herp. Rev. 27:189–190.

Graham, T. E. 1991. Life history notes: *Apalone spinifera spinifera* (eastern spiny softshell). Pattern dimorphism. Herp. Rev. 22:97.

—. 1997. Effective predator excluders for turtle nests. Herp. Rev. 28:76.

Graham, T. E., and C. B. Cobb. 1998. Sexual dimorphism of neonate eastern spiny softshells, *Apalone spinifera spinifera*. Chelonian Conserv. Biol. 3:111–112.

Graham, T. E., and A. A. Graham. 1991. *Trionyx spiniferus spiniferus* (Eastern spiny softshell). Burying behavior. Herp. Rev. 22:56–57.

—. 1997. Ecology of the eastern spiny softshell, *Apalone spinifera spinifera*, in the Lamoille River, Vermont. Chelonian Conserv. Biol. 2:363–369.

Graham, T. E., and R. W. Perkins. 1976. Growth of the common snapping turtle, *Chelydra s. serpentina*, in a polluted marsh. Bull. Maryland Herpetol. Soc. 12:123–125.

Grant, C. 1932. The herpetology of Vieques Island. J. Dept. Agriculture, Puerto Rico 16:37–39.

Gray, E. M. 1995. DNA fingerprinting reveals a lack of genetic variation in northern populations of the western pond turtle (*Clemmys marmorata*). Conserv. Biol. 9:1244–1255.

Green, D. 1991. Introduction/Establishment in Testudines: a guide to the introduction of Testudines outside of their natural range anywhere around the world. A self-published report.

Griffith, B., J. M. Scott, J. W. Carpenter, and C. Reed. 1989. Translocation as a species conservation tool: status and strategy. Science Natural History Museum, Los Angeles Co. 245:477–480.

Gritters, S. A., and L. M. Mauldin. 1994. Four years of turtle collections on Navigation Pool 13 of the upper Mississippi River. Long Term Resource Monitoring Program Special Report.; Environmental Management Technical Center, Onalaska, WI.

Groombridge, B., and L. Wright. 1982. The IUCN Amphibia—Reptilia Red Data Book: Part 1 Testudines, Crocodylia, Rhynchocephalia. Vol. 1. IUCN, Gland, Switzerland.

Gunther, A. 1864. The Reptiles of British India. London.

Gupta, A. 2002. The beleagured chelonians of northeastern India. Turtle and Tortoise Newsletter 16–17.

Gutzke, W. H. N., and D. B. Chymiy. 1988. Sensitive periods during embryology for hormonally induced sex determination in turtles. Gen. Comp. Endocrinol. 71:265–267.

Gutzke, W. H. N., and J. J . Bull. 1986. Steroid hormones reverse sex in turtles. Gen. and Comp. Endocrinol. 64:368–372.

Haitao, S. 2002. Results of turtle market surveys in Chengdu and Kunming. Turtle and Tortoise Newsletter 6:15–16.

Haitao, S., and J. F. Parham. 2001. Preliminary observations of a large turtle farm in Hainan Province, People's Republic of China. Turtle and Tortoise Newsletter 3:4–6.

Hammer, D. A. 1972. Ecological relations of waterfowl and snapping turtle populations. Ph.D. Dissertation. Utah State University, Logan, UT.

Hanski, I., and M. E. Gilpin. 1997. The metapopulation approach, its history, conceptual domain, and application to conservation, pp. 5–26. *In:* Metapopulation Biology: Ecology, Genetics, and Evolution. I. Hanski and M. E. Gilpin (eds.). Academic Press, San Diego.

Harding, J. H. 1990. A twenty year wood turtle study in Michigan: Implications for conservation, pp. 31–35. *In:* Proceedings of the First International Symposium on Turtles and Tortoises: Conservation and Captive Husbandry, Chapman University, Orange, CA.

Harding, J. H., and T. J. Bloomer. 1979. The wood turtle, *Clemmys insculpta* . . . a natural history. HERP Bull. New York Herpetol. Soc. 15:9–26.

Hardy, T. 1988. The tortoise and the scare. BioScience 38:76–69.

Harman, W. N. 1974. Snails (Mollusca: *Gastropoda*), pp. 275–312. *In:* Pollution Ecology of Freshwater Invertebrates. C. W. Hart and S. L. H. Fuller (eds.). Academic Press, New York.

Harrell, J. B., and G. L. Stringer. 1997. Feeding habits of the alligator snapping turtle (*macroclemys temminckii*) as indicated by teleostean otoliths. Herp. Rev. 28:185–187.

Harrell, J. B., C. M. Allen, and S. J. Herbert. 1996a. Movements and habitat use of subadult alligator snapping turtles (*Macroclemys temminckii*) in Louisiana. Am. Midland Naturalist 135:60–67.

Harrell, J. B., N. H. Douglas, M. M. Haraway, and R. D. Thomas. 1996b. Mating behavior in captive alligator snapping turtles (*Macroclemys temminckii*). Chelonian Conserv. and Biol. 2:101–105.

Hart, D. R. 1979. Resource partitioning amouong Louisiana turtles of the genus *Chrysemys*. Ph.D. Dissertation. Tulane University, Tulane, LA.

—. 1983. Dietary and habitat shift with size of red-eared turtles (*Pseudemys scripta*) in a southern Louisiana population. Herpetologica 39:285–290.

Harvey, P. H., and M. D. Pagel. 1991. The Comparative Method in Evolutionary Biology. Oxford University Press, Oxford.

Haskell, A., T. E. Graham, C. R. Griffin, and J. B. Hestbeck. 1996. Size related survival of headstarted redbelly turtles (*Pseudemys rubriventris*) in Massachusetts. J. Herpetol. 30:524–527.

Hedges, S. B. 1996. The origin of West Indian amphibians and reptiles, pp. 160–174. *In:* Contributions to West Indian Herpetology: a Tribute to Albert Schwartz. Vol. 12. R. Powell and R. W. Henderson (eds.). Society for the Study of Amphibians and Reptiles, Contributions to Herp. Ithaca.

Hedrick, P. W., and M. E. Gilpin. 1997. Genetic effective size of a metapopulation, pp. 164–181. *In:* Metapopulation Biology: Ecology, Genetics, and Evolution. L. Hanski and M. E. Gilpin (eds.). Academic Press, San Diego.

Hedrick, P. W., and P. S. Miller. 1992. Conservation genetics: Techniques and fundamentals. Ecol. Appl. 2:30–46.

Helwig, D. D., and M. E. Hora. 1983. Polychlorinated biphenyl, mercury, and cadmium concentrations in Minnesota snapping turtles. Bull. Eviron. Contam. Toxicol. 30:186–190.

Hemley, G., and A. Gaski. 1989. Memorandum to Chief, Office of Scientific Authority, regarding CITES proposal to list *Trachemys scripta elegans*. 2.

Hendrickson, J. R., and E. R. Alfred. 1961. Nesting populations of sea turtles on the east coast of Malaya. Bull. Raffles Mus. 26:190–196.

Hendrie, D. B. 2000. Status and conservation of tortoises and freshwater turtles in Vietnam, pp. 63–73. *In:* Asian Turtle Trade: Proceedings of a Workshop on Conservation and Trade of Freshwater Turtles and Tortoises in Asia. Vol. 2. P. P. van Dijk, B. L. Stuart, and A. G. J. Rhodin (eds.). Chelonian Research Foundation, Phnom Penh, Cambodia.

Heppell, S. S. 1997. On the importance of eggs. Marine Turtle Newsletter. 76:6–8.

—. 1998. Application of life-history theory and population model analysis to turtle conservation. Copeia 1998:367–375.

Heppell, S. S., L. B. Crowder, and D. T. Crouse. 1996. Models to evaluate headstarting as a management tool for long-lived turtles. Ecol. Appl. 6:556–565.

Herald, E. S. 1949. Effects of DDT-oil solutions upon amphibians and reptiles. Herpetologica 5:117–120.

Hewitt, J. 1935. Some new forms of batrachians and reptiles from South Africa. Rec. Albany Mus. 4:283–357.

Highfield, A. C. 1994. Tortoise Trust Guide to Tortoises and Turtles. Carapace Press, London.

Holling, H. C. 1951. Minn of the Mississippi. Houghton Mifflin Co, Boston.

Holman, J. A. 1995. Pleistocene Amphibians and Reptiles in North America. Oxford University Press, New York.

Homer, B. L., E. R. Jacobson, J. Schumacher, and G. Scherba. 1994. Chlamydiosis in mariculture-reared green sea turtles (*Chelonia mydas*). Vet. Pathol. 31:1–7.

Hoover, C. 1998. The U.S. role in the international live reptile trade: Amazon tree boas to Zululand dwarf chameleons. TRAFFIC—North America. August 1998:1–59.

Horine, B. 1996. Life's a snap for this farmer. Successful Farming. (Successful Bonus Section) Mid-March.

Horne, A. J., and C. R. Goldman. 1994. Limnology, 2nd edition. McGraw-Hill, New York.

Hsus. 1994. Live freshwater turtle and tortoise trade in the United States. Human Society of the United States/ Humane Society International.

Hughes, G. R. 1996. Nesting of the leatherback turtle (*Dermochelys coriacea*) in Tongaland, KwaZulu-Natal, South Africa, 1963–1995. Chelonian Conserv. Biol. 2:153–158.

Hulse, A. C. 1980. Notes on the occurrence of introduced turtles in Arizona. Herp. Rev. 11:16–17.

Hussain, I. S. 2003. Olive Ridley Sea Turtles in Porto-Novo, Tamil Nadu, India, with an observation of an Asian giant softshell turtle. Marine Turtle Newsletter 101:25.

Hynes, H. B. N. 1970. The Ecology of Running Waters. University of Toronto Press, Toronto, Canada.

IBAMA. 1989. Instituto Brasileiro do Meio Ambiente e dos Recursos Naturais Renovavis. Projecto Quelônios da Amazônia—10 anos. Instituto Brasileiro do Meio Ambiente e dos Recursos Naturais Renováveis, Brasilia.

IUCN. 1989. Tortoise and freshwater turtles: an action plan for their conservation. IUCN/ SSC Tortoise and Freshwater Turtle Specialist Group, Gland, Switzerland.

Iverson, J. B. 1979. Another inexpensive turtle trap. Herp. Rev. 10:55.

—. 1981. Biosystematics of the *Kinonsternon hirtipes* species group (Testudines: Kinosternidae). Tulane Stud. Botany Zool. 23:1–74.

—. 1982. Biomass in turtle populations: a neglected subject. Oecologia (Berlin) 55:69–76.

—. 1985. Checklist of the turtles of the world with english common names. Society for the Study of Amphibians and Reptiles. Herpetol. Circ. 14:1–14.

—. 1991. Patterns of survivorship in turtles (Order: Testudines). Can. J. Zool. 69:385–391.

—. 1992a. Correlates of reproductive output in turtles (Order Testudines). Herpetol. Monogr. 6:25–42.

—. 1992b. Species richness maps of the freshwater and terrestrial turtles of the world. Smithsonian Herpetological Service:18.

—. 1992c. Global correlates of species richness in turtles. Herp. J. 2:77–81.

Iverson, J. B., and J. F. Berry. 1979. The mud turtle genus *Kinosternon* in northeastern Mexico. Herp. Rev. 35:318–324.

Iverson, J. B., and A. Richmond. 1992. Checklist with Distribution Maps of the Turtles of the World. Privately printed.

Iverson, J. B., C. P. Balgooyen, K. K. Byrd, and K. K. Lyddan. 1993. Latitudinal variation, in egg and clutch size in turtles. Can. J. Zool. 71:2448–2461.

Iverson, J. B., H. Higgins, A. Sirulnick, and C. Griffiths. 1997. Local and geographic variation in the reproductive biology of the snapping turlte (*Chelydra serpentina*). Herpetologica 53:96–117.

Iverson, J. B., S. A. Thomson, and A. Georges. 2001. Validity of taxonomic changes for turtles proposed by Wells and Wellington. J. Herpetol. 35:361–368.

IWC. 1988. International Wildlife Coalition. A proposal to include the red-eared turtle *Trachemys scripta elegans* in Appendix II of the Convention on International Trade in Endangered Species of Wild Fauna and Flora. U.S. Fish and Wildlife Service, Office of the Scientific Authority.

Izadjoo, M., C. O. A. Pantoja, and R. J. Siebling. 1987. Acquisition of *Salmonella* flora by turtle hatchlings on commercial turtle farms. Can. J. Microbiol. 33:718–724.

Jackson, C. G., Jr., and J. D. Davis. 1972. A quantitative study of the courtship display of the red-eared turtle, *Chrysemys scripta elegans* (Wied). Herpetologica 28:58–64.

Jackson, D. C., J. Allen, and P. K. Strup. 1976. The contribution of non-pulmonary surfaces to CO_2 loss in 6 species of turtles at 20°C. Comp. Biochem. Physiol. 55A:243–246.

Jackson, D. R. 1986. Predataion on turtles by the American alligator: Ecological and morphological correlates. Abstract. *In:* Tortugas de Agua Dulce Neotropicales, 4–8 March, Estacion Biological "Los Tuxtlas," Veracruz.

—. 1995. Systematics of the *Pseudemys concinna-floridana* complex (Testudines: Emydidae): an alternative interpretation. Chelonian Conserv. Biol. 1:329–333.

—. 1996. Meat on the move: diet of a predatory turtle *Dierochelys reticularia* (Testudines: Emydidae). Chelonian Conserv. Biol. 2:105–108.

—. In press. *Pseudemys nelsoni* Carr 1938, Florida red-bellied turtle. *In:* The Conservation Biology of Freshwater Turtles. Vol. 3. P. C. H. Pritchard and A. G. J. Rhodin (eds.). Chelonian Research Foundation.

Jackson, J. F. 1988. Crevice occupation by musk turtles: taxonomic distribution and crevice attributes. Anim. Behav. 36:793–801.

—. 1990. Evidence for chemosensor-mediated predator avoidance in musk turtles. Copeia 1990:557–560.

Jacobson, E. R. 1993. Implications of infectious disease for captive propagation and introduction programs of threatened/endangered reptiles. J. Zool. Wildlife Med. 24:245–255.

—. 1995. Health problems of marine turtles, pp. 22–23. *In:* Proceedings: International Congress of Chelonian Conservation. B. Devaux (ed.). SOPTOM, Gonfaron, France.

—. 1996. Guest Editorial: Marine turtle farming and health issues. Marine Turtle Newsletter 72:13–15.

Jacobson, E. R., J. M. Gaskin, R. P. Shields, and F. H. White. 1979. Mycotic pneumonia in mariculture-reared green sea turtles. J. Am. Vet. Med. Assoc. 175:929–933.

Jacobson, E. R., S. L. Club, J. M. Gaskin, and C. Gardiner. 1985a. Herpesvirus-like infection in Argentine tortoises. J. Am. Vet. Med. Assoc. 187:1227–1229.

Jacobson, E. R., J. M. Gaskin, M. Roelke, E. C. Greiner, and J. Allen. 1985b. Conjunctivitis, tracheitis, and pneumonia associated with herpesvirus infection in green sea turtles. J. Am. Vet. Med. Assoc. 189:1020–1023.

Jansen, K. P. 1993. Ecology of the tropical freshwater turtle *Rhinoclemmys funerea* in Caribbean Costa Rica. M. S. Thesis. Southwest Missouri State University, Springfield, MO.

Janzen, F. J., G. L. Paukstis, and E. D. Brodie III. 1992. Observations on basking behavior of hatchling turtles in the wild. J. Herpetol. 26:217–219.

—. 1994. Vegetational cover predicts the sex ratio of hatchling turtles in natural nests. Ecology 75:1593–1599.

Janzen, F. J., and G. L. Paukstis. 1991. A preliminary test of the adaptive significance of environmental sex determination in reptiles. Evolution 45:435–441.

Janzen, F. J., J. C. Ast, and G. L. Paukstis. 1995. Influence of the hydric environment and clutch on eggs and embryos of two sympatric map turtles. Funct. Ecol. 9:913–922.

Janzen, F. J., J. K. Tucker, and G. L. Paukstis. 2000a. Experimental analysis of an early life-history stage: selection on size of hatching turtles. Ecology 81:2290–2304.

—. 2000b. Experimental analysis of an early life-history stage: avian predation selects for larger body size of hatchling turtles. J. Evol. Biol. 13:947–954.

Jaruthanin, K. 2001–2002. Salween: The dangerous river. Fish Zone (In Thai) 3:27–40.

Jenkins, M. D. 1995. Tortoises and freshwater turtles: The trade in southeast Asia. TRAFFIC International – United Kingdom, Cambridge, UK.

Jensen, J. 1998. Bait preferences of southeastern United States Coastal Plain riverine turtles: Fish or fowl? Chelonian Conserv. Biol. 3:109–111.

Jiménez, J. W., K. A. Hughes, G. Alaks, I. Graham, and R. C. Lacy. 1994. An experimental study of inbreeding depression in a natural habitat. Science Natural History Museum, Los Angeles Co. 266:271–273.

Johns, A. D. 1987. Continuing problems for Amazon river turtles. Oryx 21:25–28.

Johnson, T. 1987. The amphibians and reptiles of Missouri. Conservation Commission of the State of Missouri, Jefferson City.

Jones, R. L. 1996. Home range and seasonal movements of the turtle *Graptemys flavimaculata*. J. Herpetol. 30:376–385.

Jones, R. P. 1994. Heads up: It's turtle traffic. The Milwaukee Journal, May 27.

Jordan, D. S. 1905. The origin of species through isolation. Science 22:545–562.

Junk, W. J., and V. M. F. Da Silva. 1997. Mammals, reptiles and amphibians, pp. 409–417. *In:* The Central Amazon Floodplain. W. J. Junk (ed.). Springer Verlag, Berlin.

Kar, C. S. 1994. Crocodile census in Orissa. Crocodile Specialist Group Newsletter 13:10–11.

Kar, C. S., and R. J. Rao. 1985. A gharial (*Gavialis gangeticus*) at the Gahirmatha coast, Orissa, India. J. Bombay Natural History Soci. 82:671–674.

Kaufmann, J. H. 1986. Stomping for earthworms by wood turtles, *Clemmys insculpta*: a newly discovered foraging technique. Copeia 1986:1001–1004.

—. 1991. *Clemmys insculpta* (wood turtle): Cleaning symbiosis. Herp. Rev. 22:98.

Kaufmann, J. H., J. H. Harding, and K. N. Brewster. 1989. Worm stomping by wood turtles revisited. Bull. Chicago Herp. Soc. 24:125–126.

Kearney, P. 1972. Nocturtles of Trinidad. Int. Turtle Tortoise Soc. J. Mar.–Apr.:10–11.

Keddy, P. A. 2000. Westland Ecology: Principles and Conservation. Cambridge University Press, Cambridge, UK.

Kennedy, M. E. 1969. *Salmonella* serotypes isolated from turtle environment. Can. J. Microbiol. 15:130–132.

Kennett, R. 1992. A new trap design for catching freshwater turtles. Wildl. Res. 19:443–445.

Kennett, R., and J. Russell-Smith. 1993. Seed dispersal by freshwater turtles in northern Australia, pp. 69–70. *In:* Herpetology in Australia—A Diverse Discipline. D. Lunney and D. Ayers (ed.). Transactions of the Royal Zoological Society of New South Wales, Mosman.

Kennett, R., and O. Tory. 1996. The diet of two freshwater turtles. *Chelonia rugosa* and *Elseya dentata* (Testudines: Chelidae) from the wet-dry tropics of northern Australia. Copeia 1996:409–419.

Kennett, R., K. Christian, and D. Pritchard. 1993a. Underwater nesting by the tropical freshwater turtle, *Chelonia rugosa* (Testudinata: Chelidae). Aust. J. Zool. 41:47–52.

Kennett, R. M., A. Georges, and M. Palmer-Allen. 1993b. Early developmental arrest during immersion of eggs of a tropical freshwater turtle, *Chelonia rugosa* (Testudinata: Chelidae), from northern Australia. Aust. J. Zool. 41:37–45.

Khan bin Momin Khan, M. 1964. A note on Batagur baska (the river terrapin or tuntong). Malay Nature J. 18:184–186.

—. 1977. River terrapin. Nature Malaysiana 2:32–37.

Khan, M. A. R. 1982. Chelonians of Bangladesh and their conservation. J. Bombay Natural History Soc. 79:110–116.

—. 1987. Bangladesher Bonnyoprani. Bangla Academy, Dhaka, Bangladesh.

Kiew, B. H. 1984. Conservation status of the Malaysian fauna. IV. Turtles, terrapins, and tortoises. Malayan Nature J. 38:2–3.

Kinneary, J. J. 1996. The origin of marine turtles: a pluralistic view of evolution. Chelonian Conserv. Biol. 2:73–78.

Kitimasak, W. 2002. Natural history and captive breeding of the Siamese Narrow-Headed Softshell Turtle *Chitra chitra* Nutphanol 1986. Ph.D. Dissertation, Chulalongkorn University, Bangkok.

Kleese, W. C. 1980. Environmental effects upon herpesvirus infections in captive green sea turtles, pp. 203–210. *In:* SSAR Contributions to Herpetology. Number 1. Reproductive Biology and Diseases of Captive Reptiles. J. B. Murphy and J. T. Collins (eds.). Meseraull Printing, Inc., Lawrence, KS.

Klemens, M. W. 1993. Amphibians and Reptiles of Connecticut and Adjacent Regions. State Geological and Natural History Survey of Connecticut Bulletin. 112.

—. 2000a. From information to action: Developing more effective strategies to conserve turtles, pp. 239–258. *In:* Turtle Conservation. M. W. Klemens (ed.). Smithsonian Institution Press, Washington, D.C.

—. 2000b. Turtle Conservation, p. 334. Smithsonian Institution Press, Washington, D.C.

Klemens, M. W., and J. B. Thorbjarnarson. 1995. Reptiles as a food resource. Biodivers. Conserv. 4:281–298.

Knight, A. W., and J. W. Gibbons. 1968. Food of the painted turtle, *Chrysemys picta*, in a polluted river. Am. Naturalist 80:559–562.

Kofoid, C. A. 1903. Plankton studies. IV. The plankton of the Illinois River, 1894–1899, with introductory notes upon the hydrography of the Illinois River and its basin. Part I. Quantitative investigations and general results. Ill. Lab. Nat. Hist. Bull. 6:96–629.

Kofron, C. P. 1991. Aspects of ecology of the ringed sawback turtle, *Graptemys oculifera*. Amphibia-Reptilia 12:161–168.

Kolbe, J. J., and F. J. Janzen. 2002. Impact of nest-site selection on nest success and nest temperature in natural and disturbed habitats. Ecology 83:269–281.

Kramer, M. 1995. Home range of the Florida red-bellied turtle (*Pseudemys nelsoni*) in a Florida spring run. Copeia 1995:883–890.

Krawchuck, M. A., N. Koper, and R. J. Brooks. 1997. Observations of a possible cleaning symbiosis between painted turtles, *Chrysemys picta*, and snapping turtles, *Chelydra serpentina*, in central Ontario. Can. Field Naturalist 111:315–318.

Krebs, C. J. 1989. Ecological Methodology. Harper and Row, Publishers, New York.

Kubitzki, K., and A. Ziburski. 1994. Seed dispersal in flood plain forests of Amazonia. Biotropica 26:30–43.

Kuchling, G. 1988. Population structure, reproductive potential and increasing exploitation of the freshwater turtle *Erymnochelys madagascariensis*. Biol. Conserv. 43:107–113.

—. 1992. Distribution and status of *Erymnochelys madagascariensis* (Grandidier, 1867). Report to Conservation international and Ministere de la production Animal et des Eaux et Forets. Conservation International, Washington, D.C.

—. 1993. Possible biennial ovarian cycles of the freshwater turtle *Erymnochelys madagascariensis*. J. Herpetol. 27:470–472.

—. 1995. Turtles at a market in western Yunnan: Possible range extensions for some southern Asiatic chelonians in China and Myanmar. Chelonian Conserv. Biol. 1:223–226.

—. 1999. The Reproductive Biology of Chelonia-Zoophysiology. Springer-Verlag, New York.

—. 2003a. Status and conservation of endemic river turtles in Burma (Myanmar). Proceedings of the IUCN Turtle Survival Alliance 2003 Conference, Orlando, FL (unpaginated).

—. 2003b. A new underwater trap for catching turtles. Herp. Review 34:126–128.

Kuchling, G., and C. Blanc. In press. Madagascan big-headed turtle (*Erymnochelys madagascariensis*). *In:* The Conservation Biology of Freshwater Turtles. Vol. 3. P. C. H. Prichard and A .G. J. Rhodin (eds.). Chelonian Research Foundation.

Kuchling, G., and R. A. Mittermeier. 1993. Status and exploitation of the Madagascan big-headed turtle, *Erymnochelys madagascariensis*. Chelonian Conserv. Biol. 1:13–18.

Kushlan, J. A., and M. S. Kushlan. 1980. Everglades alligator nests: nesting site for marsh reptiles. Copeia 1980:930–932.

La Claire, L. 1996. New clues in map turtle decline. Endangered Species Bulletin XX. 15.

Lagler, K. F. 1943. Food habits and economic relations of the turtles of Michigan with special reference to fish managment. Am. Midland Naturalist 29:257–312.

Lahanas, P. N. 1982. Aspects of the life history of the southern black-knobbed sawback *Graptemys nigrinoda deticola*, Folkerts and Mount. M.S. Thesis. Auburn University, Auburn, AL.

Lamb, T., J. W. Bickham, T. B. Lyne, and J. W. Gibbons. 1995. The slider turtle as an environmental sentinel: multiple tissue assays using flow cyometric analysis. Ecotoxicology 4:5–13.

Lardie, R. L. 1976. Incubation of turtle eggs from a road kill. Bull. Oklahoma Herpetol. Soc. 1:23.

Lau, M., and S. Haitao. 2000. Conservation and trade of terrestrial and freshwater turtles and tortoises in the People's Republic of China, pp. 30–38. *In:* Asian Turtle Trade: Proceedings of a Workshop on Conservation and Trade of Freshwater Turtles and Tortoises in Asia. Vol. 2. P. P. van Dijk, B. L. Stuart, and A. G. J. Rhodin (eds.). Chelonian Research Foundation, Phnom Penh, Cambodia.

Lau, M. W. N., G. Ades, N. Goodyer, and F. S. Zou. 1996. Wildlife trade in southern China including Hong Kong and Macau. Report. Biodiversity Working Group, China Council for International Cooperation on Environment and Development project, Hong Kong.

Lau, M., B. Chan, P. Crow, and G. Ades. 2000. Trade and conservation of turtles and tortoises in the Hong Kong Special Administrative Region, People's Republic of China, pp. 39–44. *In:* Asian Turtle Trade: Proceedings of a Workshop on Conservation and Trade of Freshwater Turtles and Tortoises in Asia. Vol. 2. P. P. van Dijk, B. L. Stuart, and A. G. J. Rhodin (eds.). Chelonian Research Foundation, Lunenburg, MA.

Le Dien Duc, D., and S. Broad. 1995. Investigations in tortoise and freshwater turtle trade in Vietnam. IUCN Species Survival Commission. IUCN, Gland, Switzerland and Cambridge, UK.

Lee, J. C. 1996. The amphibians and reptiles of the Yucatan Peninsula. Cornell University Press, Ithaca, NY.

Lee, R. C. 1969. Observing the tortuga blanca. Int. Turtle Tortoise Soc. J. 3:20–26.

Legler, J. M. 1956. A social relation between snapping and painted turtles. Trans. Kansas Acad. Sci. 59:461–462.

—. 1960a. A new subspecies of slider turtle (*Pseudemys scripta*) from Coahuila, Mexico. Univ. Kansas Publ. Mus. Natur. Hist. 13:73–84.

—. 1960b. A simple and inexpensive device for trapping aquatic turtles. Proc. Utah Acad. Sci. Arts Lett. 37:63–66.

—. 1976. Feeding habits of some Australian short-necked tortoises. Victorian Naturalist 93:40–43.

—. 1978. Observations on behavior and ecology in an Australian turtle, *Chelodina expansa* (Testudines: Chelidae). Can. J. Zool. 56:2449–2453.

—. 1980. Taxonomy, distribution and ecology of freshwater turtles in the Alligator River region, Northern Territory. Report to the Supervising Scientist, Canberra.

—. 1982. The ecology of freshwater turtles in the Alligator River region. Report to the Office of the Supervising Scientist, Canberra.

—. 1985. Australian chelid turtles: Reproductive patterns in wide-ranging taxa, pp. 117–123. *In:* Biology of Australian Frogs and Reptiles. G. Grigg, R. Shine, and H. Ehmann (ed.). Royal Zoological Society of New South Wales.

—. 1989. Diet and head size in Australian chelid turtles, genus Emydura. Abstracts, pp. 1–10. *In:* Third International Congress of Vertebrate Morphology. Vol. 1. Annales de la Société Royale Zoologique de Beligque 119, University of Antwerp.

—. 1990. The genus Pseudemys in Mesoamerica: Taxonomy, distribution, and origins, pp. 82–105. *In:* Life History and Ecology of the Slider Turtle. J. W. Gibbons (ed.). Smithsonian Institution Press, Washington, D.C.

—. 1993. Morpholgy and physiology of the chelonia, pp. 108–119. *In:* Fauna of Australia. Amphibia and Reptilia. Vol. 2A. C. J. Glasby, G. J. B. Ross, and P. L. Beesley (eds.). Australian Government Publishing Service, Canberra.

Legler, J. M., and J. Cann. 1980. A new genus and species of chelid turtle from Queensland, Australia. Los Angeles Co. Mus. Nat. Sci. Contrib. Sci. 324:1–18.

—. in press. *Rheodytes leukops*, The Fitzroy River turtle. The conservation biology of freshwater turtles. *In:* The Conservation Biology and Freshwater Turtles. P. C. H. Prichard and A. G. J Rhodin (eds.). Chelonian Research Foundation.

Legler, J. M., and A. Georges. 1993a. Biogeography and phylogeny of the Chelonia, pp. 129–132. *In:* Fauna of Australia. Vol 24, Amphibia and Reptilia. C. J. Glasby, Ross, G. J. B., and Beesley, P. L. (eds.). Australian Government Publishing Service.

—. 1993b. Family Chelidae, pp. 142–152. *In:* Fauna of Australia. Amphibia and Reptilia. Vol. 2A, C. J. Glasby, G. J. B. Ross, and P. L. Beesley (eds.). Australian Government Publishing Service, Canberra.

Legler, J. M., and R. M. Winokur. 1979. Unusual neck tubercles in an Australian turtle, *Elseya Latisternum*. Herpetologica 35:325–329.

Leopold, L. B., M. B. Wolman, and J. P. Miller. 1964. Fluvial Processes in Geomorphology. W. H. Freeman, San Francisco.

Levell, J. P. 1995. A Field Guide to Reptiles and the Law. Serpents Tale Natural History Book Distributors, Excelsior, MN.

Licata, L., and X. Elguezabal. 1997. Management plan for the giant Amazonian turtle, *Podocnemis expansa*, in De La Tortuga Arrau Wildlife Refuge, Orinoco River, Venezuela, pp. 171–173. *In:* Proceedings: Conservation, Restoration, and Management of Tortoises and Turtles: An International Conference. J. van Abbema (ed.), State University of New York, Purchase, NY.

Liem, D. S. 1983. Survey and management of wildlife resources along the Purari River, pp. 269–282. *In:* The Purari-Tropical environment of a high rainfall river basin. T. Petr (ed.). Dr W. Junk Publishers, The Hague.

Lim, K. K. P., and F. L. K. Lim. 1992. A Guide to the Amphibians and Reptiles of Singapore. Singapore Science Centre.

Lindeman, P. V. 1996. Distribtuion, relative abundance, and basking ecology of the razor-back musk turtle, *Kinosternon carinatum*, in the Pearl and Pascagoula River drainages. Herpetol. Nat. Hist. 4:23–34.

—. 1997. A comparative spotting-scope study of the distribution and relative abundance of river cooters (*Pseudemys concinna*) in western Kentucky and southern Mississippi. Chelonian Conserv. Biol. 2:378–383.

—. 1999. Surveys of basking map turtles, *Graptemys* ssp. in three river drainages and the importance of deadwood abundance. Biol. Conserv. 88:33–42.

—. 2000. Resource use of five sympatric turtle species: effects of competition, phylogeny, and morphology. Can. J. Zool. 78:992–1108.

—. 2001a. A contrast in the basking habits of the sympatric trionychid turtles *Apalone mutica* and *A. spinifera*. Herpetol. Natural Hist. 8:87–89.

—. 2001b. Notes on nesting by the smooth softshell turtle (*Apalone mutica*) in a river impoundment in Western Kentucky. J. Ky. Acad. Sci. 62:117–120.

—. 2001c. Turtle fauna of the Upper Tradewater River near Dawson Springs, Kentucky. J. Ky. Acad. Sci. 62:121–124.

Lindeman, P. V., and M. J. Sharkey. 2001. Comparative analyses of functional relationships in the evolution of trophic morphology in the map turtles (Emydidae: *Graptemys*). Herpetologica 57:313–318.

Loch, J. H. 1950. Notes on the Perak River turtle. Malay Nature J. 5:157–160.

Lock, M. A., and D. D. Williams. 1981. Prespectives in Running Water Ecology. Plenum Press, New York.

Loehr, V. J. T. 2002. Population characteristics and activity patterns of the Namaqualand speckled padloper (*Homopus signatus signatus*) in the early spring. J. Herpetol. 36:378–389.

Lotze, H. U. 1963. Kleine Mitteilungen. Aquarien Terr. 16:220–221.

Loveridge, A. H. 1941. Revision of the African terrapin of the family Pelomedusidae. Bull. Mus. Comp. Zool. Harvard College 88:468–524.

Loveridge, A. H., and E. E. Williams. 1957. Revision of the African tortoise and turtles of the suborder Cryptodira. Bull. Mus. Comp. Zool. Harvard College 115:163–557.

Lovich, J. E. 1994. Biodiversity and zoogeography of non-marine turtles in Southeast Asia, pp. 380–391. *In:* Biological Diversity: Problems and Challenges. S. K. Majumdar, F. J. Brenner, J. E. Lovich, J. F. Schalles, and E. W. Miller (eds.). Pennsylvania Academy of Science.

—. 1996. Possible demographic and ecologic consequences of sex ratio manipulation in turtles. Chelonian Conserv. Biol. 2:114–117.

Lovich, J. E., and J. W. Gibbons. 1997. Conservation of covert species: Protecting species we don't even know, pp. 426–429. *In:* Conservation Restoration and Management of

Turtles and Tortoises: An International Conference. J. van Abbema (ed.), State University of New York, Purchase, NY.

Lovich, J. E., C. J. McCoy, and W. R. Garstka. 1990. The development and significance of melanism in the slider turtle, pp. 233–254. *In:* The Life History and Ecology of the Slider Turtle. J. W. Gibbons (ed.). Smithsonian Institution Press, Washington, D.C.

Lovich, J. E., A. D. Tucker, D. E. Kling, J. W. Gibbons, and T. D. Zimmerman. 1991. Behavior of hatchling diamondback terrapins released in a South Carolina salt marsh. Herp. Rev. 22:81–83.

Lovich, J. E., S. W. Gotte, C. H. Ernst, J. C. Harshbarger, A. F. Laemmerzahl, and J. W. Gibbons. 1996. Prevalence and histopathology of shell disease in turtles from Lake Blackshear, Georgia. J. Wildlife Dis. 32:259–265.

Lovich, J. E., R. A. Mittermeier, P. C. H. Pritchard, A. G. J. Rhodin, and J. W. Gibbons. 2000. Powdermill Conference: Trouble for the world's turtles. Turtle and Tortoise Newsletter 16–17.

Luiselli, L. 1998. Food habits of the pelomedusid turtle *Pelusios castaneus castaneus* in south-eastern Nigeria. Chelonian Conserv. Biol. 3:106–107.

Luiselli, L., M. Capula, D. Capizzi, E. Fillippi, V. T. Jesus, and C. Anibaldi. 1997. Problems for conservation of pond turtles (*Emys obicularis*) in central Italy: Is the introduced red-eared turtle (*Trachemys scripta*) a serious threat? Chelonian Conserv. Biol. 2:417–419.

Lunney, D., and D. Ayers. 1993. Herpetology in Australia: A Diverse Discipline, p. 414. Transactions of the Royal Zoological Society of New South Wales, Mosman.

M.W.W.G. 1986. The value of Malaysian wetlands and their role in national development: A special report to the minister of Science, Technology, and Environment. Malaysian Wetland Working Group, Department of Wildlife and National Parks, Peninsular Malaysia, Malaysia.

Macarthur, R. H., and E. O. Wilson. 1967. The Theory of Island Biogeography. Princeton University Press, New York.

Macarthur, R. H., and R. Levins. 1964. Competition, habitat selection, and character displacement in a patchy environment. Proc. Natl. Acad. Sci. USA 51:1207–1210.

—. 1967. The limiting similarity, convergence, and divergence of coexisting species. Am. Naturalist 101:377–385.

MacCulloch, R. D., and D. M. Gordon. 1978. A simple trap for basking turtles. Herp. Rev. 9:133.

MacCulloch, R. D., and D. M. Secoy. 1983. Movement in a river population of *Chrysems picta bellii* in southern Saskatchewan. J. Herpetol. 17:283–285.

Magnusson, W. E. 2002. Commercial extinction exists and is often a conservation objective. Turtle and Tortoise Newsletter 5:11.

Magnusson, W. E., A. C. De Lima, V. L. De Costa, and R. C. Vogt. 1997. Home range of the turtle, *Phrynops rufipes*, in an isolated reserve in central Amazonia, Brazil. Chelonian Conserv. Biol. 2:494–499.

Mahmoud, I. Y. 1968a. Feeding behavior in kinosternid turtles of Oklahoma. Southwestern Naturalist 14:31–66.

—. 1968b. Feeding behavior in kinosternid turtles. Herpetologica 24:300–305.

—. 1969. Comparative ecology of the kinosternid turtles of Oklahoma. Southwestern Naturalist 14:31–66.

Maier, T. J., M. N. Marchand, R. M. Degraaf, and J. A. Litvaitis. 2002. A subterranean camera trigger for identifying predators excavating turtles nests. Herp. Rev. 33:284–287.

Maitland, P. S. 1990. Biology of Fresh Waters, 2nd edition. Chapman & Hall, New York.

Manchester, D. 1982. Red-eared sliders in Pennsylvania. Testudo 2:27–30.

Mansergh, I., G. Davey, and P. Robertson. 1993. Reptiles and amphibians in Victoria – legislation, pp. 373–376. *In:* Herpetology in Australia: A Diverse Discipline. D. Lunney, and D. Ayers (eds.). Royal Zoological Society of New South Wales, Mosman.

Mao, S. H. 1971. Turtles of Taiwan. The Commercial Press, Ltd, Tapai, Taiwan.

Marchand, L. J. 1942. A contibution to a knowledge of the natural history of certain fresh-water turtles. M.S. Thesis. University of Florida, Gainesville.

—. 1945. The individual range of some Florida turtles. Copeia 1945:191–192.

Martín, J., and P. López. 1990. Amphibians and reptiles as prey of birds in Southwestern Europe. Smithsonian Herpetol. Inform. Serv. 82:1–24.

Martin, P. S. 1966. Africa and Pleistocene overkill. Nature 212:339–342.

—. 1967. Prehistoric overkill, pp. 75–120. *In:* Pleistocene Extinctions: The Search for a Cause. P. S. Martin and H. E. Wright, Jr. (eds.). Yale University Press, New Haven, CT.

—. 1984. Prehistoric overkill, pp. 354–403. *In:* Quaternary Extinctions: A Prehistoric Revolution. P. S. Martin and R. G. Klein (eds.). University of Arizona Press, Tucson.

Martuscelli, P. 1995. *Hydromedusa tectifera* (South American snake-necked turtle). Predation. Herpetol. Rev. 26:34–35.

Mascort, R. 1997. An overview of a threatened population of the European pond turtle, *Emys orbicularis*, pp. 312–313. *In:* Proceedings: Conservation, Restoration, and Management of Tortoises and Turtles: An International Conference. J. van Abbema (ed.). State University of New York, Purchase, NY.

Maxwell, F. D. 1911. Reports on Inland and Sea Fisheries in the Thongwa, Myaungmya, and Bassein Districts and the Turtle Banks of the Irrawaddy Division. Govt. Printing Office, Rangoon.

McCloud, K. 1985. Letter to the editor. Herpetol. Rev. 16:33.

McCord, W. P., and P. C. H. Pritchard. 2002. A review of the softshell turtles of the genus *Chitra*, with the description of new taxa from Myanmar and Indonesia (Java). Hamadryad 27:11–56.

McCord, W. P., O. S. G. Pauwels, R. Bour, F. Chérot, J. Iverson, P. C. H. Prichard, K. Thirakhupt, W. Kitimasak and T. Bundhitwongrut. 2003. *Chitra burmanica* sensu Jaruthanin, 2002 (Testudines: Trionychidae): An unavailable name. Hamadryad 27:214–216.

McCoy, C. J., and J. Lovich. In press a. *Graptemys ernsti*, Escambia map turtle. *In:* The Conservation Biology of Freshwater Turtles. Vol. 3. P. C. H. Pritchard and A. Rhodin (eds.). Chelonian Research Foundation.

McCoy, C. J., and J. Lovich. In press b. *Graptemys gibbonsi*, Pascagoula map turtle, Conservation of Freshwater turtles. IUCN Species Survival Commission.

McCoy, C. J., and R. C. Vogt. 1980a. Distribution and population status of the ringed sawback *Graptemys oculifera* Baur in Mississippi and Louisiana. Final Report. U.S. Fish and Wildlife Services.

—. 1980b. Distribution and population status of the black-knobbed sawback *Graptemys nigrinoda* in Alabama and Mississippi. Final Report. U.S. Fish and Wildlife Services.

McDiarmid, R. W., G. T. Bancroft, and J. S. Godley. 1983. Large-scale operations management test of use of the white amur for control of problem aquatic plants, Reports 2 and 3 , First and second year poststocking results; Vol. 5: The herpetofauna of Lake Conway, Florida: Community analysis. U.S. Army Engineer Waterways Experiment Station, Ce, Vicksburg, MS.

McDougal, J. 2000. Conservation of tortoises and terrestrial turtles, pp. 180–206. *In*: Turtle Conservation. M. W. Klemens (ed.). Smithsonian Institution Press, Washington, D.C.

McDowell, S. B. 1964. Partition of the genus *Clemmys* and related problems in the taxonomy of the aquatic Testudinidae. Proc. Zool. Soc. London. 143:239–279.

McKeown, S. 1978. Hawaiian reptiles and amphibians. Oriental Publ. Co., Honolulu.

McKeown, S., and R. G. Webb. 1982. Softshell turtles in Hawaii. J. Herpetol. 16:107–111.

Medem, F. 1960. Datos zoogeograficos y ecologicos sobre los Crocodylia y Testudinata del Alto Caqueta y Rio Caguan. Caldasia 10:329–353.

—. 1962. La distribicion geográfica y ecologia de los Crocodylia y Testudinata en el Departmento del Choco. Rev. Acad. Colomb. Cien. Ex. Fis. Nat. 11:279–303.

—. 1969. Estudios adicionales sobre los Crocodylia y Testudinata del alto Caquetá y Rio Caguán. Caldasia 10:329–353.

—. 1977. Contribucion al conocimiento sobre la taxonomia, distribucion geografica y ecologia de la tortuga "Bache" (*Chelydra serpentina acutirostris*). Caldasia 12:41–101.

Meffe, G. K. 1992. Techno-arrogance and halfway technologies: Salmon hatcheries on the Pacific Coast of North America. Conserv. Biol. 6:350–354.

Meffe, G. K., and C. R. Carroll. 1994. Principles of Conservation Biology. Sinauer Associates, Inc., Sunderland, MA.

Meffe, G. K., C. R. Carroll, and Contributors. 1997. Principles of Conservation Biology, 2nd edition. Sinauer Associates, Inc., Sunderland, MA.

Meshaka, W. E. 1988. *Pseudemsy nelsoni* (Florida red-bellied slider). Mutualism. Herpetol. Rev. 19:88.

Metrailler, S., and G. Le Gratiet. 1996. Tortues continentales de Guyane française, Continental Turtles of the French Guiana. Privately printed by Sébastien Métrailler.

Meyers-Schöne, L., and B. T. Walton. 1994. Turtles as monitors of chemical contaminants in the environment, pp. 93–153. *In:* Reviews of Environmental Contamination and Toxicology. Vol. 135. G. W. Ware (ed.). Springer-Verlag, New York.

Meylan, A. B., and D. Ehrenfeld. 2000. Conservation of marine turtles, pp. 96–125. *In:* Turtle Conservation. M. W. Klemens (ed.). Smithsonian Institution Press, Washington, D.C.

Meylan, P. A. and P. P. van Dijk. In press. *Nilssonia formosa* (Gray, 1869a) Burmese peacock softshell turtle. *In:* The Conservation Biology of Freshwater Turtles. P. C. H. Pritchard and A. G. J. Rhodin (eds.). Chelonian Research Foundation.

Meylan, P. A., C. A. Stevens, M. E. Barnwell, and E. D. Dohm. 1992. Observations on the turtle community of Rainbow Run, Marion Co., Florida. Florida Scientist 55:219–228.

Miller, J. D. 1985. Embryology of marine turtles, pp. 269–328. *In:* Biology of the Reptilia. Vol. 14. C. Gans, F. Billett, and P. F. A. Maderson (eds.). John Wiley, New York.

Miller, K., G. C. Packard, and M. J. Packard. 1987. Hydric conditions during incubation influence locomotor performance of hatchling snapping turtles. J. Exp. Biol. 127:401–412.

Miller, R. R. 1946. The probable origin of the softshelled turtle in the Colorado River Basin. Copeia 1946:46.

Mills, H. B., W. C. Starrett, and F. C. Bellrose. 1966. Man's effect on the fish and wildlife of the Illinois River. Ill. Nat. Hist. Surv. Biol. Notes 57:24.

Minckley, W. L. 1966. Coyote predation on aquatic turtles. J. Mammal. 47:137.

Minton, S. A., Jr. 1966. A contribution to the herpetology of West Pakistan. Bull. Amer. Mus. Natur. Hist. 134:27–184.

Minton, S. A., and Minton, M. R. 1973. Giant reptiles. Charles Scribner's Sons, New York.

Mitchell, B. L. 1946. A naturalist in Nyasaland. Nayasaland Agriculture Q. J. 6:1–47.

Mitchell, J. C. 1994. The Reptiles. Smithsonian Institution Press, Washington, D.C.

Mitchell, J. C., and M. W. Klemens. 2000. Primary and secondary effects of habitat alteration, pp. 5–32. *In:* Turtle Conservation. M. W. Klemens (ed.). Smithsonian Institution Press, Washington, D.C.

Mitsukuri, K. 1905. The cultivation of marine and freshwater animals in Japan. The snapping turtle, or softshelled tortoise, "suppon." Bull. U.S. Bur. Fish. 24:260–266.

Mittermeier, R. A. 1970. Turtles in Central American markets. Int. Turtle Tortoise Soc. J. 4:20–26.

—. 1971. Status—The market in So. E. Mexico. Int. Turtle Tortoise Soc. J. 5:15–19, 36–38.

—. 1975. A turtle in every pot. Animal Kingdom. April/May:9–14.

—. 1978. South America's river turtles: saving them by use. Oryx 14:233–230.

Mittermeier, R. A., and R. A. Wilson. 1974. Redescription of *Podocnemis erythrocephala* (Spix, 1824), an Amazonian pelomedusid turtle. Papeis Avulsos de Zoologia 28:147–162.

Moll, D. 1976a. Environmental Influence on growth rate in the Ouachita map turtle, *Graptemys Pseudogeographica ouachitensis.* Herpetologica 32:439–143.

—. 1976b. Food and feeding strategies of the Ouachita map turtle (*Graptemys pseudogeographica ouachitensis*). Am. Midland Naturalist 96:478–482.

—. 1980. Dirty river turtles. Nat. Hist. 89:42–49.

—. 1977. Ecological investigations of turtles in a polluted ecosystem: the Central Illinois River and adjacent flood plain lakes. Ph.D. Dissertation. Illinois State University, Normal, IL.

—. 1985. The trophic ecology of aquatic turtles: investigations of the relationship of prey availability with dietary overlap and specialization. Assoc. for the Study of Reptiles and Amphibians J. 2:1–22.

—. 1986. The distribution, status, and level of exploitation of the freshwater turtle, *Dermatemys mawei* in Belize, Central America. Biol. Conserv. 35:87–96.

—. 1989. Food and feeding behavior of the turtle, *Dermatemys mawei*, in Belize. J. Herpetol.. 23:445–447.

—. 1990. Population sizes and foraging ecology in a tropical freshwater stream turtle community. J. Herpetol. 24:48–53.

—. 1994. The ecology of sea beach nesting in slider turtles (*Trachemys scripta venusta*) from Caribbean Costa Rica. Chelonian Conserv. Biol. 1:107–116.

—. 1995. Conservation and management of river turtles: A review of methodology and techniques, pp. 290–294. *In:* Proceedings: International Congress of Chelonian Conservation July, 1995. B. Devaux (ed.). SOPTOM, Gonfaron, France.

—. 1997a. Ecological characteristics of the snapping turtle, *Chelydra serpentina acutirostris* in Costa Rica. Herpetology '97: Abstracts, p. 145. *In:* Third World Congress of Herpetology 2–10 August 1997, Prague, Czech Republic.

—. 1997b. Ecological characteristics of the Torguguero, Costa Rica, non-marine turtle community, p. 214. *In:* ASIH/HL/SSAR/AFS-ELHS/AES/GIS Joint Meetings, 26 June–2 July, 1997, Seattle, WA.

Moll, D., and E. Moll. 1990. The slider turtle in the Neotropics: Adaptations of a temperate species to a tropical environment, pp. 152–161. *In:* Life History and Ecology of the Slider Turtle. J. W. Gibbons (ed.). Smithsonian Institution Press, Washington, D.C.

Moll, D., and K. P. Jansen. 1995. Evidence of a role in seed dispersal by two tropical herbivorous turtles. Biotropica 27:121–127.

Moll, D., and M. W. Klemens. 1996. Ecological characteristics of the pancake tortoise (*Malacochersus tornieri*) in Tanzania. Chelonian Conserv. Biol. 2:26–35.

Moll, D., K. A. Buhlmann, and J. W. Gibbons. in press. *Trachemys scripta* (Schoepff, 1792): Slider turtle. *In:* The Conservation Biology of Freshwater Turtles. P. C. H. Pritchard and A. G. J. Rhodin (eds.). Chelonian Research Foundation.

Moll, E. O. 1976a. West Malaysian turtles: Utilization and conservation. Herpetol. Rev. 7:163–166.

—. 1976b. Ecology and management of the tuntong (*Batagur baska*) in Malaysia, p. 30. Game Department of West Malaysia, Kuala Lumpur.

—. 1976c. Ecology and management of the tuntong (*Batagur baska*) in Malaysia. A report submitted to the Game Department of West Malaysia, August 1976. Kuala Lumpur.

—. 1978a. Drumming along the Perak. National History 87:36–42.

—. 1978b. Report of research on the distribution, ecology, and management of coastal nesting turtles in Trengganu, West Malaysia. Submitted to Terengganu Dept. of Fisheries, Kuala Terengganu.

—. 1979. Reproductive cycles and adaptations, pp. 305–331. *In:* Turtles: Perspectives and Research. M. Harless and H. Morlock (eds.). John Wiley and Sons, Inc., New York.

—. 1980a. Natural history of the river terrapin, *Batagur baska* (Gray) in Malaysia (Testudines: Emydidae). Malaysian J. Sci. 6:23–62.

—. 1980b. Tuntong laut: the river turtle that goes to sea. Nature Malaysiana 5:16–21.

—. 1982. Freshwater turtle trade: Drugs. IUCN/SSC Freshwater Chelonian Specialist Group Newsletter 3:1–2.

—. 1983. IUCN/SSC Meeting-Malaysia. IUCN/SSC Freshwater Chelonian Specialist Group Newsletter 5:37–40.

—. 1984a. River terrapin recovery plan for Malaysia. J. Wildlife Parks III:37–46.

—. 1984b. Regional News: Neotropical - Caribbean Region. IUCN/SSC Freshwater Chelonian Group Newsletter 6:7–8.

—. 1984c. K L tortoise nibbles wrong bait. IUCN/SSC Freshwater Chelonian Group Newsletter 6:9.

—. 1985a. Comment: Sexually dimorphic plastral kinesis—The forgotten papers. Herp. Rev. 16:16.

—. 1985b. Freshwater turtles. Sanctuary V:49–59, 66.

—. 1985c. Estuarine turtles of tropical Asia: Status and management, pp. 214–226. *In:* Symp. Endangered Marine Animals and Marine Parks. Vol. 1.

—. 1986. Survey of the freshwater turtles of India Part I: The genus *Kachuga*. J. Bombay Natural History Soc. 83:538–552.

—. 1987a. Survey of the freshwater turtles of India, Part II: the genus *Kachuga*. J. Bombay Natural History Soc. 84:7–25.

—. 1987b. The river terrapin conservation program in Malaysia: follow-up study and evaluation. Submitted to Dept. of Wildlife and National Parks, Kuala Lumpur.

—. 1989. Malaysia's efforts in the conservation of the river terrapin, pp. 173–176. *In:* International Conference on Natural Parks and Protected Areas. Dept. of Wildlife and National Parks, Peninsular Malaysia, Kuala Lumpur, Malaysia.

—. 1990a. India's freshwater turtle resource with recommendations for management, pp. 501–515. *In:* Conservation in Developing Countries: Problems and Prospects. J. C. Daniel, and J. S. Serrao (eds.). Oxford University Press, Oxford, UK.

—. 1990b. Final Report: Status and management of the river terrapin (*Batagur baska*) in tropical Asia. World Wide Fund for Nature, WWF3901/Asia, Gland, Switzerland.

—. 1995. The turtle *Trachemys scripta* and the pet trade. Aliens, IUCN Invasive Species Specialist Group Newsletter 2:3.

—. 1997. Effects of habitat alteration on river turtles of tropical Asia with emphasis on sand mining and dams, pp. 37–41. *In:* Proceedings: Conservation, Restoration, and Management of Tortoises and Turtles: An International Conference. J. van Abbema (ed.), State University of New York, Purchase, NY.

Moll, E. O., and J. M. Legler. 1971. The life history of a neotropical slider turtle, *Pseudemys scripta* (Schoepff), in Panama. Los Angeles Co. Museum Sci. Bull. 11:1–102.

Moll, E. O., and D. Moll. 2000. Conservation of river turtles, pp. 126–155. *In:* Turtle Conservation. M. W. Klemens (ed.). Smithsonian Institution Press, Washington, D.C.

Moll, E. O., and Mohd Khan bin Momin Khan. 1990. Turtles of Taman Negara. J. Wildlife and Parks X:135–138.

Moll, E. O., K. E. Matson, and E. B. Krehbiel. 1981. Sexual and seasonal dichromatism in the Asian river turtle *Callagur borneoensis*. Herpetologica 37:181–194.

Moodie, K. B., and T. R. Van Devender. 1979. Extinction and extirpation in the herpetofauna of the Southern High Plains with emphasis on *Geochelone wilsonii* (Testudinae). Herpetologica 35:198–206.

Mooney, H. A., S. P. Hamburg, and J. A. Drake. 1986. The invasions of plants and animals into California, pp. 250–272. *In:* The Ecology of Biological Invasions of North America and Hawaii. H. A. Mooney, and J. A. Drake (eds.). Springer-Verlag, New York.

Mora, J. M., and A. N. Ugalde. 1991. A note on the status and exploitation of *Pseudemys scripta emolli* (Reptilia: Emydidae) in northern Costa Rica. Bull. Chicago Herpetol. Soc. 26:111.

Morreale, S. J., G. J. Ruiz, J. R. Spotila, and E. A. Standora. 1982. Temperature-dependent sex determination: Current practices threaten conservation of sea turtles. Science Natural History Museum, Los Angeles Co. 216:1245–1247.

Morris, K. A., G. C. Packard, T. J. Boardman, G. L. Paukstis, and M. J. Packard. 1983. Effect of the hydric environment on growth of embyronic snapping turtles (*Chelydra serpentina*). Herpetologica 39:272–285.

Morrison, W. S. 1915. Renzada District. Burma Gazetteer. Government Printing Office, Rangoon.

Mortimer, J. A. 1988. Management options for sea turtles: re-evaluating priorities. Florida Defenders of the Environment 25:1–4.

—. 1989. Threats to the survival of Terengganu's sea turtles and a review of management options. Turtle Sanctuary Advisory Council of Terengganu. Produced under WWF Project 3868.

—. 1992. The threatened turtles of the Rio Trombetas, pp. 270–271. *In:* Amazon Wildlife, Produced by H. U. Bernard. Insight Guides: APA Publications (HK) Ltd. Singapore.

—. 1995a. Teaching critical concepts for the conservation of sea turtles. Marine Turtle Newsletter 71:1–4.

—. 1995b. Headstarting as a managing tool, pp. 613–615. *In:* Biology and Conservation of Sea Turtles. K. A. Bjorndal (ed.). Smithsonian Institution Press, Washington, D.C.

Mount, R. H. 1981. The status of the flattened musk turtle, *Sternotherus minor depressus*. Tinkle and Webb. Contract Report 14-16-0004-80-096. U.S. Fish and Wildlife Service.

Mrosovsky, N. 1981. Editorial. Marine Turtle Newsletter 19:1–2.

—. 1982. Sex ratio bias in hatchling sea turtles from artificially incubated eggs. Biol. Conserv. 23:309–314.

—. 1983. Conserving Sea Turtles. The British Herpetological Society c/o The Zoological Society of London, Regent's Park, London.

Mrosovsky, N., and M. H. Godfrey. 1995. Manipulating sex ratios: Turtle speed ahead! Chelonian Conserv. Biol. 1:238–240.

Muir, J. H. 1984. Commercial exploitation of *Graptemys* and *Sternotherus* turtles. Bull. Chicago Herp. Soc. 19:98–100.

Murphy, J. B., and J. T. Collins. 1983. A Review of the Diseases and Treatments of Captive Turtles. AMS Publishing, Lawrence, KS.

N.R.C. 1992. Restoration of aquatic ecosystems: science, technology, and public policy. National Research Council, National Academy Press, Washington D.C.

Nair, P. N. R., and M. Badrudeen. 1975. On the occurrence of the softshelled turtle, *Pelochelys bibroni* (Owen), in marine environment. Ind J. Fisheries 22 (1/2):270–274.

Neill, W. T. 1958. The occurrence of amphibians and reptiles in saltwater areas and a bibliography. Bull. Marine Sci. Gulf and Caribbean 8:1–97.

Netting, M. G. 1944. The spineless soft-shelled turtle, *Amyda mutica* (Le Sueur), in Pennsylvania. Annals of the Carnegie Museum 30:85–88.

Newbery, R. 1984. The American red-eared terrapin in South Africa. African Wildlife 38:186–189.

Nicholls, R. E. 1977. The Running Press Book of Turtles. Running Press, Philadelphia.

Normile, D. 2003. The new face of traditional Chinese medicine. Science 299:188–190.

Nutaphand, W. 1979. The Turtles of Thailand. Siamfarm Zoological Gardens, Bangkok.

N.Y.T.T.S. 2002. Asian Scholarship Program Summer 2002 Scholarship Interns. New York Turtle and Tortoise Society Web Site.

Obbard, M. E. 1983. Population ecology of the common snapping turtle *Chelydra serpentina*, in north-central Ontario. Ph.D. Dissertation. University of Guelph, Guelph, Ontario.

Obbard, M. E., and R. J. Brooks. 1980. Nesting migrations of the snapping turtle (*Chelydra serpentina*). Herpetologica. 36:158–162.

—. 1981. Fate of overwintered clutches of the common snapping turtle (*Chelydra serpentina*) in Algonquin Park, Ontario. Can. Field Naturalists 93:350–352.

Obst, F. J. 1986. Turtles, Tortoises and Terrapins. St. Martin's Press, New York.

Odum, H. T. 1957. Trophic structure and productivity of Silver Springs, Florida. Ecol. Monogr. 27:55–112.

Ojasti, J. 1967. Consideraciones sobre la ecologia y conservacion de la tortuga *Podocnemis expansa* (Chelonia, Pelomedusidae). Atas Simp. Sobre Biota Amazonica 7:201–206.

Olaffson, P. G., A. M. Bryan, B. Bush, and W. Stone. 1983. Snapping turtle: a biological screen for PCB's. Chemosphere 12:1525–1532.

Oliver, R. C. D. 1979. Wildlife conservation and management in Bangladesh. Report to the FAO, Rome.

Oruc, A., F. Demirayak, and G. Sat. 1996. Fishery in the eastern Mediterranean and its impact on sea turtles, p. 17. Report for WWF International and DHKD. Coastal Management Section Report No. 96/1, Istanbul.

Owen, J. G., and J. R. Dixon. 1989. An ecogeographic analysis of the herpetofauna of Texas. Southwestern Naturalist 34:165–180.

Packard, G. C. 1999. Water relations of chelonian eggs and embroys: is wetter better? Am. Zool. 39:289–303.

Packard, G. C., and M. J. Packard. 1995. The basis for cold tolerance in hatchling painted turtles (*Chrysemys picta*). Physiol. Zool. 68:129–148.

—. 1997. Overwintering strategies of neonatal turtles, p. 157. *In:* Abstracts of the 3rd World Congress of Herpetology, Prague.

Packard, G. C., M. J. Packard, and T. J. Boardman. 1981a. Patterns and possible significance of water exchange by flexible-shelled eggs of painted turtles (*Chrysemys picta*). Physiol. Zool. 54:165–178.

Packard, G. C., M. J. Packard, T. J. Boardman, and M. D. Ashen. 1981b. Possible adaptive value of water exchanges in flexible-shelled eggs of turtles. Science Natural History Museum, Los Angeles Co. 213:471–473.

Packard, G. C., M. J. Packard, T. J. Boardman, K. A. Morris, and R. D. Shuman. 1983. Influence of water exchanges by flexible-shelled eggs of painted turtles *Chrysemys picta* on metabolism and growth of embryos. Physiol. Zool. 56:217–230.

Packard, G. C., M. J. Packard, K. Miller, and T. J. Boardman. 1988. Effects of temperature and moisture during incubation on carcass composition of hatchling snapping turtles (*Chelydra serpentina*). J. Comp. Physiol. B. 158:117–125.

Packard, G. C., J. W. Lang, L. D. Lohmiller, and M. J. Packard. 1997a. Cold tolerance in hatchling painted turtles (*Chrysemys picta*): supercooling or tolerance for freezing? Physiol. Zool. 70:670–678.

Packard, G. C., S. L. Fasano, M. B. Attaway, L. D. Lohmiller, and T. L. Lynch. 1997b. Thermal environment for overwintering hatchlings of the painted turtle (*Chrysemys picta*). Can. J. Zool. 75:401–406.

Packard, M. J., and G. C. Packard. 1986. Effect of water balance on growth and calcium mobilization of embryonic painted turtles (*Chrysemys picta*). Physiol. Zool. 59:398–405.

Páez, V. P. 1995. The Conservation and Nesting Ecology of the Endangered Yellow-Spotted Amazonian Turtle, *Podocnemis unifilis*. Ph.D. Dissertation. Ohio University, Miami.

Páez, V. P., and B. C. Bock. 1997. Nesting ecology of the Yellow-Spotted river turtle in the Columbian Amazon, pp. 219–224. *In:* Proceedings: Conservation, Restoration, and Management of Tortoises and Turtles: An International Conference. J. van Abbema (ed.). State University of New York, Purchase, NY.

—. 1998. Temperature effect on incubation period in the yellow-spotted river turtle, *Podocnemis unifilis*, in the Colombian Amazon. Chelonian Conservation and Biology 3:31–36.

Palmer, T. 1994. Lifelines: The Case for River Conservation. Island Press, Washington D.C.

Pappas, M. J., and B. J. Brecke. 1992. Habitat selection of juvenile Blanding's turtles, *Emydoidea blandingii*. J. Herpetol. 26:233–234.

Pardo, H. A. 1969. Contribucion al conocimiento de la morfologia, ecologia, comportamiento y distribucion geografica de *Podocnemis vogli*, Testudinata (Pelomedusidae). Rev. Acad. Colomb. Cienc. Exactas Fis. Nat. 13:303–326.

Parmalee, P. W. 1989. Muskrat predation on softshell turtles. J. Tennessee Acad. Sci. 64:225–227.

Parmenter, R. R. 1980. Effects of food availability and water temperature on the feeding ecology of pond sliders (*Chrysemys scripta*). Copeia 1980:503–514.

Parmenter, R. R., and H. W. Avery. 1990. The feeding ecology of the slider turtle, pp. 257–266. *In:* Life History and Ecology of the Slider Turtle. J. W. Gibbons (ed.). Smithsonian Institution Press, Washington, D.C.

Parsons, J. J. 1962. The Green Turtle and Man. University of Florida Press, Gainesville, FL.

Paukstis, F. L., and F. J. Janzen. 1990. Sex determination in reptiles: Summary of effects of constant temperatures of incubation on sex ratios of offspring. Smithsonian Herpetological Service 83:1–8.

Payne, A. I. 1986. The Ecology of Tropical Lakes and Rivers. John Wiley and Sons, Chichester, UK.

Pearce, F. 1995. The biggest dam in the world. New Scientist 145:25–29.

Penn, G. H. 1950. Utilization of crawfishes by cold-blooded vertebrates in the eastern United States. Am. Midland Naturalist 44:643–658.

Perez-Eman, J. L., and A. Paolillo O. 1997. Diet of the pelomedusid turtle *Peltocephalus dumerilianus* in the Venezuelan Amazon. J. Herpetol. 31:173–179.

Petokas, P. J., and M. M. Alexander. 1979. A new trap for basking turtles. Herpetol. Rev. 10:90.

Pezzuti, J. C. B., and R. C. Vogt. 1999. Nesting ecology of *Podocnemis sextuberculata* (Testudines, Pelomedusidae) in the Japura River, Amazonas, Brazil. Chelonian Conserv. Biol. 3:419–424.

Philippart, J. C. 1995. Is captive breeding an effective solution for the preservation of endemic species? Biol. Conserv. 72:291–295.

Phillips, C. A., W. W. Dimmick, and J. L. Carr. 1996. Conservation genetics of the common snapping turtle (*Chelydra serpentina*). Conserv. Biol. 10:397–405.

Pianka, E. R. 1977. Reptile species diversity, pp. 1–34. *In:* Biology of the Reptilia. C. Gans and D. W. Tinkle (eds.). Academic Press Inc., New York.

—. 1994. Evolutionary Ecology. HarperCollins College Publishers, New York.

Pieau, C. 1971. Sur la proportion sexuelle chez les embryons de deux Cheloniens (Testudo graeca L. et *Emys orbicularis* L.) issus d'eoufs incubes artificiellement. C. R. Hebd. Seanc. Acad. Sci, Paris 272:3071–3074.

—. 1974. Differenciation du sexe en fonction de la temperature chez les embryons *d'Emys orbicularis L.* (Chelonien) effets des hormones sexuelles. Ann. Embryol. Morphol. 7:365–394.

Pierce, L. 1992. Diet content and overlap of six species of turtles among the Wabash River. M.S. Thesis. Eastern Illinois Univ., Charleston.

Platt, S. G., and L. W. Fontenot. 1992. The red-eared slider, *Trachemys scripta elegans* (Weid) in South Korea. Bull. Chicago Herp. Soc. 27:113–114.

Platt, S. G., Kalyar, and W. K. Ko. 2000. Exploitation and conservation status of tortoises and freshwater turtles in Myanmar, pp. 95–100. *In:* Asian Turtle Trade: Proceedings of a Workshop on Conservation and Trade of Freshwater Turtles and Tortoises in Asia. Vol. 2. P. P. van Dijk, B. L. Stuart, and A. G. J., Rhodin (eds.). Chelonian Research Foundation, Phnom Penh, Cambodia.

Plummer, M. V. 1976. Some aspects of nesting success in the turtle, *Trionyx muticus.* Herpetologica 32:353–359.

—. 1977. Activity, habitat, and population structure in the turtle *Trionyx muticus.* Copeia 1977:431–440.

—. 1979. Collection and marking. *In:* Turtles: Perspectives and Research. M. Harless and H. Morlock (eds.). John Wiley and Sons, New York.

Plummer, M. V., and J. C. Burnley. 1997. Behavior, hibernacula, and thermal relations of softshell turtles (*Trionyx spiniferus*) overwintering in a small stream. Chelonian Conserv. Biol. 2:489–493.

Plummer, M. V., and D. B. Farrar. 1981. Sexual dietary differences in a population of *Trionys muticus.* J. Herpetol. 15:175–179.

Plummer, M. V. and H. W. Shirer. 1975. Movement patterns in a river population of the softshell turtle, *Trionyx muticus.* Univ. Kansas Mus. Nat. Hist. Occ. Papers 43:1–26.

Plummer, M. V., N. E. Mills, and S. L. Allen. 1997. Activity, habitat, and movement patterns of softshell turtles (*Trionyx spiniferus*) in a small stream. Chelonian Conserv. Biol. 2:514–520.

Pluto, T. G., and Bellis, E. D. 1986. Habitat utilization by the turtle, *Graptemys geographica*, along a river. J. Herpetol. 20:22–31.

—. 1988. Seasonal and annual movements of riverine map turtles, *Graptemys geographica*. J. Herpetol. 22:152–158.

Polisar, J. 1990a. Exploitation and Management of the Central American River Turtle, *Dermatemys mawei*, in Belize. Tortoises & Turtles 5:12–13.

—. 1990b. The reproductive biology and conservation of the Central American River turtle *Dermatemys mawii* in Belize. Belize Audubon Society, Belize City.

—. 1991. An inventory of the turtle fauna of the Rio Bravo Conservation and Management Area. Final Report. Programme for Belize, Belize City, Belize.

—. 1992. Reproductive biology and exploitation of the Central American river turtle *Dermatemys mawii* in Belize. M. S. Thesis. University of Florida, Gainesville, FL.

—. 1994. New legislation for the protection and management of *Dermatemys mawii* in Belize, Central America. Herpetol. Rev. 25:47–49.

—. 1995. River turtle reproductive demography and exploitation patterns in Belize: Implications for management. Vida Silvestre Neotropical 4:10–19.

—. 1996. Reproductive biology of a flood-season nesting freshwater turtle of the northern neotropics: *Dermatemys mawii* in Belize. Chelonian Conserv. Biol. 2:13–25.

—. 1997. Effects of exploitation on "*Dermatemys mawii*" populations in northern Belize and conservation strategies for rural riverside villages, pp. 441–443. *In:* Proceedings: Conservation, Restoration, and Management of Tortoises and Turtles: An International Conference. J. van Abbema (ed.), State University of New York, Purchase, NY.

Polisar, J., and R. H. Horwich. 1994. Conservation of the large, economically important river turtle *Dermatemys mawii* in Belize. Conserv. Biol. 8:338–342.

Pope, C. H. 1935. The Reptiles of China. Turtles, Crocodilians, Snakes, Lizards. Amer. Mus. Nat. Hist New York, Nat. Hist. Central Asia, New York.

—. 1939. Turtles of the United States and Canada. Alfred A. Knopf, New York.

Porej, D. 2001. Herpetofauna of the Muthurajawela Marsh, Sri Lanka, with notes on natural history. Herpetological Natural History 8:27–35.

Preston, F. W. 1960. Time and space and the variation of species. Ecology 41:611–627.

—. 1962a. The canonical distribution of commonness and rarity: Part I. Ecology 43:185–215.

—. 1962b. The canonical distribution of commonness and rarity: Part II. Ecology 43:410–432.

—. 1968. On modeling islands. Ecology 40:611–627.

Priest, T. E., and C. E. Franklin. 2002. Effect of water temperature and oxygen levels on the diving behavior of two freshwater turtles: *Rheodytes leukops* and *Emydura macquarii*. J. Herpetol. 36:555–561.

Pritchard, P. C. H. 1967. Living Turtles of the World. T.F.H. Publications, Inc., Neptune, NJ.

—. 1979. Encyclopedia of Turtles. T.F.H. Publications, Inc., Neptune, NJ.

—. 1980a. Record size turtles from Florida and South America. Chelonologica 1:113–123.

—. 1980b. The conservation of sea turtles: Practices and problems. Am. Zool. 20:609–617.

—. 1984. Piscivory in turtles and evolution of the long-necked Chelidae. Symp. Zool. Soc. London 52:87–110.

—. 1986. *Podocnemis expansa* in Brazil; *Podocnemis expansa* in Venezuela . . . and in Peru. SSC Freshwater Turtle Group Newsletter 1–2.

—. 1989. The Alligator Snapping Turtle: Biology and Conservation. Milwaukee Public Museum, Milwaukee.
—. 1993a. A ranching project for freshwater turtles in Costa Rica. Chelonian Conserv. Biol. 1:48.
—. 1993b. Carapacial pankinesis in the Malayan softshell turtle, *Dogania subplana*. Chelonian Conserv. Biol. 1:31–36.
—. 1994. Comments on *Chinemys palaennamitica* and certain other chelonian taxa based upon material from archaeological sites. Chelonian Conserv. Biol. 1:131–137.
—. 2001. Observations on body size, sympatry, and niche divergence in softshell turtles (*Trionychidae*). Chelonian Conserv. Biol. 4:5–27.
Pritchard, P. C. H., and W. Greenhood. 1968. The sun and the turtle. Int. Turtle Tortoise. Soc. J. 2:20–25, 34.
Pritchard, P. C. H., and P. Trebbau. 1984. The Turtles of Venezuela. Society for the Study of Amphibians and Reptiles, Ithaca, NY.
Pritchard, P. C. H., P. Bacon, F. Berry, A. Carr, J. Fletemeyer, R. Gallagher, S. Hopkins, R. Lankford, R. Marquez M., L. Ogren, W. Pringle, Jr., H. Reichart, and R. Witham. 1983. Manual of Sea Turtle Research and Conservation Techniques. K. A. Bjorndal and G. H. Balazs (eds.). Center for Environmental Education, Washington, D.C.
Quaid, L. 2002. Missouri River's loss of life alarming, study finds. *Springfield News-Leader*, January 10, Springfield, MO.
Quinn, A. J., and J. L. Christiansen. 1972. The relationship between pond bottom type and growth of western painted turtles *Chrysemys picta belli* in Iowa, a preliminary report. Proc. Iowa Acad. Sci. 78:67–69.
Rahbek, C. 1993. Captive breeding—a useful tool in the preservation of biodiversity? Biodiversity Conserv. 2:426–437.
Ramirez, M. 1956. Estudio biológico de la tortuga arrau del Orinoco Venezuela. El Agricultor Venezolano 190:44–63.
Ramo, C. 1982. Biologia del galapago (*Podocnemis vogli* Muller, 1935) en el Hato "El Frio" Llanos de Apure (Venezuela). Revista de Vertebrados de la Estacion Biologica de Doñana (Consejo Superior de Investigaciones Cientificas), Sevilla, Spain.
Ranson, B. 2002. Newsnotes: Importation of red-eared sliders banned in Korea. Turtle and Tortoise Newsletter 25.
Rao, R. J. 1986. Freshwater turtle conservation in National Chambal Sanctuary. Tiger Paper 13:28–29.
—. 1991. Conservation management of freshwater turtles in the National Chambal Sanctuary. J. Ecol. Soc. 4:43–53.
Rashid, S. M. A. 1987. The *Aspideretes nigricans* mystery. Br. Herpetol. Soc. Bull. 34:42–43.
Rashid, S. M. A., and I. R. Swingland. 1990. Interim report on the freshwater turtle trade in Bangladesh. Asiatic Herpetol. Res. 3:123–128.
Rashid, S. M. A., and S. M. Munjurul Hannan Khan. 2000. Trade and conservation status of freshwater turtles and tortoises in Bangladesh, pp. 77–85. *In*: Asian Turtle Trade: Proceedings of a Workshop on Conservation and Trade of Freshwater Turtles and Tortoises in Asia. Vol. 2. P. P. van Dijk, B. L. Stuart, and A. G. J. Rhodin (eds.). Chelonian Research Foundation, Phnom Penh, Cambodia.
Ream, C., and R. Ream. 1966. The influence of sampling methods on the estimation of population structure in painted turtles. Am. Midland Naturalist 7:325–338.
Reese, D. A. 1996. Comparative demography and habitat use of western pond turtles in northern California: the effects of damming and related alterations. Ph.D. Dissertation. University of California, Berkeley, CA.

Reese, D. A., and H. H. Welsh. 1997. Use of terrestrial habitat by western pond turtles, *Clemmys marmorata:* Implications for management, pp. 352–357. *In:* Proceedings: Conservation, Restoration, and Managemetn of Tortoises and Turtles: an International Conference. J. van Abbema (ed.). New York Turtle and Tortoise Society, Purchase, NY.

Reichart, H. A. 1995. Farming and ranching as a strategy for sea turtle conservation, pp. 465–471. *In:* Biology and Conservation of Sea Turtles, Revised Edition. K. A. Bjorndal (ed.). Smithsonian Institution Press, Washington, D.C.

Reinert, H. K. 1991. Translocation as a conservation strategy for amphibians and reptiles: some comments, concerns, and observations. Herpetologica 47:357–363.

Rensch, B. 1959. Evolution above the Species Level. Columbia University Press, New York.

Rhodin, A. G. J. 1992. Chelonian zooarchaeology of Eastern New England: turtle bone remains from Cedar Swamp and other prehistoric sites. Bull. Mass. Archaeol. Soc. 53:21–30.

—. 1995. Archaeological turtle bone remains from Concord Shell Heap. Bull. Mass. Archeol. Soc. 56:71–82.

Rhodin, A. G. J., and V. R. Genorupa. 2000. Conservation status of freshwater turtles in Papua, New Guinea, pp. 129–136. *In:* Asian Turtle Trade: Proceedings of a Workshop on Conservation and Trade of Freshwater Turtles and Tortoises in Asia. Vol. 2. P. P. van Dijk, B. L. Stuart, and A. G. J. Rhodin (eds.). Chelonian Research Foundation, Phnom Penh, Cambodia.

Rhodin, A. G. J., F. Medem, and R. A. Mittermeier. 1981. The occurrence of neustophagia among podocnemine turtles. Br. *J.* Herpetol. 6:175–176.

Rhodin, A. G. J., R. A. Mittermeier, and P. M. Hall. 1993. Distribution, osteology, and natural history of the Asian giant softshell turtle, *Pelochelys bibroni,* in Papua, New Guinea. Chelonian Conserv. Biol. 1:19–30.

Richter, B. D., A. P. Braun, M. A. Mendelson, and L. L. Master. 1997. Threats to imperiled freshwater fauna. Conserv. Biol. 11:1081–1093.

Roberts, T. R. 1973. Ecology of fishes in the Amazon and Congo basins, pp. 239–254. *In:* Tropical Forest Ecosystems in Africa and South America: Comparative Review. B. J. Meggers, E. S. Ayers, and W. D. Duckworth (eds.). Smithsonian Institution Press, Washington, D.C.

Robinson, K. M., and G. G. Murphy. 1975. A new method for trapping softshell turtles. Herpetol. Rev. 6:111.

Rochat, K. C., N. N. Deane, and A. M. Erasmus. 1962. Foods and feeding: water tortoise *Pelomedusa subrufa.* Lammergeyer 2:69.

Roff, W. R. 1967. Stories and Sketches by Sir Frank Swettenham. Oxford in Asia Historical Reprints. Oxford University Press, Kuala Lumpur.

Rogers, J. S. 1976. Species density and taxonomic diversity of Texas amphibians and reptiles. Syst. Zool. 25:26–40.

Rogner, M. 1995. Schildkröten 1. Heidi Rogner, Hürtgenwald, Germany.

—. 1996. Schildkröten 2. Heidi Rogner, Hürtgenwald, Germany.

Roosenburg, W. M., and K. C. Kelley. 1996. The effect of egg size and incubation temperature on growth in the turtle, *Malaclemys terrapin.* J. Herpetol. 30:198–204.

Rose, M. R., F. Parker, and A. Rhodin. 1982. New Guinea plateless turtle or pitted shell turtle (Fly River or pig-nosed turtle), pp. 243–246. *In:* The IUCN Amphibia—Reptilia Red Data Book: Part 1: Testudines, Crocodylia, Rhynchocephalia. B. Groombridge and L. Wright (eds.). IUCN, Gland, Switzerland.

Ross, D. A., and J. E. Lovich. 1992. Does the color pattern of two species of turtles imitate duckweed? J. Pennsylvania Acad. Sci. 66:39–42.

Ross, J. P. 1998. Crocodiles, Status Survey and Conservation Plan. IUCN Gland, Switzerland.

Rowe, J. W. 1992. Dietary habits of the Blanding's turtle (*Emydoidea blandingii*) in northeastern Illinois. J. Herpetol 26:111–114.

Roze, J. A. 1964. Pilgrim of the river. Natural History 73:35–41.

Ruby, D. E., J. R. Spotila, S. K. Martin, and S. J. Kemp. 1994. Behavioral responses to barriers by desert tortoises: Implications for wildlife management. Herpetol. Monogr. 8:144–160.

Rüppel, E. 1835. Neue Wirbelthiere zu der Fauna von Abyssinien gehörig. III Amphibien. Frankfurt a.M.

Saharia, V. B. 1982. Wildlife in India. Natraj Publishers, Dehra Dun, India.

St. Cricq, L. 1874. A journey across South America from the Pacific Ocean to the Atlantic Ocean. Half-Volume III. Blackie and Son, London.

Salter, R. E. 1983. Summary of currently available information on internationally threatened wildlife species in Burma. Nature Conservation and National Parks Project Burma., p. 76. Food and Agriculture Organization of the United Nations (Fo:Bur /80/006, Field Document 7/83), Rangoon.

Salwasser, H. 1990. Sustainability as a conservation paradigm. Conserv. Biol. 4:213–216.

Salzberg, A. 1998a. Chelonian conservation news. Chelonian Conserv. Biol. 3:147–150.

—. 1998b. Preliminary report on the status and origins of turtles sold in Ca food markets with observations on the environmental impact of the trade. Herpetol. Digest. 1.

—. 2001. Turtles for sale. Turtle and Tortoise Newsletter 22.

Samedi, and D. T. Iskandar. 2000. Freshwater turtle and tortoise conservation and utilization in Indonesia, pp. 106–111. *In:* Asian Turtle Trade: Proceedings of a Workshop on Conservatoin and Trade of Freshwater Turtles and Tortoises in Asia. Vol. 2. P. P. van Dijk, B. L. Stuart, and A. G. J. Rhodin (eds.). Chelonian Research Foundation, Phnom Penh, Cambodia.

Schall, J. J., and E. R. Pianka. 1977. Species densities of reptiles and amphibians on the Iberian Peninsula. Doþana, Acta Vertebrata 4:27–34.

—. 1978. Geographical trends in numbers of species. Science Natural History Museum, Los Angeles Co. 201:679–686.

Schodde, R., I. Mason, and T. O. Wolfe. 1973. Further records of the pitted-shelled turtle (*Carettochelys insulpta*) from Australia. Trans. Roy. Soc. South Aust. 96:115–117.

Schoener, T. W. 1977. Competition and the niche, pp. 35–136. *In:* Biology of the Reptilia. C. Gans and D. W. Tinkle (eds.). Academic Press Inc., New York.

Schultze-Westrum, T. 1995. The Papua turtle from New Guinea. Bull. Chicago Herpetol. Soc. 30:189–192.

Schwartz, A., and R. W. Henderson. 1991. Amphibians and Reptiles of the West Indies: Descriptions, Distributions, and Natural History. University of Florida Press, Gainesville.

Scott, N. J. 1982. Herpetological Communities. Wildlife Research Report 13. U.S. Dept. of the Interior, Fish and Wildlife Service, Washington, D.C.

Seidel, M. E. 1988. Revision of the West Indian emydid turtles (Testudines). American Museum Novitates 2918:1–41.

—. 1994. Morphometric analysis and taxonomy of cooter and red-bellied turtles in the North American genus *Pseudemys* (Emydidae). Chelonian Conserv. Biol. 1:117–130.

—. 1995. How many species of cooter turtles and where is the scientific evidence?: a reply to Jackson. Chelonian Conserv. Biol. 1:333–336.

—. 1996. Current status and biogeography of the West Indian turtles in the genus *Trachemys* (Emydidae), pp. 160–174. *In:* Contributions to West Indian Herpetology: A Tribute to Albert Schwartz. Vol. 12. R. Powell, and Henderson, R. W. (eds.). Society for the Study of Amphibians and Reptiles, Contributions to Herpetology, Ithaca, NY.

Seidel, M. E., and W. M. Palmer. 1991. Morphological variation in turtles of the genus *Pseudemys* (Testudines: Emydidae) from central Atlantic drainages. Brimleyana 17:105–135.

Seigel, R. A. 1980. Predation by raccoons on diamondback terrapins, *Malaclemys terrapin tequesta*. J. Herpetol. 14:87–89.

Seigel, R. A., and R. J. Brauman. 1994. Food habits of the yellow-blotched map turtle (*Graptemys flavimaculata*), p. 20. Mississippi Mus. Nat. Sci. Tech. Rprt. 28.

Seigel, R. A., and C. K. Dodd, Jr. 2000. Manipulation of turtle populations for conservation: halfway technologies or viable options? pp. 218–238. *In:* Turtle Conservation. M. W. Klemens (ed.). Smithsonian Institution Press, Washington, D.C.

Seigel, R. A., and J. W. Gibbons. 1995. Workshop on the ecology, status, and management of the diamondback terrapin (*Malaclemys terrapin*), Savannah River Ecology Laboratory, 2 August 1994: Final results and recommendations. Chelonian Conserv. Biol. 1:240–243.

Semlitsch, R. D., and J. W. Gibbons. 1989. Lack of largemouth bass predation on hatchling turtles (*Trachemys scripta*). Copeia 1989:1030–1031.

Seuss, D. 1958. Yertle the Turtle and Other Stories by Dr. Seuss. Random House, New York.

Shafer, C. L. 1990. Nature reserves: Island theory and conservation practice. Smithsonian Institution Press, Washington, D.C.

Sharma, D. S. K. 1995. Conservation and management project for endangered estuarine turtles in peninsular Malaysia. *In:* Proceedings: International Congress of Chelonian Conservation. B. Devaux (ed.). SOPTOM, Gonfaron, France.

—. 1999. Tortoise and freshwater turtle trade and utilization in peninsular Malaysia. TRAFFIC—Southeast Asia:39.

Sharma, D. S. K., and O. B. Tisen. 2000. Freshwater turtle and tortoise utilization and conservation status in Malaysia, pp. 120–128. *In:* Asian Turtle Trade: Proceedings of a Workshop on Conservation and Trade of Freshwater Turtles and Tortoises in Asia. Vol. 2. P. P. van Dijk, B. L. Stuart, and A. G. J. Rhodin (eds.). Chelonian Research Foundation, Phnom Penh, Cambodia.

Shaver, D. J. 1996. Head-started Kemp's Ridley turtles nest in Texas. Marine Turtle Newsletter 74:5–7.

Shaver, D. J., and M. R. Fletcher. 1992. Letters: Kemp's Ridley sea turtles. Science Natural History Museum, Los Angeles Co. 257:465–466.

Shealy, R. M. 1976. The natural history of the Alabama map turtle, *Graptemys pulchra* Baur, in Alabama. Bull. Florida St. Mus., Biol. Sci. 21:47–111.

Shepherd, C. R. 2000. Export of live freshwater turtles and tortoises from North Sumatra and Riau, Indonesia: a case study, pp. 112–119. *In:* Asian Turtle Trade: Proceedings of a Workshop on Conservation and Trade of Freshwater Turtles and Tortoises in Asia. Vol. 2. P. P. van Dijk, B. L. Stuart, and A. G. J. Rhodin (eds.). Chelonian Research Foundation, Phnom Penh, Cambodia.

Shipman, P. A. 1993. Natural history of the alligator snapping turtle (*Macroclemys temminckii*) in Kansas. Kansas Herpetol. Soc. Newsletter. 93:14–17.

Shively, S. H., and J. F. Jackson. 1985. Factors limiting the upstream distribution of the Sabine Map turtle. Am. Midland Naturalist 114:292–303.

Shoumatoff, A. 1978. The Rivers Amazon. Sierra Club Books, San Francisco, CA.

Shrestha, T. K. 1997. Status, biology, conservation, and management of tortoises and turtles in the Himalayan foothills of Nepal. Proceedings 11–16 July, pp. 278–286. *In:* Conservation, Restoration, and Management of Tortoises and Turtles: An International Conference. J. van Abbema (ed.). State University of New York, Purchase, NY.

Simberloff, D. 1988. The contribution of population and community biology to conservation science. Ann. Rev. Ecol. Systematics 5:473–511.

Simpson, G. G. 1953. The Major Features of Evolution. Columbia University Press, New York.

—. 1964. Species density of North American recent mammals. Systematic Zool. 13:57–73.

Singh, G. R. 2000. Ganges soft-shell turtle *Aspideretus gangeticus* predating on Nilgai *Boselaphus tragocamelus* in Keoladeo National Park, Bharatpur, Rajasthan. J. Bombay Natural History Soc. 97:285.

Sinha, R. K. 1995. Commercial Exploitation of freshwater turtle resource in the middle Ganges river system in India, pp. 14–20. *In:* Proceedings: International Congress of Chelonian Conservation; July 6–10. B. Devaux (ed.). SOPTOM, Gonfaron, France.

Siow, K. T., and E. O. Moll. 1982. Status and conservation of estuarine and sea turtles in West Malaysian water, pp. 339–348. *In:* Biology and Conservation of Sea Turtles. K. A. Bjorndal (ed.). Smithsonian Institution Press, Washington, D.C.

Sites, J. W. J., N. N. Fitz Simmons, N. J. D. Silva, Jr., and V. H. Cantarelli. 1999. Conservation genetics of the giant Amazon river turtle (*Podocnemis expansa*; Pelomedusidae)—Inferences from two classes of molecular markers. Chelonian Conserv. Biol. 3:454–463.

Sloan, K. N., and J. E. Lovich. 1995. Exploitation of the alligator snapping turtle, *Macroclemys temminckii*, in Louisiana: a case study. Chelonian Conserv. Biol. 1:221–222.

Sloan, K. N., and D. Taylor. 1987. Habitats and movements of adult alligator snapping turtles in northeast Louisiana. Proc. Ann. Conf. Southeast. Assoc. Fish. Wildl. Agencies 41:343–348.

Sloan, K. N., K. A. Buhlmann, and J. E. Lovich. 1996. Stomach contents of commercially harvested adult alligator snapping turtles, *Macroclemys temminckii*. Chelonian Conserv. Biol. 2:96–99.

Smith, H. M., and R. B. Smith. 1979. Synopsis of the herpetofauna of Mexico. Vol. 6. Guide to Mexican turtles. John Johnson, North Bennington, VT.

Smith, H. M., K. L. Williams, and E. O. Moll. 1963. Herpetological explorations on the Rio Conchos, Chihuahua, Mexico. Herpetologica 19:205–215.

Smith, M. A. 1931. The fauna of British India, including Ceylon and Burma. Reptilia and Amphibia. Vol. 1. Loricata and Testudines. Taylor and Francis, London.

Smith, N. J. H. 1974. Destructive exploitation of the South American river turtle. Yearbook, Assoc. Pacific Coast Geographers 36:85–102.

—. 1975. Destructive exploitation of the South American river turtle. (Reprint Edition of 1974, article above). Chelonia 2:1–9.

—. 1979. Aquatic turtles of Amazonia: An endangered resource. Biol. Conserv. 16:165–178.

Snyder, M. F. R., S. R. Derrickson, S. R. Bessinger, J. W. Wiley, T. B. Smith, W. D. Toone, and B. Miller. 1996. Limitations of captive breeding in endangered species recovery. Conserv. Biol. 10:338–348.

Soares, G. S. 1995. Conservation and management of *Podocnemis* in the Guapore River, Rondonia, Brazil, pp. 287–288. *In:* Proceedings: International Congress of Chelonian Conservation; July 6–10, 1995. B. Devaux (ed.). SOPTOM, Gonfaron, France.

Soini, P. 1995. Investigaciones en la Estacion Biologica Cahuana. Reporte Pacaya-Samiria. Universidad Nacional Agraria La Molina, Molina.

Soulé, M. E. 1980. Thresholds for survival: Maintaining fitness and evolutionary potential, pp. 151–169. *In:* Conservation Biology: An Evolutionary-Ecological Perspective. M. E. Soulé and B. A. Wilcox (eds.). Sinauer Associates, Sunderland, MA.

Soulé, M. E., and B. A. Wilcox. 1980. Conservation biology: Its scope and its challenge, pp. 1–8. *In:* Conservation Biology: An Evolutionary-Ecological Perspective. M. E. Soulé and B. A. Wilcox (eds.). Sinauer Associates, Sunderland, MA.

Souza, F. L., and A. S. Abe. 1995. Observations on feeding habits of *Hydromedusa maximiliani* (Testudines: Chelidae) in southeastern Brazil. Chelonian Conserv. Biol. 1:320–322.

—. 1997. Population structure, activity, and conservation of the Neotropical freshwater turtle, *Hydromedusa maximiliani*, in Brazil. Chelonian Conserv. Biol. 2:521–525.

—. 1998. Resource partitioning by the Neotropical freshwater turtle, *Hydromedusa maximiliani*. J. Herpetol. 32:106–112.

—. 2000. Feeding ecology, density and biomass of the freshwater turtle, *Phrynops geoffroanus*, inhabiting a polluted urban river in south-eastern Brazil. J. Zool. Soc., London. 252:437–446.

Souza, R. R. D., and Vogt, R. C. 1994. Incubation temperature influences sex and hatchling size in the neotropical turtle *Podocnemis unifilis*. J. Herpetol. 28:453–464.

Sparks, R. E., J. C. Nelson, and Y. Yin. 1998. Naturalization of the Flood Regime in Regulated Rivers: the case of the upper Mississippi River. BioScience 48:706–720.

Spotila, J. R., E. A. Standora, S. J. Morreale, G. J. Ruiz, and C. Puccia. 1983. Methodology for the study of temperature related phenomena affecting sea turtle eggs. U.S.F.W.S. Endangered Species Report No. 11.

Spruce, R. 1908. Notes of a botanist on the Amazon and Andes, New York.

Stancyk, S. E. 1982. Non-human predators and their control, pp. 139–152. *In:* Biology and Conservation of Sea Turtles. K. A. Bjorndal (ed.). Smithsonian Institution Press, Washington, D.C.

Standing, K. L., T. B. Herman, and I. P. Morrison. 2000. Predation of neonate Blanding's turtles (*Emydoidea blandingii*) by short-tailed shrews (*Blarina brevicauda*). Chelonian Conserv. Biol. 3:658–660.

Stapp, W. B., M. M. Cromwell, and A. Wals. 1995. A global rivers environmental education network, pp. 177–197. *In:* Conserving Wildlife, International Education and Communication Approaches. S. K. Jacobson (ed.). Columbia University Press, New York.

Starrett, W. C. 1971. A Survey of the Mussels (Unionacea) of the Illinois River: a Polluted Stream. Illinois Natural History Surv. Bull. 30:265–403.

—. 1972. Man and the Illinois River. *In:* River Ecology and Man. R. T. Oglesby, C. A. Carlson, and J. A. McCann (eds.). Academic Press, New York.

Stebbins, G. L. 1950. Variation and Evolution in Plants. Columbia University Press, New York.

Stein, R. J., W. K. Eames, and D. C. Parris. 1980. *Chrysemys scripta elegans* (red-eared slider). Herpetol. Rev. 11:115.

Stewart, J. H. 1988. A recovery plan for the ringed sawback turtle *Graptemys oculifera*. U.S. Fish and Wildlife Service, Southeast Region.

Stiner, M. C., N. D. Munro, T. A. Surovell, E. Tchernov, and O. Bar-Yosef. 1999. Paleolithic population growth pulses evidenced by small animal exploitation. Science 283:190–194.

Stone, W. B., E. Kiviat, and S. A. Butkas. 1980. Toxicants in snapping turtles (*Chelydra serpentina*). N.Y. Fish Game J. 27:39–62.

Stuart, B. L., and R. J. Timmins. 2000. Conservation status and trade of turtles in Laos, pp. 58–62. *In:* Asian Turtle Trade: Proceedings of a Workshop on Conservation and Trade of Freshwater Turtles and Tortoises in Asia. Vol. 2. P. P. van Dijk, B. L. Stuart, and A. G. J. Rhodin (eds.). Chelonian Research Foundation, Phnom Penh, Cambodia.

Stuart, B. L., R. J. Timmins, D. B. Hendrie, L. Sopha, C. Sophat, H. Piseth, H. Kimchay, T. S. Tana, P. L. Hour, C. Thach, J. Compton, and R. Holloway. 2000. Turtle trade in Indochina: Regional summary (Cambodia, Laos, and Vietnam). Chelonian Res. Monogr. 2:74–76.

Summers, A. 2003. Open wide (and fast). Natural History 12:38–39.

Swettenham, F. 1895. With a casting-net, pp. 138–147. *In:* Stories and Sketches by Sir Frank Swettenham. W. R. Roff (ed.). Blackie & Son, London.

Swingland, I. R., and M. W. Klemens. 1989. The conservation biology of tortoises., pp. 1–204. Vol. 5. Occas. Pap. Interatl. Union Cons. Nature Nat. Res., Spec. Srv. Comm.

Tana, T. S., P. L. Hour, C. Thach, and L. Sopha. 2000. Overview of turtle trade in Cambodia. With contributions from C. Sophat, H. Piseth, and H. Kimchay, pp. 55–57. *In:* Asian Turtle Trade: Proceedings of a Workshop on Conservation and Trade of Freshwater Turtles and Tortoises in Asia. Vol. 2. P. P. van Dijk, B. L. Stuart, and A. G. J. Rhodin (eds.). Chelonian Research Foundation, Phnom Penh, Cambodia.

Taskavak, E. 1995. Threats to survival of Euphrates soft-shelled turtle *Rafetus euphraticus* in southeastern Anatolia, Turkey. *In:* Proceedings: International Congress of Chelonian Conservation, July 6–10, 1995. B. Devaux (ed.). SOPTOM, Gonfaron, France.

Taskavak, E., and M. K. Atatur. 1998. Distribution and habitats of the Euphrates softshell turtle, *Rafetus euphraticus*, in Southeastern Anatolia, Turkey, with observations on biology and factors endangering its survival. Chelonian Conserv. Biol. 3:20–30.

Taskavak, E., M. J. Reimann, and W. N. Polder. 1999. First record of the Nile softshelled turtle, *Trionyx triunguis*, from Kos Island, Greece, with comments on its occurrence in the eastern Mediterranean. Chelonian Conserv. Biol. 3:510–512.

Taubes, G. A. 1992a. A dubious battle to save the Kemp's Ridley sea turtle. Science 256:614–616.

—. 1992b. Letters: Kemp's Ridley sea turtles. Science 257:466–467.

Tauxe, R., J. Rigau-Perez, J. Wells, and P. Blake. 1985. Turtle associated salmonellosis in Puerto Rico. Hazards of the global turtle trade. J. Am. Med. Assoc. 254:237–239.

TCF. 2002. A global action plan for conservation of tortoises and freshwater turtles. Strategy and funding prospectus 2002–2007. K. A. Buhlmann, R. Hudson, and A. G. J. Rhodin (eds.). Conservation International and Chelonian Research Foundation, Washington, D.C.

Telecky, T. M. 2001. United States import and export of live turtles and tortoises. Turtle and Tortoise Newsletter 5–13.

Templeton, A. R. 1990. The role of genetics in captive breeding and reintroduction for species conservation. Endangered Species Update 8:14–17.

—. 1996. Translocation in conservation., pp. 315–325. *In:* Biodiversity in Managed Landscapes: Theory and Practice. R. C. Szaro and D. W. Johnson (eds.). Oxford University Press, New York.

Templeton, A. R., H. Hemmer, G. Mace, U.S. Seal, W. M. Shields, and D. S. Woodruff. 1986. Local adaptation, coadaptation, and population boundaries. Zool. Biol. 5:115–125.

Theobald, W. 1868a. Catalogue of reptiles in the Museum of the Asiatic Society of Bengal. J. Asiatic Soc. Bengal. Extra number:1–88.

—. 1868b. Catalogue of the reptiles of British Burma, embracing the provinces of Pegu, Martaban, and Tenasserim; with descriptions of new or little-known species. J. Linn. Soc. Zool. 1868:4–67.

Thirakhupt, K., and P. P. Van Dijk. 1995 "1994". Species diversity and conservation of turtles of western Thailand. Nat. Hist. Bull. Siam Soc. 42:207–259.

—. 1997. The turtles of Western Thailand: Pushed to the edge by progress, pp. 272–277. *In:* Proceedings: Conservation, Restoration and Management of Tortoises and Turtles: An International Conference. J. van Abbema (ed.). New York Turtle and Tortoise Society and the Wcs Turtle Recovery Program, Purchase, NY.

Thomas, L. 1974. The Lives of a Cell: Notes of a Biology Watcher. The Viking Press, New York.

Thomas, R. B. 1993. Growth, diet, and reproduction of the red-eared slider (*Trachemys scripta*) inhabiting a reservoir receiving a cold effluent. M. S. Thesis. Southwest Missouri State University, Springfield, MO.

Thomas, R. B., and W. S. Parker. 2000. Intrasexual variations in overland movements of slider turtles (*Trachemys scripta*). J. Herpetol. 34:469–472.

Thomas, R. B., D. Moll, and J. Steiert. 1994. Evidence of a symbiotic relationship between cellulolytic bacteria and a freshwater herbivorous turtle. Southwestern Naturalist 39:386–388.

Thomas, R. B., N. Vogrin, and R. Altig. 1999. Sexual and seasonal differences in behavior of *Trachemys scripta* (Testudines: Emydidae). J. Herpetol. 33:511–515.

Thompson, C. M., and R. E. Sparks. 1977. The Asiatic Clam, *Corbicula manilensis* in the Illinois River. Illinois Natural History Surv. Bull. 91:34–36.

Thompson, M. B. 1983. Populations of the Murray River tortoise, *Emydura* (Chelodina): the effect of egg predation by the red fox, *Vulpes vulpes*. Aust. Wildl. Res. 11:491–499.

—. 1993. Hypothetical considerations of the biomass of chelid tortoises in the River Murray and the possible influences of predation by introduced foxes, pp. 219–224. *In:* Herpetology in Australia—A Diverse Discipline. D. Lunney and D. Ayers (eds.). Transaction of the Royal Society of New South Wales, Mosman.

Thomson, S. 1996. Field keys of the turtles of the Northern Territory. Applied Ecology Research Group. Belconnen Act 2616. University of Canberra.

Thomson, S., R. Kennett, and A. Georges, 2000. A new species of long-necked turtle (Testudines: Chelidae) from the Arnhem Land Plateau, Northern Territory, Australia. Chelonian Conserv. Biol. 3:675–685.

Thorbjarnarson, J. B., N. Perez, and T. Escalona. 1993. Nesting of *Podocnemis unifilis* in the Capanaparo rier, Venezuela. J. Herpetol. 27:344–347.

Thorbjarnarson, J., C. J. Lagueux, D. Bolze, M. W. Klemens, and A. B. Meylan. 2000. Human use of turtles: a worldlwide perspective, pp. 33–84. *In:* Turtle Conservation. M. W. Klemens (ed.). Smithsonian Institution Press, Washington, D.C.

Thornhill, G. M. 1982. Comparative reproduction of the turtle, *Chrysemys scripta elegans*, in heated and natural lakes. J. Herpetol. 16:347–353.

Timken, R. L. 1968. *Graptemys pseudogeographica* in the upper Missouri River of the north-central United States. J. Herpetol. 1:76–82.

Tinkle, D. W. 1959. The relation of the fall line to the distribution and abundance of turtles. Copeia 1959:167–170.

TRAFFIC. 1997. Turtles protected. TRAFFIC Int. Bull. 16:86.

Trauth, S. E., J. D. Wilhide, and A. Holt. 1998. Population structure and movement patterns of alligator snapping turtles (*Macroclemys temminckii*) in Northeastern Arkansas. Chelonian Conserv. Biol. 3:64–70.

Tucker, A. D., and K. N. Sloan. 1997. Growth and reproductive estimates from alligator snapping turtles, *Macroclemys temminckii*, taken by commercial harvest in Louisiana. Chelonian Conserv. Biol. 2:587–592.

Tucker, A. D., N. N. Fitz Simmons, and J. W. Gibbons. 1995. Resource partioning by the estuarine turtle *Malaclemys terrpain*: Trophic, spatial, and temporal foraging constraints. Herpetologica 51:167–181.

Tucker, J. K. 1995. Salvage of eggs from road-killed red-eared sliders, *Trachemys scripta elegans*. Chelonian Conserv. Biol. 1:317–318.

—. 1997. Natural history notes on nesting, nests, and hatchling emergence in the red-eared slider turtle, *Trachemys scripta elegans*, in west-Central Illinois. Illinois Natural History Surv. Biol. Notes 140:13.

—. 1999. Environmental correlates of hatchling emergence in the red-eared turtle, *Trachemys scripta elegans*, in Illinois. Chelonian Conserv. Biol. 3:401–406.

—. 2000a. Body size and migration of hatchling turtles: inter- and intraspecific comparisons. J. Herpetol. 34:541–546.

—. 2000b. Annual variation in hatchling size in the red-eared slider turtle (*Trachemys scripta elegans*). Herpetologica 56:8–13.

Tucker, J. K., and D. Moll. 1997. Growth, reproduction, and survivorship in the red-eared turtle, *Trachemys scripta elegans*, in Illinois with conservation implications. Chelonian Conserv. Biol. 2:352–357.

Tucker, J. K., and G. C. Packard. 1998. Overwinter survival by hatchling sliders (*Trachemys scripta*) in west-central Illinois. J. Herpetol. 32:431–434.

Tucker, J. K., and G. L. Paukstis. 1999. Post-hatching substrate moisture and overwintering hatchling turtles. J. Herpetol. 33:608–615.

Tucker, J. K., R. J. Maher, and C. H. Theiling. 1995a. Year-to-year variation in growth in the red-eared turtle, *Trachemys scripta elegans*. Herpetologica 51:354–358.

—. 1995b. Melanism in the red-eared turtle (*Trachemys scripta elegans*). J. Herpetol. 29:291–296.

Tucker, J. K., F. J. Janzen, and G. L. Paukstis. 1997. Response of embryos of the red-eared turtle (*Trachemys scripta elegans*) to experimental exposure to water-saturated substrates. Chelonian Conserv. Biol. 2:345–351.

Tucker, J. K., G. L. Paukstis, and F. J. Janzen. 1998a. Annual and local variation in reproduction in the red-eared slider, *Trachemys scripta elegans*. J. Herpetol. 32:515–526.

Tucker, J. K., N. I. Filoramo, G. L. Paukstis, and F. J. Janzen. 1998b. Response of red-eared slider, *Trachemys scripta elegans*, eggs to slightly differing water potentials. J. Herpetol. 32:124–128.

Tucker, J. K., S. A. Gritters, and R. A. Hrabic. 1998c. Turtle Communities: an analysis of aquatic species in the Upper Mississippi River system. Unpublished report. National Biological Service Environmental Management Technical Center, Onalaska, WI.

Tucker, J. K., D. Moll, and E. O. Moll. 1999. Growth of slider turtles (*Trachemys scripta*) from temperate and tropical populations. Chelonian Conserv. Biol. 3:464–467.

Turner, L. K. 1995. Reproduction and diet of *Pseudemys concinna* inhabiting a cold water reservoir in southwest Missouri. M.S. Thesis. Southwest Missouri State University, Springfield, MO.

U.S.F.W.S. 1988. Ringed sawback turtle recovery plan. U.S. Fish and Wildlife Service, Jackson, MS.

—. 1989. Alabama red-bellied turtle recovery plan. U.S. Fish and Wildlife Service, Jackson, MS.

—. 1990. Flattened musk turtle recovery plan. U.S. Fish and Wildlife Service, Jackson, MS.

—. 1993. Yellow-blotched map turtle (*Graptemys flavimaculata*). U.S. Fish and Wildlife Service, Jackson, MS.

Valenzuela, N. 2001a. Maternal effects on life-history traits in the Amazonian giant river turtle *Podocnemis expansa*. J. Herpetol. 35:368–378.

—. 2001b. Constant, shift, and natural temperature effects on sex determination in *Podocnemis expansa* turtles. Ecology 82:3010–3024.

Valenzuela, N., R. Botero, and E. Martinez. 1997. Field study of sex determination in *Podocnemis expansa* from Colombian Amazonia. Herpetol. 53:390–398.

Van Abbema, J. 1997. Proceedings: Conservation, Restoration, and Management of Tortoises and Turtles – An International Conference. State University of New York, Purchase, NY.

Van De Bunt, P. 1990. Tortoise Exploitation in Sumatra. Tortoises & Turtles 5:14–15.

Van Dijk, P. P. 1990. Hard-shelled eggs may need special care during incubation. Tortoises & Turtles 5:16–17.

—. 1994. Report on a visit to Myanmar. An unpublished report to the American Museum of Natural History and the IUCN/SSC Turtle Recovery Program of the Tortoise and Freshwater Turtle Specialist Group.

—. 1997. Turtle conservation in Myanmar: Past, present and future, pp. 265–271. *In:* Proceedings: Conservation, Restoration, and Management of Tortoises and Turtles: An International Conference. J. van Abbema (ed.). State University of New York, Purchase, NY.

—. 2000. The status of turtles in Asia, pp. 15–23. *In:* Asian Turtle Trade: Proceedings of a Workshop on Conservation and Trade of Freshwater Turtles and Tortoises in Asia. Vol. 2. P. P. van Dijk, B. L. Stuart, and A. G. J. Rhodin (eds.). Chelonian Research Foundation, Phnom Penh, Cambodia.

Van Dijk, P. P., and T. Palasuwan. 2000. Conservation status, trade, and management of tortoises and freshwater turtles in Thailand, pp. 137–144. *In:* Asian Turtle Trade: Proceedings of a Workshop on Conservation and Trade of Freshwater Turtles and Tortoises in Asia. Vol. 2. P. P. van Dijk, B. L. Stuart, and A. G. J. Rhodin (eds.). Chelonian Research Foundation, Phnom Penh, Cambodia.

Van Dijk, P. P., and Thirakhupt. 1995. Southeast Asian *Chitra* – from description to extinction in 15 years? *In:* Proceedings: International Congress of Chelonian Conservation. B. Devaux (ed.). SOPTOM, Gonfaron, France.

Van Dijk, P. P., B. L. Stuart, and A. G. J. Rhodin. 2000a. Asian Turtle Trade: Proceedings of a Workshop on Conservation and Trade of Freshwater Turtles and Tortoises in Asia, p. 164 p. *In:* Chelonian Research Monographs. Vol. 2. P. C. H. Prichard, B. L. Stuart, and A .G. J. Rhodin (ed.). Chelonian Research Foundation, Phnom Penh, Cambodia.

Van Dijk, P. P., D. T. Iskandar, T. Palasuwan, A. G. J. Rhodin, Samedi, D. S. K. Sharma, C. R. Shepherd, O. B. Tisen, and V. R. Genorupa. 2000b. Turtle trade in Southeast Asia: Regional summary (Indonesia, Malaysia, Papua New Guinea, and Thailand). pp. 145–147. *In:* Asian Turtle Trade: Proceedings of a Workshop on Conservation and Trade of

Freshwater Turtles and Tortoises in Asia. Vol. 2. P. P. van Dijk, B. L. Stuart, and A. G. J. Rhodin (eds.). Chelonian Research Foundation, Phnom Penh, Cambodia.

Vance, J. M. 1994. Plugging up the drain. Missouri Conservationist, August, 1994:22–27.

Vandewalle, T. J. and J. L. Christiansen. 1993. The impact of stream modification on the distribution of riverine turtles in Iowa, An unpublished report. The Iowa Department of Natural Resources and The U.S. Army Corps of Engineers, Des Moines, IA.

Varghese, G., and G. T. Tonapi. 1986. Observations on the identity of some Indian freshwater turtles and their feeding habits. Biol. Conserv. 37:87–92.

Vasudevan, K. 1998. Reproductive ecology of the Indian softshell turtle, *Aspideretes gangeticus*, in Northern India. Chelonian Conserv. Biol. 3:96–99.

Velasco, A. M. 1995. Reproductive biology of *Podocnemis unifilis* and *P. expansa* in the Cuyabeno National Park, Ecudor. Abstract, p. 46. *In:* Proceedings: International Congress of Chelonian Conservation July 6–10, 1995. B. Devaux (ed.). SOPTOM, Gonfaron, France.

Vermersch, T. G. 1992. Lizards and Turtles of South-Central Texas. Eakin Press, Austin, TX.

Vijaya, J. 1982a. *Pelochelys bibroni* from the Gahirmata coast. Orissa Hamadryad. 7:17.

—. 1982b. The flapshell turtles. Hornbill 56:71–82.

Villiers, A. 1957. Tortues marines du Senegal. Science et Nature 21:17–23.

—. 1958. Tortues et crocodiles de l'Afrique Noire Française. Inst. Franç. D Afrique Noire Initiat. Afriq. 15:1–354.

—. 1962. West African tortoises, turtles, and terrapins. African Wildlife 39–52.

Vogt, R. C. 1979. Cleaning/feeding symbiosis between grackles (*Quiscalus*: Icteridae) and map turtles (*Graptemys*: Emydidae). The Auk 96:608–609.

—. 1980a. Natural history of the map turtles *Graptemys pseudogeographica* and *G. ouachitensis* in Wisconsin. Tulane Studies Zool. Botany 22:17–48.

—. 1980b. New methods for trapping aquatic turtles. Copeia 1980:368–371.

—. 1981a. Food partitioning in three sympatric species of map turtle, genus *Graptemys* (Testudinata, Emydidae). Am. Midlands Naturalist 105:102–111.

—. 1981b. Turtle egg (*Graptemys*: Emydidae) infestation by fly larvae. Copeia. 1981:457–459.

—. 1981c. Graptemys versa, pp. 1–2. Vol. Rept. 280. Catalog. Am. Amphib. Report.

—. 1981d. Natural History of Amphibians and Reptiles of Wisconsin. Milwaukee Public Museum, Milwaukee, WI.

—. 1989. Information on biological supply house collecting of turtles in Wisconsin. In litt. to Dr. Charles Dane, Chief, Office of Scientific Authority, USFWS.

—. 1990. Reproductive parameters of *Trachemys scripta venusta* in southern Mexico, pp. 162–168. *In:* Life History and Ecology of the Slider Turtle. J. W. Gibbons (ed.). Smithsonian Institution Press, Washington, D.C.

—. 1991. Effects of estradiol on sex determination in turtles. (Abstract). Annual meeting of the Herpetologists' Leaque and the Society for the Study of Amphibians and Reptiles, Penn. State University, PA.

—. 1992. Ecologia reproductiva de uma comunidade de quelonios da Amazonia. (Abstract). Congresso de Zoologia de America Latina, Belem, Brazil.

—. 1993. Systematics of the false map turtles (*Graptemys pseudogeographica* complex: Reptilia, Testudines, Emydidae). Ann. Carnegie Mus. 62:1–46.

—. 1994. Temperature controlled sex determination as a tool for turtle conservation. Chelonian Conserv. Biol. 1:159–162.

—. 1995. Brazilian freshwater turtles, to eat or not to eat? Save them by eating them? pp. 279–280. *In:* Proceedings: International Congress of Chelonian Conservation. B. Devaux (ed.). SOPTOM (Station d'observation et de protection des tortues des maures), Gonfaron, France.

—. 1997. Freshwater turtle nesting and reproduction in the neotropics. Abstract, p. 301. ASIH/HL/SSAR/AFS-ELHS/AES/GIS Joint Meetings, Seattle, WA.

—. In press. *Podocnemis expansa* (Schweigger, 1812) Tartaruga, Arrau, Giant South American River Turtle. *In:* The Conservation Biology of Freshwater Turtles. Vol. 3. P. C. H. Pritchard and A. G. J. Rhodin (eds.). Chelonian Research Foundation.

Vogt, R. C., and J. J. Bull. 1982. Genetic sex determination in the spiny softshell *Trionyx spiniferus* (Testudines: Trionychidae). Copeia 1982:699–700.

—. 1984. Ecology of hatchling sex ratio in map turtles. Ecology. 65:582–587.

Vogt, R. C., and O. Flores-Villela. 1986. Determinacion del sexo en tortugas por la temperatura de incubacion de los huevos. Ciencia 37:21–32.

Vogt, R. C., and O. Flores-Villela. 1992a. Effects of incubation temperature on sex determination in a community of neotropical freshwater turtles in southern Mexico. Herpetologica 48:265–270.

Vogt, R. C., and O. A. Flores-Villela. 1992b. Aspectos de la ecología de la tortuga blanca (*Dermatemys mawii*) en la Reserva la Biósfera Montes Azules, pp. 221–231. *In:* Reserva de la Biósfera Montes Azules, Selva Lacandona: Investigación para su conservación. Vol. 1. M. A. Vásquez-Sáchez and M. A. Ramos (eds.). Publ. Esp. Ecosfera.

Vogt, R. C., and S. Guzman-Guzman. 1988. Food partitioning in a Neotropical freshwater turtle community. Copeia 1988:37–47.

Vogt, R. C., and P. Soini. in press. *Podocnemis unifilis* Troschel, 1848 Tracaja, Terecay, Yellow-spotted Amazon River Turtle. *In:* The Conservation Biology of Freshwater Turtles. Vol. 3. P. C. H. Pritchard and A. G. J. Rhodin (eds.). Chelonian Research Foundation.

Vogt, R. C., and J. L. Villarreal-Benitez. 1993. Abstract: Species abundance and biomass distributions in freshwater turtles. Estación de Biología Tropical Los Tuxtlas, Instituto de Biología, San Andrés Tuxtla, Varacruz, Mexico.

—. 1997. Species abundance and biomass distributions in freshwater turtles, pp. 210–218. *In:* Proceedings: Conservation, Restoration, and Management of Tortoises and Turtles: An International Conference 11–16 July. J. van Abbema and M. W. Klemens (eds.), State University of New York, Purchase, NY.

Vogt, R. C., J. R. Polisar, and D. Moll. In press. *Dermatemys mawii* Gray, 1847. *In:* The Conservation Biology of Freshwater Turtles. P. C. H. Pritchard and A. G. J. Rhodin (eds.). Chelonian Research Foundation.

Vogt, R. C., V. H. Cantarelli, and A. G. D. Carvalho. 1994. Reproduction of the cabecudo, *Peltocephalus dumerilianus*, in the Biological Reserve of Rio Trombetas, Para, Brazil. Chelonian Conserv. Biol. 1:145–148.

Von Hildebrand, P., C. Saenz, M. C. Penuela, and C. Caro. 1988. Biologia reproductiva y manejo de la tortuga charapa (*Podocnemis expansa*) en el bajo Rio Caqueta. Columbia Amazonica 3:89–112.

Von Humboldt, A. 1819. Relation Historique du Voyage aux Régions Équinoxiales du Nouveau Continent, Fait en 1799–1804, Maze, Paris.

—. 1852. Personal Narrative of Travels to the Equinoctial Regions of America 1799–1804. H. G. Bohn, London.

Vrba, E. S. 1980. Evolution, species, and fossils: How does life evolve? South African J. Sci. 76:61–84.

—. 1984. Evolutionary pattern and process in the sister group Alcelaphini-Aepycerotini (Mammalia: *Bovidae*), pp. 62–79. *In:* Living Fossils. N. E. Eldredge, and S. M. Stanley (eds.). Springer-Verlag, New York.

—. 1987. Ecology in relation to speciation rates: Some case histories of Miocene to Recent mammal clades. Evol. Ecol. 1:283–300.

—. 1992. Mammals as a key to evolutionary theory. J. Mamm. 73:1–28.

Wade, S. E., and C. E. Gifford. 1965. A preliminary study of the turtle population of a northern Indiana lake. Proc. Indiana Acad. Sci. 74:371–374.

Wagner, M. 1868. Die Darwin'sche Theorie und das Migrationsgesetz der Organismen. Duncker and Humblot, Leipzig.

Wahlquist, H. 1970. Sawbacks of the Gulf Coast. Int. Turtle Tort. Soc. J. 4:10–13, 28.

Walker, W. F. 1979. Locomotion, pp. 435–454. *In:* Turtles Perspectives and Research. M. Harless and H. Morlock (eds.). John Wiley, New York.

Wallace, A. R. 1855. On the law which has regulated the introduction of new species. Annals and Magazine of Natural History. Vol. 16.

—. 1876. The Geographical Distribution of Animals. Harper, New York.

Warwick, C. 1985a. Terrapin farming in the United States. RSPCA (Royal Society for the Prevention of Cruelty to Animals) Today 49:24–25.

—. 1985b. The trade in red-eared terrapins. Animals Int. 16:8–9.

—. 1986. Red-eared terrapin farms and conservation. Oryx 20:237–240.

Warwick, C., and C. Steedman. 1988. Report on the use of red-eared turtles (*Trachemys scripta elegans*) as a food source utilized by man. People's Trust for Endangered Species, Surrey, UK.

Warwick, C., C. Steedman, and T. Holford. 1990. Ecological Implications of the red-eared turtle trade. Texas J. Sci. 42:419–422.

Watermolen, D. J., and J. F. Amrhein. 1995. Mercury levels in snapping turtles, *Chelydra serpentina*, from Wisconsin. Bull. Chicago Herp. Soc. 30:230–233.

Waters, J. C. 1974. The biological significance of the basking habit in the black-knobbed sawback, *Graptemys nigrinoda* Cagle. M.S. Thesis. Auburn University, Auburn, AL.

WCED. 1987. Our Common Future. World Commission on Environment and Development. Oxford University Press, Oxford, UK.

Weaver, J. 1973. Profits, politics, and *Podocnemis*. Int. Turt. Tort. Soc. J. 7:10–15.

Webb, R. G. 1961. Observations on the life histories of turtles (genus *Pseudemys* and *Graptemys*) in Lake Texoma, Oklahoma. Am. Midlands Naturalist 65:17–48.

—. 1995. Redescription and neotype designation of *Pelochelys bibroni* from southern New Guinea (Testudines: Trionychidae). Chelonian Conserv. Biol. 1:301–310.

Welcomme, R. L. 1979. Fisheries Ecology of Floodplain Rivers. Longman Group Limited, London.

Whitaker, R. 1982. Bangladesh: A general survey. Hamadryad 7:4–12.

—. 1983. Bangladesh: a general survey. Hornbill 7:11–13.

—. 1997. Turtle rearing in village ponds, pp. 106–108. *In:* Proceedings: Conservation, Restoration, and, Management of Tortoises, and Turtles: An International Conference: 11–16 July 1993. J. van Abbema (ed.), State University of New York, Purchase, NY.

Whitaker, R., and H. Andrews. 1997. Captive breeding of Indian turtles and tortoises at the Centre for Herpetology/Madras Crocodile Bank, pp. 166–170. *In:* Proceedings: Conservation, Restoration, and, Management, of Tortoises, and Turtles: An International Conference: 11–16 July 1993. J. van Abbema (ed.), State University of New York, Purchase, NY.

White, D., Jr., and D. Moll. 1992. Restricted diet of the common map turtle *Graptemys geographica* in a Missouri stream. Southwest. Naturalist 37:317–318.

White, P., and T. Flannery. 1992. The impact of people on the Pacific world, pp. 1–8. *In:* The Native Lands: Prehistory and Environmental Change in Australia and the Southwest Pacific. J. Dodson (ed.). Longman Cheshire, Melbourne.

Wibbels, T. 1992. Letters: Kemp's Ridley sea turtles. Science Natural History Museum, Los Angeles Co. 257:465.

Wibbels, T., F. C. Killebrew, and D. Crews. 1991. Sex determination in Cagle's map turtle: implications for evolution, development, and conservation. Can. J. Zool. 69:2693–2696.

Wilbur, H. M., and P. J. Morin. 1988. Life history evolution in turtles, pp. 387–439. *In:* Biology of the Reptilia. C. Gans and R. B. Huey (eds.). Alan R. Liss, Inc., New York.

Wilcox, B. A. 1984. In situ conservation of genetic resources: Determinants of minimum area requirements, pp. 639–666. *In:* National Parks, Conservation, and Development: The Role of Protected Areas in Sustaining Society. J. A. McNeely and K. R. Miller (eds.). Smithsonian Institution Press, Washington D.C.

Wiley, J. W., and F. E. Lohrer. 1973. Additional records of non-fish prey taken by osprey. Wilson Bull. 85:468–470.

Willers, B. 1994. Sustainable development: a new world deception. Conserv. Biol. 8:1146–1148.

Williams, K. L., H. M. Smith, and P. S. Chrapliwy. 1960. Turtles and lizards from northern Mexico. Trans. Illinois St. Acad. Sci. 53:36–45.

Williams, T. A., and J. L. Christiansen. 1981. The niches of two sympatric softshell turtles, *Trionyx muticus* and *Trionyx spiniferus*, in Iowa. J. Herpetol. 15:303–308.

Williams, T. W. 1996. What good is a wetland? Audubon (November–December):43–100.

Williamson, M. H. 1981. Island Populations. Oxford University Press, Oxford, UK.

Wilson, L. D., and L. Porras. 1983. The ecological impact of man on the South Florida herpetofauna. Univ. Kansas Mus. Nat. Hist., spec. publ. 9:1–89.

Wilson, T. B. 1999. Reproduction and movement patterns of the common snapping turtles (*Chelydra serpentina*) inhabiting a hypothermic reservoir in southwest Missouri. M.S. Thesis. Southwest Missouri State University, Springfield, MO.

Witherington, B. E. 1995. Hatchling orientation, pp. 577–578. *In:* Biology and Conservation of Sea Turtles, revised ed. K. A. Bjorndal (ed.). Smithsonian Institution Press, Washington, D.C.

Woinarski, J. C. Z. 1993. The status of herpetofauna and herpetology in the Northern Territory, pp. 81–106. *In:* Herpetology in Australia: A Diverse Discipline. D. Lunney, and D. Ayers (eds.). Transactions of the Royal Zoological Society of New South Wales, Mosman.

Wood, R. C. 1997. The impact of commercial crab traps on northern diamondback terrapins, *Malaclemys terrapin terrapin*, pp. 21–27. *In:* Proceedings: Conservation, Restoration, and Management of Tortoises and Turtles: An International Conference. J. van Abbema (ed.). State University of New York, Purchase, NY.

Wood, R. C., and R. Herlands. 1995. Terrapins, tires, and traps: Conservation of the northern Diamonback Terrpin (*Malaclemys terrapin terrapin*) on the Cape May Peninsulua, New Jersey, USA, pp. 254–258. *In:* Proceedings: International Congress of Chelonian Conservation, July 6–10, 1995. B. Devaux (ed.). SOPTOM, Gonfaron, France.

—. 1997. Turtles and tires: The impact of roadkills on northern diamondback Terrapin, *Malaclemys terrapin terrapin*, populations on the Cape May Peninsula, southern New Jersey, USA, pp. 46–53. *In:* Proceedings: Conservation, Restoration, and Management of Tortoises and Turtles: An International Conference: 11–16 July. J. van Abbema (ed.), State University of New York, Purchase, NY.

Woody, J. B. 1990. Guest editorial: Is 'headstarting' a reasonable conservation measure? On the surface, yes; in reality, no. Marine Turtle Newsletter 50:8–11.

—. 1991. Guest editorial: It's time to stop headstarting Kemp's ridley. Marine Turtle Newsletter 55:7–8.

Worthington, E. B. 1929. A report on the fishing survey of Lakes Albert and Kioga, Cambridge, UK.

Wright, J. 1995. Herping in the great white north. Reptiles, June:94–99.

Xianlin, M., Z. Zhihua, and B. L. Stuart. 2002. Recent actions by the People's Republic of China to better control international trade of turtles. Turtle and Tortoise Newsletter 15–16.

Yawetz, A., I. Sidis, and A. Gasith. 1983. Metabolism of parathion and brain cholinesterase inhibition in Aroclor 1254-treated and untreated Caspian terrapin (*Mauremys caspica rivulata* Emydidae, Chelonia) in comparison with two species of wild birds. Comp. Biochem. Physiol. 75C:377–382.

Ypsilanti, A. 1993. Turtles days of August, Thousands of hatchlings seized. The New York Turtle and Tortoise Society (NYTTS) NewsNotes IV:16.

Zhao, E. M. 1995. Chelonian fauna and their conservation in China. *In:* Proceedings: International Congress of Chelonian Conservation. B. Devaux (ed.). SOPTOM, Gonfaron, France.

—. 1998. Chinese Red Data Book of Endangered Animals: Amphibia and Reptilia. Vol. 3, Beijing.

Zhao, E. M., and K. Adler. 1993. Herpetology of China. Society for the Study of Amphibians and Reptiles. Vol. 10 Oxford, OH.

Zhou, J., and T. Zhou. 1992. Chinese Chelonians Illustrated. Jiangsu Science & Technology Publishing House, Nanjing.

Zimmerman, L. C., and C. R. Tracy. 1989. Interactions between the environment and herbivory in reptiles. Physiol. Zool. 62:374–409.

Zug, G. R. 1966. The penial morphology and the relationships of cryptodiran turtles. Occ. Pap. Mus. Zool. Univ. Michigan 647:1–24.

—. 1971. Buoyancy, locomotion, morphology of pelvic girdle and hindlimb, and systematics of Cryptodiran turtles. Misc. Publ. Mus. Zool. Univ. Michigan 142:1–98.

Species Index

Printed and bound in the UK by
CPI Antony Rowe, Eastbourne

The manufacturer's authorised representative in the EU for product safety is Oxford University Press España S.A. of El Parque Empresarial San Fernando de Henares, Avenida de Castilla, 2 - 28830 Madrid (www.oup.es/en or product.safety@oup.com). OUP España S.A. also acts as importer into Spain of products made by the manufacturer.
Printed and bound by CPI Group (UK) Ltd, Croydon, CR0 4YY
12/09/2026
02215931-0002